Libuše Hannah Vepřek
At the Edge of AI

Science Studies

Libuše Hannah Vepřek is a postdoctoral researcher at the Ludwig Uhland Institute for Historical and Cultural Anthropology at Eberhard Karls Universität Tübingen. The cultural anthropologist and computer scientist completed her doctorate at Ludwig-Maximilians-Universität München in the context of the "Playing *in the Loop*" project (2021-2024) funded by the German Research Foundation. Her main research areas are digital anthropology, anthropology of technology, science and technology studies, moral anthropology and ethics of technology, and digital methods.

Libuše Hannah Vepřek

At the Edge of AI

Human Computation Systems and
Their Intraverting Relations

This text is a revised version of the author's dissertation in European Ethnology and Cultural Analysis at the Ludwig-Maximilians-Universität München, 2023.

The publication of this book was funded by the LMU Open Access Fonds.
The research was funded as part of the research project "Playing in the Loop: New Human-Software Relations in Human Computation Systems and their Impacts on the Spheres of Everyday Life" funded by the German Research Foundation (DFG 464513114). Parts of the field research were funded by the DAAD program "IFI – Internationale Forschungsaufenthalte für Informatikerinnen und Informatiker (Doktoranden), 2019-2022 (57515303)".

Bibliographic information published by the Deutsche Nationalbibliothek
The Deutsche Nationalbibliothek lists this publication in the Deutsche Nationalbibliografie; detailed bibliographic data are available in the Internet

This work is licensed under the Creative Commons Attribution 4.0 (BY) license, which means that the text may be remixed, transformed and built upon and be copied and redistributed in any medium or format even commercially, provided credit is given to the author.
Creative Commons license terms for re-use do not apply to any content (such as graphs, figures, photos, excerpts, etc.) not original to the Open Access publication and further permission may be required from the rights holder. The obligation to research and clear permission lies solely with the party re-using the material.

First published in 2024 by transcript Verlag, Bielefeld
© **Libuše Hannah Vepřek**

Cover layout: Maria Arndt, Bielefeld
Cover illustration: created with the assistance of DALL•E 2 (2023-03-08)
Print-ISBN: 978-3-8376-7228-2
PDF-ISBN: 978-3-8394-7228-6
ISSN of series: 2703-1543
eISSN of series: 2703-1551

Contents

Acknowledgments 7

List of Figures 9

List of Abbreviations 11

1 Introduction: "We're Doing Something Completely New" 13

2 Approaching Human Computation-Based Citizen Science Analytically 33
Crowdsourcing and Crowdworking 34
Citizen Science (Games) and the Entanglements of Play, Work, and Science 36
Sociotechnical Systems and the Study of Algorithms, Computer Code,
and Artificial Intelligence 40
Infrastructures and Infrastructuring 48
A Theoretical Framework for Analyzing Emerging Hybrid Systems 50

3 Methodology: Encountering Human Computation Ethnographically 79
Praxiographically Inspired Co-Laborative Ethnography 80
Constructivist Grounded Theory 83
Doing Research On, With, and Among Researchers and Developers 85
A Toolkit of Methods for Emerging Hybrid Systems 90

4 Envisioning and Designing the Future 101
Human-In-The-Loop Imaginaries 105
Weaving Together the Imaginaries 127
Imagining as Practice: Infrastructuring and Experimentation 129
Between Counter-Imaginary and Infrastructuring 136

5 Multiple Meanings and Everyday Negotiations: Play/Science Entanglements ... 139
A Snapshot of Foldit 140
A Snapshot of Stall Catchers 143

Contributing to Cope With Everyday Life 145
A Phenomenon Between Play and Science 153
Emerging Spaces in Play/Science Entanglements and Frictions 179

6 Intraversions: Human-Technology Relations in Flux 183
Never Obsolete: Intraversions of Participant-Technology Relations in Foldit 186
"First Use No Humans:" Intraversions of Participant-Technology Relations in Stall Catchers 198
Reconfigurations of Participant-Technology Relations 209
Extending the Loop: Intraversions of Researcher-Technology Relations 210
Moving Forward in Concert 236
Contingent, Imagined, and Emergent Intraversions 243

7 Building Trust in and With Human Computation 247
Trust as an Analytical Concept... 248
Trust as a Sociomaterial Practice 252
The Question of Trust and Proprietary Software 269
Distributed Trust 270

8 Conclusions 273
Engaging at the Edge of Artificial Intelligence 276
Beyond the Edge 278

Glossary... 281

References... 283

Acknowledgments

From my early exploratory research to the submission of my dissertation and its transformation into this book, I have been fortunate to have been accompanied by many people who have inspired and made this work possible.

First and foremost, I would like to extend my heartfelt gratitude to my research partners across the different projects. Their openness, generosity of time, and invaluable exchange of knowledge, thoughts, and reflections have been essential in shaping the outcomes of this endeavor. I am sincerely thankful to those who remain anonymous in this work but who have read, reviewed, and provided feedback on my work, and who have supported me with unwavering friendship beyond this research. My special thanks go to Pietro Michelucci, François Bry, Chris Schaffer, and Nozomi Nishimura for warmly welcoming and supporting me at their esteemed institutions and laboratories.

I am indebted to my doctoral supervisor, Johannes Moser, for his invaluable support and guidance. His mentorship has been instrumental not only during my doctoral studies but also throughout my entire academic journey. In addition, I am grateful to my second doctoral supervisor, Christoph Bareither, whose constructive feedback has always challenged me to delve deeper in my thinking and research.

My debt to my colleagues is no less significant, and while I cannot name everyone individually, I would like to express my gratitude to my colleagues at the Institute for European Ethnology and Cultural Analysis at LMU Munich.

Over the years, I have had the privilege of participating in various working groups that have not only been a source of inspiration but have also provided invaluable support and opportunities for collaboration that I deeply appreciate. I would like to thank all members of the Code Ethnography Collective, the studiolab, the digital anthropology lab, and the informal but very well-established digidoc working group. Their knowledge sharing and insightful discussions have greatly influenced the development and pursuit of this project.

I am particularly grateful to Rebecca Carlson for all the late-night meetings and exchanges across time zones and oceans. Her guidance has been incredibly valuable to me. I would also like to sincerely thank Lina Franken and Sarah Thanner for the hours of discussions in Tokyo, Madrid, and Hamburg that helped to advance my analysis. I am also thankful to Alina Becker, Anne Dippel, Dennis Eckhardt, Ruth Dorothea Eggel,

Laura Gozzer, Felicitas Muth, Petra Schmidt, Pia Schramm, Roman Tischberger, and Leonie Thal, who contributed ideas and suggestions during conversations or in response to presentations. My gratitude also extends to the members of the international PhD program "Transformations in European Societies" and to scholars at Cornell University who shared their expertise and perspectives in our dialogues. This list is undoubtedly incomplete, and I apologize for any unintentional omissions, while expressing my sincere appreciation to all who have supported me. Thanks to all of them, this work has become what it is, while any remaining mistakes are, of course, my own.

I would like to thank the German Research Foundation for the financial funding that made my research possible. Their support of our research project "Playing *in the Loop*: New Human–Software Relations in Human Computation Systems and their Impacts on the Spheres of Everyday Life" (DFG – 464513114) was instrumental. In addition, I also wish to thank the DAAD program "IFI – Internationale Forschungsaufenthalte für Informatikerinnen und Informatiker (Doktoranden), 2019–2022 (57515303)," which provided essential support for parts of my field research. Acknowledgment is also due to Vanessa Fuller for her exceptional assistance in editing key parts of this book.

Furthermore, I am genuinely thankful to my friends, who always provide me with the necessary energy when I lack it and remind me from time to time to look to the side. It remains for me to express my heartfelt gratitude to my family, especially to Maritza Vepřek-Heijman, Stan Vepřek, and Nynke Vepřek for their endless love. I am particularly grateful to my mother, who dedicates herself daily to care-taking responsibilities and is still always ready to lend an ear and support me, no matter how irrelevant my thoughts may be. This work is dedicated to her.

Finally, I would like to thank Philipp Dowling, whose love and unwavering support have carried me throughout the research and writing process. Thank you for being by my side every step of the way.

List of Figures

Figure 1:	Stall Catchers' main UI	99
Figure 2:	Puppies in Stall Catchers. Error page	133
Figure 3:	Foldit overview UI after login	141
Figure 4:	Foldit main game UI	142
Figure 5:	Stall Catchers' main UI with the "virtual microscope"	145
Figure 6:	Screenshot of a Foldit GUI recipe	191
Figure 7:	Final leaderboards of the April 2021 Catchathon	204
Figure 8:	Data transformation in summary performed in the laboratory's data pipeline	211
Figure 9:	Simplified stages of the dataflow. Green refers to stages performed at the laboratory, light blue refers to the processing stage at the Human Computation Institute, and blue refers to the Stall Catchers platform	214
Figure 10:	ARTigo Play mode "annotate image regions based on tags"	241
Figure 11:	Excerpt of the Stall Catchers source code with the saveNextMovie function	264

List of Abbreviations

AI	Artificial intelligence
AGI	Artificial general intelligence
ANT	Actor–network theory
CCS	Critical code studies
CNN	Convolutional neural network
CS	Citizen science
DL	Deep learning
GUI	Graphical user interface
HC	Human computation
HI	Hybrid intelligence
I/O	Input/output
LLM	Large language model
ML	Machine learning
NLP	Natural language processing
PI	Principal Investigator
SQL	Structured query language
STS	Science and technology studies
UI	User interface
VDC	Voluntary distributed computing

1 Introduction: "We're Doing Something Completely New"

On April 29, 2022, the Human Computation Institute hosted a live event for the final hour of their so-called Catchathon. For this special 24-hour event, the institute invited participants, schoolchildren, libraries, and the general public to join a timed competition taking place within the human computation (HC)–based citizen science (CS) game Stall Catchers. The participants' task consisted of analyzing Alzheimer's disease research data presented as short video sequences in a gamified setting on the Stall Catchers platform. In doing so, they checked the blood vessels depicted in the videos for blockages, annotating them as either "flowing," or "stalled" if they detected a blockage or "stall" in the blood flow. Within this competition format, teams of participants competed against each other aiming to annotate the highest number of research videos by the end of the competition. In this specific Catchathon, they also competed against the artificial intelligence (AI) bot GAIA.

During the final-hour event hosted on Zoom, joined by several classes of students from Miami along with other participants from around the world, Pietro Michelucci, the director of the institute and Stall Catchers project lead, summarized this Catchathon competition: "GAIA is fast, but not quite as skillful as" (Human Computation Institute 2021, 32:28–32:31) Stall Catchers' best human participants. While the institute had previously organized several such Catchathons, this competition was unique: "We're doing something completely new" (Human Computation Institute 2021, 17:36–17:40). In addition to the human crowd participating in the competition, the "intelligent bot," as it was called on the institute's blog (Egle [Seplute] 2021c), or the "artificial intelligence agent" (Human Computation Institute 2021, 18:15–18:17) GAIA analyzed research videos alongside human participants. Not knowing how human participants would respond to and engage with the bot and how these new participant–bot relations would unfold, the Stall Catchers team had worked hard on building a "bot-wrapper" to introduce what appeared to be the very first "world's citizen science bot" (Human Computation Institute 2021, 18:10–18:12).[1] Named after the primordial goddess in Greek mythology, the personifica-

---

1   This included the development of an application programming interface (API) to allow the machine learning (ML) model to communicate with the Stall Catchers platform and bot–user profiles

tion of Earth, GAIA was trained on human participants' annotation data and built using techniques from deep learning (DL) employed to play Stall Catchers.

Michelucci assessed GAIA's performance as good, although not exceeding some human participants' skill levels. This assessment seemingly also reflected some relief after one participant referred to GAIA as a "pesky bot" (Egle [Seplute] 2021c) capable of annotating research videos all night, while most of the human participants slept. The institute's blog post summarizing the final statistics of the Catchathon eventually stated that, even though it was a "close race," "it's 1 for supercatchers, 0 for GAIA [robot emoji], eh? [emoji with a winking face and tongue sticking out] Just kidding—we're all in this together, GAIA [robot emoji] included, and hopefully she will help us analyze data faster in the future!" (Egle [Seplute] 2021c). The experiment, therefore, succeeded, demonstrating that a bot such as GAIA could be successfully introduced to the platform without completely outpacing and beating human participants. Instead, this introduction stimulated competition.

Stall Catchers, aimed at advancing the Alzheimer's disease research conducted at the Schaffer–Nishimura Lab in biomedical engineering at Cornell University, was created to solve a specific data analysis problem that could not be solved by the laboratory's researchers nor via computational methods alone. The introduction of GAIA into Stall Catchers kicked off the Human Computation Institute's ongoing research into how human participants and AI bots could be combined to not only speed up analysis, but also ensure the scientifically required quality of crowdsourced answers (or, simply, crowd answers). Even without AI bots, Stall Catchers relies on a complex "wisdom-of-the-crowd" approach (Surowiecki 2005) to calculate crowd answers from individual participant's video annotations. This approach itself relies on nontrivial human–technology,[2] human–software or human–algorithm relations, which together form the core of the HC system Stall Catchers.

The Human Computation Institute followed an HC approach, combining humans and computers in new ways to solve the data analysis problem. While participant–software relations had thus far relied on humans performing the actual data analysis with algorithms consigned to evaluating and combining human inputs, the introduction of AI bots to the regular Stall Catchers game would alter these relations. Creating partnerships between humans and AI bots (Vaicaityte 2021a), for instance, would redistribute human and software roles and shift the respective responsibilities in this sociotechnical system. These human–technology relations, however, do not simply rely upon and come into being based solely on the designers' and developers' imaginations, decisions, and implementations. Instead, they depend just as much upon participants' active engagement and adoptions, alongside technological affordances (Gibson 1977; 1979; Bareither 2020a) and action potentials, which only become actualized through usage and practice (Beck 1997). Together, these human and nonhuman actors form dynamic and contingent

---

allowing ML models to become part of the game. This also included adapting the platform and improving its resilience to anticipate increased traffic during the event.

2   Human–technology relations here do not suggest any order between humans and technology but represent different relations, such as human–human, technology–human or human–technology–human relations.

relations which together create the HC-based CS system. This first encounter between human participants and AI bots serves as a salient and illustrative moment in Stall Catchers' evolution, given that it introduces new forward movements within its participant–AI relations. I call these movements involving the redistribution of agency, shifts in the role assignments of subjects and objects, and reconfigurations of tasks and practices between human and nonhuman actors *intraversions*.

Fundamentally, this book concerns HC-based CS projects such as Stall Catchers as sociotechnical assemblages and the human–technology relations unfolding within them and simultaneously forming them. The assemblage concept was originally proposed by philosopher Gilles Deleuze and psychoanalyst Félix Guattari (2013), and subsequently further defined and developed by different scholars with various foci and without always strictly following Deleuze and Guattari's thinking (e.g., Law 2004; DeLanda 2006; 2016; Ong and Collier 2005; Brenner, Madden, and Wachsmuth 2011; Buchanan 2015). While I discuss the concept in detail in Chapter 2, assemblages can be broadly defined as temporally consistent and volatile compositions of heterogeneous elements, such as human and nonhuman actors, and their relations, temporarily coming together and forming certain configurations from which assemblages emerge that go beyond the sum of the individual elements (Welz 2021a, 161). I analyze how HC systems in the field of CS are formed in the interplay of different human—such as developers, scientists, and participants—and nonhuman (or more-than-human)[3] actors to determine how they are 1) imagined and developed as new forms of *hybrid intelligence* (HI) and 2), at the same time, negotiated in everyday life and ethical practice in the entanglements of play and science. 3) I investigate the role of trust in the continuous formation processes. Building upon this, I focus on how human–technology relations unfold within this complex interplay and these negotiations, as well as how they continuously transform through future thinking, everyday adaptations, and failures. Thus, I follow an analytical approach that focuses on the becoming (Hultin 2019) of human–technology relations and HC-based CS assem-

---

[3] With the aims of moving past the divide between nature and culture, decentering the human, and "get[ting] non-humans to speak as more than spokespersons for human interests" (Latimer and Miele 2013, 7), ecological, anthropological, science and technology studies (STS), feminist, and other programs turned to naturecultures and more-than-human approaches (cf., e.g., Gesing et al. 2019; Welz 2021b). The philosopher, anthropologist, and sociologist Bruno Latour has aptly shown in his book *We Have Never Been Modern*, aiming to reconnect nature and culture, that the divide is a modern invention that cannot hold (1993). I discuss actor–network theory's (ANT) role in these discussions and its symmetric ontology in Chapter 2. Taking it a step further, the biologist, philosopher of science, and feminist theorist Donna Haraway, in *The Companion Species Manifesto*, describes the aim of the manifesto's agenda as follows: "Cyborgs and companion species each bring together the human and non-human, the organic and technological, carbon and silicon, freedom and structure, history and myth, the rich and the poor, the state and the subject, diversity and depletion, modernity and postmodernity, and nature and culture in unexpected ways" (2003, 4). Today, these pathbreaking conceptual shifts have gained much support across research programs, as stated above. In this work, I discuss humans and nonhumans, or human and nonhuman actors, and further specify, whenever possible, who and what are acting and engaging with each other in which relation.

blages that, with the concept of intraversions (see below), considers both instantaneous situations and historical becoming.[4]

As the example of the Catchathon has already shown, different human actors—including the institute's designers and developers, the volunteer participants, and the biomedical researchers—come together and contribute to form the assemblage. They do so not only by engaging in relations with other humans, but also (indirectly or directly, depending on their role) with nonhuman actors. I consider nonhuman actors as including materialities, such as infrastructure and microscopes, along with other entities such as data, algorithms, user interfaces (UIs), AI bots, and mice. I refer to nonhuman actors stressing that they are neither neutral nor passive objects only acted upon. Nevertheless, and following physicist and feminist theorist Karen Barad's understanding (1996, 181), I consider human and nonhuman agency asymmetrical. The different but interwoven human–technology relations resulting from the engagement of human and nonhuman actors continuously lead to the becoming of Stall Catchers, which, in turn, (re)configures the relations embedded within it. This becoming of the assemblage is simultaneously situated in a space which is both productive and features tensions because of the different affordances, expectations, and goals associated with play and science in CS games. Additionally, various processes impact the assemblages, bringing them closer together or tearing them apart. One example of such processes affecting HC-based CS assemblages, which I observed during my research, is trust. As my analysis shows, trust and trust-building mechanisms emerge and must be adapted alongside the intraverting relations in HC-based CS.

Thus, I employ the concept of *intraversions* to describe how human–technology relations in HC-based CS projects unfold in everyday life and develop continuously over time. As a concept, intraversions refer to the processual forward movements and shifts within relations between humans and technology. These movements and shifts result from the introduction of new computational capabilities and through the potential arising from existing relations directly forming based on human actors' practices or algorithmic and material affordances. Various forms of reconfigurations occur along the processual forward movements within human–technology relations. These include 1) shifts in the role assignments of subjects and objects—or, more precisely, in the distributed agency across the different actors—, which can never be fully attributed to one side or another; or 2) redistributions of tasks or practices. Intraversions take place along two dimensions: via instantaneous interactions—or intraactions (Barad 1996)—and via gradual temporal developments, thereby justifying why they must always be analyzed in and across time. Following this understanding, power dynamics are also not fixed in time, but change with these reconfigurations.

---

[4] Emergence-theoretic approaches, like ANT, which emphasize specific moments, have faced criticism for overlooking the historical and societal embeddedness of phenomena (Hinrichs, Röthl, and Seifert 2021, 93; Wietschorke 2021, 57). While other scholars have stressed that ANT does, in fact, include processualism as one of its theoretical dimensions (Belliger and Krieger 2006, 24), using the concept intraversions, I aim to provide an analytical and heuristic tool that overcomes the risk of neglecting these formative factors.

Intraversion, in contrast to inversion, does not merely mean that the exact opposite of what previously existed emerges (this may be possible, but is not a necessary or defining characteristic). Instead, intraversions describe cyclical modifications that build upon previous instances of relations and are, thus, always connected to the past while generating something new. If assigned subject/object positions, for instance, at some point flip back to a previous constellation, they are, nevertheless, not the same as they were previously. A simplified general example is as follows: An AI model was first used to analyze specific data and the human participants' task was to review the result, followed by a new task distribution where the human's task was to analyze the data while the result was reviewed by a computational model (because the previous AI model did not perform sufficiently well on the analysis), in order to ultimately retrain the AI model to take over the analysis once again; the relation emerging from this constellation differs from its first iteration. This is because the AI model and human participant are no longer the same in this new relation (e.g., the AI model *learns* from the human participants and the participants change their practices in working with an *improved* AI model). In this example, tasks and subjectivities are not merely swapped, but transformed or generated anew.

The concept of intraversions is conceptualized by building upon and drawing from a combination of existing theoretical approaches, which I outline in detail in Chapter 2, to derive an analytical and heuristic tool that contributes a specific focus on changing human–technology relations in HC systems. I employ Barad's understanding of intraactions which, instead of referring to interactions between fixed and independent entities, moves beyond such dichotomies (1996, 179). According to this understanding, humans and technologies can never be considered independent of each other, but instead are co-constituted in relations. Accordingly, responsibilities and power are understood in the Foucauldian sense as things that do not belong to one party or actor, but are distributed and move across these relations (Foucault 1998).

The notion of intraversions emerged during the analysis of my empirical material of HC-based CS when I attempted to understand how human–technology relations in the projects I studied evolve and unfold in everyday life. The fact that they continuously change became clear quite early: HC-based CS projects are created to solve a specific (scientific) problem that cannot be solved with current AI capabilities only, relying instead on new interplays between humans and computational or AI entities to do so. Thus, these projects necessarily must evolve alongside AI developments to remain at the edge of AI and scientific problem solving. They, and specifically their human–technology relations, must stay (and are intentionally pushed) open for future tweaking and changes. The relations' intraversions are, therefore, imagined by different actors, particularly by designers and developers. However, intraversions are also material, situational, and contingent, and guided by the encounter of different actors and a continual attempt to structure these human–technology relations differently. From this emerges resistance, counteractions, and failings, along with variously ascribed meanings;[5] intraverting relations, then, are multiples (Mol 2002b).

---

[5] I am interested in the everyday meanings actors ascribe to the HC-based CS systems studied when referring to "meaning" (Beck 1997, 14).

In this work, I analyze intraversions of human–technology relations in HC-based CS projects designed as games or so-called "games with a purpose" (GWAPs) (Von Ahn 2005). The intraversions observed are, thus, specific to the context of HC and the play/science "interferences" (Dippel and Fizek 2017a; 2019) in which they emerge. Returning to the AI bot example, the possibility of including AI bots in the Stall Catchers game only emerged from previous participant–software relations, which, at some point, enabled the training of AI models on data generated by participants. Subsequently, when wrapped in a bot participant on Stall Catchers (identifiable for participants by its username and bot icon on the leaderboard), AI models became new subjects in the game. Instead of human participants' performances being merely evaluated by computational tools in the background, software is now perceived as another fellow participant or, as I show in Chapter 6, as a competitor.

In addition to examining the intricate entanglements of play and science, analyzing human–technology relations in HC-based CS assemblages necessitates a comprehensive understanding of the broader field of HC and its overarching goals and visions. I, therefore, now turn to a brief overview of the emergence and historical development of HC. This overview is intended to provide a foundational understanding of the field before I shift to its current state of the art and how it relates to other areas of AI research.

According to computer scientists Alexander Quinn and Benjamin Bederson (2011), HC can be situated in the history of AI research beginning from the 1950s. They refer to mathematician, logician, computer scientist, and cryptanalyst Alan Turing's article "Computing Machinery and Intelligence" (1950) and computer scientist and psychologist Joseph Carl Robnett Licklider's "Man–Computer Symbiosis" (1960) as examples of the interconnectedness between HC and machine computation (Quinn and Bederson 2011, 1403). However, scholars only began thoroughly investigating the idea in the early twenty-first century (Quinn and Bederson 2011, 1403). One of the first HC systems was the "completely automated public Turing test to tell computers and humans apart" (CAPTCHA), developed in 2000 by computer scientist Luis Von Ahn and colleagues at Carnegie Mellon University in Pittsburgh (Von Ahn 2005). CAPTCHA was created as a security mechanism on the Internet to block programs and bots from accessing platforms and Internet services by asking users to recognize and correctly type a distorted word. This task proved simple for humans, while it was unsolvable for computer programs. Note that humans, here, are understood as seeing and without visual impairment.

CAPTCHAs were also considered free cognitive workers (Aytes 2012, 79). In 2010, about 200 million such CAPTCHAs were completed daily worldwide, which, at ten seconds per CAPTCHA, equated to about 500,000 hours of human work per day (Von Ahn 2010). Von Ahn explained in a presentation at the US National Science Foundation that this figure prompted him to think about how CAPTCHA could be used for something "good;" after all, it is not only valuable human time, but also a valuable activity computers could not simply take over (Von Ahn 2010). As a result, scientific research at Carnegie Mellon University gave rise to the reCAPTCHA project, subsequently purchased by Google (Google, n.d.), which now used these access restrictions to digitize books and train AI using data. By presenting people with images or words, one part of which is machine "known" or readable with the other part remaining unknown, reCAPTCHA checks whether it is a human completing the reCAPTCHA. At the same time, it "learns" a new word. The same task is presented to several people and their answers are combined to ensure that the word or image is correctly annotated or recognized. The paradox of reCAPTCHA is that a computer program initially creates a task for humans that it cannot solve itself, but simultaneously checks whether humans solve the task correctly.

However, these relations between humans and software are not fixed, because even reCAPTCHAs are constantly changing with the emergence of new machine solutions and user tactics aimed at automatically bypassing such tests.[6] I refer to such changes as *intraversions*. While Internet users contributed to reCAPTCHA because they had no choice, Von Ahn soon developed the first computer games building upon the HC approach in his doctoral thesis in 2005, introducing GWAPs (2005). In his dissertation, Von Ahn likely first defined[7] the term *human computation*[8] in its current understanding in HC research as "a paradigm for utilizing human processing power to solve problems that computers cannot yet solve" (2005, 3). In a later book Von Ahn published together with Edith Law, they further specify HC as "a new and evolving research area that centers around harnessing human intelligence to solve computational problems [...] that are beyond the scope of existing Artificial Intelligence (AI) algorithms" (Law and Von Ahn 2011, xv). Compared to Von Ahn's first definition, they now specifically situate HC within AI research, using a definition that can be understood as minimal, and upon which most HC proponents appear to agree. At the core of HC research lies the combination of humans—or to be more precise, human *intelligence* (see above) or *cognition* (Michelucci et al. 2015, 2)—and computational systems.

Aiming to define and distinguish HC from other concepts that build upon a "wisdom-of-the-crowd" approach like Wikipedia, Law and Von Ahn revisit computer science understandings of computation and algorithm (Law 2011; Law and Von Ahn 2011). Building on the perception of computation as "the process of mapping of some input representation to some output representation using a explicit, finite set of instructions (i.e., an

---

6 See, for example, the list of practices to bypass CAPTCHAs in the digital publication "HackTricks" by Carlos Polop (n.d.). On a website regarding CAPTCHA by Carnegie Mellon University, CAPTCHAs are described as a win–win situation since, even if they were broken by malicious users, this would have the advantage of solving an AI problem: "CAPTCHA tests are based on open problems in artificial intelligence (AI): decoding images of distorted text, for instance, is well beyond the capabilities of modern computers. Therefore, CAPTCHAs also offer well-defined challenges for the AI community, and induce security researchers, as well as otherwise malicious programmers, to work on advancing the field of AI. CAPTCHAs are, thus, a win–win situation: either a CAPTCHA is not broken and there is a way to differentiate humans from computers, or the CAPTCHA is broken and an AI problem is solved" (Carnegie Mellon University, n.d.). From the perspective of HC, subversive or malicious practices were, thus, considered an AI advancement rather than a problem. However, this understanding might not be shared by service providers on the Internet who rely on CAPTCHA as a security measure.

7 Computer scientist Edith Law refers to 2006 as the year in which HC was first coined (2011).

8 The term "human computation" forms an interesting return to the term "computer" as used beginning in the 1600s, where "computers" referred to humans performing calculations (Grier 2013). In the twentieth century, "human computers" were mostly women, who "were the computational processors behind everything" (Gray and Suri 2019, 52), including, for example, supporting the US in World War II and in space exploration (Light 1999; Holt 2016).

algorithm)" (Law 2011, 2, emphasis i.o.), Law and Von Ahn specify HC as "intelligent systems that explicitly organize[ ] human efforts to carry out the process of computation – whether it be performing the basic operations, or taking charge of the control process itself (e.g., specifying what operations need to be performed and in what order)" (Law 2011, 2). In addition, they build upon "explicit control" as an important element of HC systems. Explicit control here refers to the notion that the computation is a direct result of a predetermined algorithm, controlled by either humans or computers (Law 2011, 3). Applying this definition, Law and Von Ahn frame HC based on the ideas of, on the one hand, people being "engaged to perform meaningful tasks through some other activities that they are already deeply interested in (e.g., playing games, signing up for email accounts)" (Law 2011, 1), such as in GWAPs or in reCAPTCHA. However, I would argue, people are probably engaged in the latter example because they cannot avoid it. On the other hand, computations are always controlled to accurately and efficiently solve a problem addressed (Law 2011, 1). While an understanding of HC systems as "purposeful" is shared as a key concept to HC by various advocates, including Michelucci (e.g., 2013d, 84), the purpose does not necessarily refer to an individual's enjoyment, but can instead refer to results "that derive from collective behavior or interactions, such as the advancement of science that results from citizen science projects" (Michelucci 2013d, 84).

There have also been numerous attempts to differentiate HC from other terms like crowdsourcing, human-based computation, organismic or social computing, and collective intelligence (for different taxonomies, see, for example, Quinn and Bederson 2011; Michelucci 2013d; Newman 2014). At times, these terms are used synonymously, while at other times controversy arises, illustrating the fuzzy concept of HC and the numerous attempts to delineate the boundaries of this emerging field.

While the first years of HC were characterized by individual researchers, the first Human Computation Workshop (HCOMP 2009) took place in Paris, France, in 2009, bringing together "a wide variety of perspectives" (Ipeirotis, Chandrasekar, and Bennett 2009) from different disciplines (Quinn and Bederson 2011, 1403). Less than ten years after the first mention of HC in Von Ahn's doctoral thesis, researchers from the fields of AI, art, genetic algorithms, cryptography, and human–computer interaction—each field itself describing interdisciplinary fields—contributed to HC research (Quinn and Bederson 2011, 1403). The Association for the Advancement of Artificial Intelligence (AAAI) Conference on Human Computation and Crowdsourcing (HCOMP) was first organized in 2013 in Palm Springs, California, as a new space to bring together these different disciplines and researchers in a recurring format, which has since taken place annually.[9] The first conference covered topics and research ranging from human–computer interaction to cognitive psychology, economics, and various fields of AI. While the overlap with AI research is rather broad, according to the co-chairs, HCOMP extends beyond AI research. "[H]uman computation promises to play an important role in research on principles of artificial intelligence as well as in the engineering of systems that can take

---

9   However, critical scholars, science and technology studies researchers, and cultural anthropologists, for example, seem to be missing or, at least, do not yet appear represented within this interdisciplinary conference.

advantage of the (changing) complementarities of human and machine intellect" (Hartman and Horvitz 2013, xi). Pointing to the interdisciplinary nature of the new research area of HC, the co-chairs aimed to highlight the context, field, and attention HC gained over the years. Despite the domination of hard science perspectives, varied disciplinary perspectives and approaches to knowledge production come together in the discourse on HC. Conceivably, the diversity reflected in the various definitions of HC is, to some extent, explained by these varied traditions and epistemologies.

Following the publication of Von Ahn and Law's book *Human Computation* in 2011, the field witnessed another significant contribution with the publication of *Handbook of Human Computation*. Edited by Michelucci and published in 2013, this handbook presented a more extensive and broader approach to the field of HC (Michelucci 2013a). One specific aim of the handbook was to further broaden the interdisciplinarity of HC. This, for example, manifests itself in the preface written by cultural anthropologist Mary Catherine Bateson. The handbook includes chapters from both scientists and practitioners as well as visionaries in the field, offering the most comprehensive collection of different perspectives on HC to date. Only one year later, Michelucci established the transdisciplinary *Human Computation Journal*, the first journal dedicated specifically to HC, which further contributes to the HC discourse by continuing to publish research in the field, thereby accompanying and steering the developments in HC research (Michelucci and Gadiraju, n.d.).

Beyond organizing conferences dedicated to HC and the publications mentioned, the field appears to have gained further attention in recent years with the term "hybrid intelligence." In the HI literature, combining the skills of humans and machines or AI aims not only "to collectively achieve superior results" (Dellermann, Calma, et al. 2019, 276), but to also ensure that both "continuously improve by learning from each other" (Dellermann, Calma, et al. 2019, 274). Not only have research institutes named after HI been founded over the years (e.g., Elmann 2022; The Hybrid Intelligence Centre, n.d.) and the first HI conferences organized (Humane AI Net; The Hybrid Intelligence Centre, n.d.), but startups and companies such as McKinsey also seem to claim the term for their approach to AI, understanding HI as the "future of artificial intelligence at McKinsey" (McKinsey 2022). While definitions for HC and HI may vary and their research agendas sometimes focus on different aspects, considerable overlap remains in the understanding of HC and HI. In addition, researchers in these fields often employ similar approaches. Consequently, given these commonalities, I often discuss HC and HI together within my research (and "HC" can mostly be read as "HC and HI"); however, in general, they should not be considered as completely synonymous.[10]

If we now look at existing HC systems, in a broad sense they appear in various fields of everyday life, such as within access control systems for web services in the case of reCaptcha or in crowdworking platforms such as Amazon Mechanical Turk (Amazon Mechanical Turk, Inc., n.d.) and Clickworker (Clickworker GmbH, n.d.), where humans are monetarily compensated for completing so-called microtasks (Gray and Suri 2019). Microtasks can be understood as a "contemporary instantiation of piecework" (Alkhatib,

---

10   A more detailed definition of HI and how its development as a research field relates to HC is provided in Chapter 4.

Bernstein, and Levi 2017, 4609). HC systems can also be found in digital CS projects, such as Stall Catchers in which participants voluntarily contribute and where the form of involvement primarily relies on initialization from researchers, HC designers, and developers.

In general, the development of HC systems closely aligns with that of AI. In more recent years, AI has witnessed significant achievements, providing numerous examples of machines outperforming humans given their incomparable speed, accuracy, and tremendous memory. A recent example from natural language processing (NLP) lies in OpenAI's ChatGPT (OpenAI, n.d.) (along with the subsequent surge of models it inspired, including Anthropic's Claude [Anthropic PBC, n.d.] and Meta's LLaMa [Meta 2023.; *cf.* Touvron et al. 2023]). ChatGPT is a conversational model that builds upon DL and reinforcement learning from human feedback to generate outputs often indistinguishable from human responses, usually capable of performing complex, advanced tasks such as writing code, creating poetry or solving reasoning problems. Other domains include OpenAI's DALL-E (OpenAI, n.d.; *cf.* Ramesh et al. 2021; 2022) and CompVis Ludwig Maximilian University (LMU)'s Stable Diffusion developed together with Stability.AI (Rombach et al. 2022). Both of these represent DL models that generate digital images from natural language prompts. In addition, DeepMind's AlphaFold (EMBL-EBI, n.d.; *cf.* Jumper et al. 2021), is a high-performance AI system that can predict three-dimensional (3D) protein structures from amino acid sequences.

Despite these advancements, however, AI-based computer algorithms and models often still face fundamental limitations in tasks such as mathematical reasoning and some comparatively basic problems easily solved by humans like planning and creative thinking (Bry, Schefels, and Wieser 2018; Bubeck et al. 2023). Furthermore, the limitations of current AI systems repeatedly become apparent when the promises of new approaches remain unfulfilled or AI systems demonstrate their ability to cause real societal harm, such as in the field of law enforcement with the example of predictive policing (e.g., Brayne 2017; Ferguson 2017; McDaniel and Pease 2021). In addition, even where AI systems appear to solve complex problems with a high accuracy (and, in fact, may do so), the problem remains that most AI systems, especially the most *successful* ones often based on DL, are effectively black boxes whose outputs cannot be easily explained and whose accuracy is difficult to verify. Today's large language models (LLMs) are specifically known to perform well, producing *confident* responses, while unreliably discerning true facts from plausible fiction in doing so. Research directions in the field of AI generally aim to develop a strictly computational AI, often in pursuit of *strong AI* and *artificial general intelligence* (AGI). Such developments, at some point, are expected to achieve *human-like intelligence*, or at least *weak AI*, which aims to develop systems with superhuman performance targeting specific, albeit limited tasks. By contrast, HC pursues the goal of combining the respective strengths of humans and machines to realize unprecedented capabilities (Michelucci and Simperl 2014, 1). Human Computation is guided by the idea that combining humans and machines can solve complex problems for which no solutions currently exist with either *merely* computational or *merely* manual human approaches.

In the Stall Catchers example, the problem to solve was the data analysis problem that biomedical researchers at the Schaffer–Nishimura Lab faced in their Alzheimer's disease research. Building on a biomedical understanding of Alzheimer's disease based on the understanding of neuropathological changes in the brain as causing disease—an approach medical anthropologist Margaret Lock refers to as "localization theory" (2013)—researchers study the reasons for the decreased blood flow in Alzheimer's disease using genetically engineered mice.[11] In their previous work, they found that blockages or stalls in capillaries, the smallest blood vessels, occurred in mice with Alzheimer's disease ten times more often than in mice without the disease, generally leading to a 30 percent reduction in the brain's blood flow (Egle [Seplute] 2018; cf. Bracko et al. 2019; Ali et al. 2021). This decreased blood flow, also associated with Alzheimer's disease in humans, could also lead to an accumulation of amyloid beta, which is likely partly responsible for characteristic Alzheimer's symptoms. To understand the reduced blood flow better and how it can be ameliorated, researchers take *in vivo* images from the brains of mice using highly advanced fluorescence microscopy techniques.[12] The images acquired must be thoroughly analyzed individually, which, given the amount of data generated, is a tedious and time-consuming task. Eventually, this process led to a backlog of data to be analyzed, substantially slowing down the research process.

The laboratory's attempt to automate the image analysis using ML algorithms was unsuccessful because no model at that time achieved the data quality and accuracy required. This problem provided the perfect opportunity to build an HC-based CS project for cognitive scientist, mathematical psychologist, and founder of the Human Computation Institute Michelucci, via which to explore and develop "novel methods leveraging the

---

11   I do not specifically focus on mice or build upon the interdisciplinary field of human–animal studies (DeMello 2021) in my research, even though such studies play a fundamental role in Alzheimer's disease research conducted at the biomedical laboratory and, thus, in the Stall Catchers project. In fact, Stall Catchers would not exist without mice. However, such a focus lies beyond the scope of this study due to my research interest, which focuses on HC, and specifically on HC-based CS games and their human–technology relations, which do not necessarily include animal research. Furthermore, to ensure the feasibility of my research, I needed to draw boundaries around the Stall Catchers assemblage studied. In this research, mice are, therefore, mostly present in the form of digital research data. Mouse models, nevertheless, played an important role during my fieldwork at the laboratory, since most research practices related to Stall Catchers involved mice, including caring for mice, performing surgeries on them, doing experiments with them, and, in the end, also euthanizing mice. Conducting participant observation on these practices was challenging for me and it became a topic of conversation with various members of the laboratory and the Human Computation Institute. Research with animals is controversially discussed in the public and scientific discourses, and my research partners were aware of that. They cared for the animals, at times beyond the requirements in various guidelines and ethical regulations.

12   Fluorescence microscopy is a technique in which expressed fluorescent proteins or administered small molecule fluorophores are excited with a specific wavelength; during recovery into their energy ground state, they emit a photon of a defined higher wavelength. Two-photon imaging was most commonly used for *in vivo* mouse studies (Denk, Strickler, and Webb 1990; Palikaras and Tavernarakis 2015). In two-photon imaging, two photons of lower energy—for example, near-infrared light—are used to achieve the excitation of fluorophores. The use of low-energy near-infrared light and a good tissue penetration allows for the imaging of a thick specimen and even living tissue. Researchers at the Schaffer–Nishimura Lab primarily used two-photon microscopy for Alzheimer's disease research, although the laboratory also had a microscope for three-photon microscopy.

complementary strengths of networked humans and machines" (Human Computation Institute, n.d.). Coincidentally, Michelucci was searching for a problem to solve using HC at that time. Researchers hoped that this new project could solve their data analysis problem by combining a crowd of volunteer participants performing analytical tasks alongside computer algorithms that tracked and evaluated each individual participant's contribution and calculated and finalized crowd answers by combining them. Stall Catchers is, thus, an example of how HC systems rely on humans and algorithms to jointly tackle problems neither can easily solve on their own. They do so by delegating specific computational steps or tasks to humans "in the loop." These computational tasks can range from classification tasks, as is the case in Stall Catchers, to taking over complex design tasks (e.g., Center for Game Science [University of Washington] et al., n.d.a). Human computation systems, therefore, can assume various configurations regarding how humans are invited to contribute.

Starting from current scientific and AI problems, HC remains at the edge of AI and, as such, must continuously adapt to new developments in AI, while simultaneously also influencing these developments (albeit indirectly). Human Computation, like HI, not only begins with current AI problems, but also actively distances itself from other AI research approaches: "Research in the field of Artificial Intelligence seeks to model and emulate human intelligence using a machine. Research in human computation leverages *actual* human intelligence to perform computationally-difficult tasks" (Crouser, Hescott, and Chang 2014, 48). Representatives of HC even view it as a *better* alternative for the development of a "superior intelligence" (Michelucci 2016, 5). Due to this ethical framing of HC in the field of HC-based CS, I consider the development of HC systems "ethical projects" (Ege and Moser 2021a), since they are "future-oriented undertakings" (Ege and Moser 2021a, 7) that strive for "better" human–AI systems. The concepts of "sociotechnical imaginaries" (Jasanoff and Kim 2015), and philosopher of ethics and technology Steven Dorrestijn's "subjectivation and technical mediation" (2012a), which he develops following philosopher Michel Foucault, provide further helpful theoretical approaches to analyze how HC-based CS systems are imagined and how human actors relate to, shape, and are shaped by them.

As "laboratories" for exploring new human–technology relations of the future, HC-based CS forms a particularly fruitful research field for cultural anthropological analysis. In particular, designers and developers of these sociotechnical systems do not pursue the goal of developing a system that at some point is complete, but instead focus on tackling specific scientific problems and move on to new challenges once a solution is found. In this sense, human–technology relations, as well as the sociotechnical systems themselves and their purposes remain open and continuously changing. Projects range from astronomy and biochemistry to flood prediction and art history.[13] In general, CS is commonly understood as "the active engagement of the general public in scientific research tasks" (Vohland et al. 2021, 1). The term itself has become an umbrella (Wiggins and Wilbanks 2019, 5) for various kinds of public involvement in scientific projects,

---

13 See, for example, Stardust@home (Westphal et al. 2005; Stardust@home, n.d.), Foldit (Center for Game Science [University of Washington] et al., n.d.a), UpRiver (Suarez 2015), and ARTigo (Ludwig-Maximilians-Universität n.d.).

allowing "citizen" scientists without professional training to collaborate with academic researchers in various ways and at all stages of the scientific process, primarily contributing to data collection and data analysis.[14] In my research, I focus on CS initiated by professional scientists themselves, and more specifically on online and digital CS, to which participants can contribute using their own computers or mobile devices. Most projects in the field of HC-based CS are designed as GWAPs, in which "players perform a useful computation as a side effect of enjoyable game play" (Von Ahn and Dabbish 2008, 61). This "useful computation" often both directly serves to advance the scientific research behind the game and the development of AI models to solve the underlying problems, while simultaneously allowing participants to contribute to scientific research and enjoy the games. Following the term "playbour," first coined by game researcher Julian Kücklich (2005), such platforms serve not only as laboratories for new human–technology relations, but also as playgrounds, in which humans and algorithms fuse into "playbouring cyborgs" (Dippel and Fizek 2017b).

Following sociologist Pierre Bourdieu's understanding of social space and different "social fields" (1985), play, science, and work can be understood as social fields each following their own specific logics. In the example of HC-based CS games, the "interferences" (Dippel and Fizek 2017a; 2019) of the different fields form a productive space in which HC systems unfold. This space, however, is not without friction as different field logics merge and, at times, conflict with one another. Transferring the term from physics, historian and cultural anthropologist Anne Dippel and media and game scholar Sonia Fizek use "interferences" to describe "the overlay" of work and play in the digital sphere and specifically in "[c]itizen science games as new modes of work/play in the digital age" (2019, 263). While Dippel and Fizek talk about "work/play" interferences in CS games, I focus on "science/play" interferences, which, of course, are a form of "work/play" interferences, because science is often understood as the counterpart (and sometimes even opposite) to play in my examples.

The science/play entanglements allow volunteer participants to engage in scientific projects in an enjoyable way, while (partly unresolvable) tensions and frictions also pervade the sociotechnical assemblages. By moving between these fields of everyday life, HC-based CS often uncovers unquestioned ascriptions of the entangled fields. Moreover,

---

14   The terms "citizen science" and "citizen scientists" remain controversial due to the meanings and exclusions they carry. For example, the term "citizen," according to the *Cambridge Academic Content Dictionary*, refers to "a person who was born in a particular country and has certain rights or has been given certain rights because of having lived there" (Cambridge University Press n.d.). Various alternatives have been proposed, such as *community science, participatory research* or *open science*, each including their own problems. For an overview of the different considerations, ongoing discussions, and alternatives, see, for example, Eitzel et al. (2017). Moreover, the mode of involvement of citizen scientists is much discussed in the CS literature. At its most simplified, while CS proponents claim that it makes science more democratic by opening up the production of knowledge to society, others understand CS as a "renewed approach to exploit citizens by making them work for free" (Vohland et al. 2021, 2), thereby criticizing CS projects for both reproducing hierarchies and exploiting volunteers without involving them in the actual knowledge production step (Vepřek 2021b). While I consider these questions throughout my research, they do not form the focus of my research interest. In this work, I use CS as a term from the field itself, choosing the term "participants" for "citizen scientists" (*cf.* Chapter 3).

the sociotechnical systems must not only be scientifically sound and technically functioning, but also engaging and enjoyable for participants. Considering *how* play and science interfere is crucial to understanding HC-based CS as assemblages and their continuously changing human–technology relations, since these interferences create specific affordances and open up new action potentials, which, if activated, can lead to new intraversions.

As I learned during my fieldwork, the scientific data analysis problem challenging the work of Alzheimer's disease researchers presented not only an interesting problem for the Human Computation Institute in Ithaca, NY to solve employing HC, but, at the same time, introduced productive constraints into the otherwise infinite space of possibilities (fieldnote Oct. 20, 2022). These constraints both facilitated building an HC system and set the direction for how the Human Computation Institute would build HC-based CS systems introducing "path dependencies," thereby guiding the further evolution of such systems and rendering some developments more likely than others (Klausner et al. 2015; De Munck 2022).

The project could build upon existing HC-based CS platforms, which, similar to the scientific constraints, introduced further path dependencies. However, the development of Stall Catchers—and HC-based CS systems more broadly—requires enormous effort in exploring the possibilities of human–technology relations, creating infrastructures, and developing algorithms. This is because the scientific problems tackled lie at the very frontiers of both AI and science.

The development of HC-based CS is shaped by visions of future human–technology relations alongside normative assumptions regarding how humans and machines *should* work together and how projects *should* unfold. More precisely, the envisioned human–technology relations of the future refer to participant–technology relations and questions regarding how to include unpaid and untrained volunteer participants in algorithmic systems and scientific research. Building upon the understanding that HC systems "all serve a purpose" (Michelucci 2013b, xxxvii), the Human Computation Institute considers it unethical to ask humans to perform a task that a computer can solve. This implies a necessity to evolve and adapt as soon as computational solutions advance and move on to new problems requiring human input. However, as I demonstrate in this work, such programmed inscriptions and a project's purpose and meaning by design are not uncontested. Instead, they are frequently challenged in their everyday unfolding by various human actors, such as participants and scientists, and materialities such as technical possibilities and breakdowns. Together, these continuously reshape the sociotechnical assemblage of HC-based CS projects.

Some participants of Stall Catchers, for example, objected to the project being labelled a game because that label did not properly reflect their motivation to contribute nor how they perceived their engagement.[15] They did not simply contribute to *just any* scientific research project, but had a personal connection to and were in a direct or indirect way affected by Alzheimer's disease. Participating in Stall Catchers, as I argue in Chapter 5, can be interpreted as a form of coping with everyday life marked

---

15 For this reason, I use the term "participants" throughout this work—rather than, for example, "players"— to refer to those who voluntarily engage in these projects as part of a crowd.

by this disease.[16] Analyzing participants' perspectives, their motivations to contribute to HC-based CS, and the meanings they ascribe to their engagement is important to understand how HC-based CS projects—initially imagined and designed by researchers and developers—are (re)negotiated in practice. These negotiations contribute to the formation of the assemblage and influence how human–technology relations in these systems unfold and intravert.

Thus, by considering not only the visions and design of HC-based CS, but also the motivations of the actors involved, "serendipitous discover[ies]" (Schaffer in Human Computation Institute 2018, 00:33), timing (Mousavi Baygi, Introna, and Hultin 2021), breakdown (e.g., Larkin 2008),[17] and the unruliness of nonhuman actors, it becomes possible to understand how the sociotechnical assemblage evolves. Moreover, and returning to the introductory example, it helps to analyze how, for example, the introduction of AI bots in Stall Catchers changes the participant–technology relations informing the calculation of "cyborg answers," as Michelucci described them (Human Computation Institute 2021, 22:25–22:26).

Human computation, and specifically HC-based CS as a research field and subject, has thus far primarily been addressed by computer science, information science, and related fields with a focus on quantitative and standardized analysis.[18] These studies do not centrally focus on the social, cultural or semiotic dimensions, which always represent a part of and form sociotechnical systems. As Barad aptly states, "[P]henomena are the place where matter and meaning meet" (1996, 185); I consider this to include the sociocultural sphere. However, HC systems do not merely imagine possible future combinations of humans and technology in order to move beyond today's AI capabilities and human abilities. As laboratories for such new combinations, they already impact, create, and change our everyday lives today. Moreover, given the rapid developments in the field of HC, which increasingly inform and contribute to AI discourse, it is important to analyze and critically engage with these developments. Through my research, I aim to contribute to a digital anthropological and science and technology studies (STS) understanding of HC and its human–technology relations. I also attempt to inform the development of HC systems by including perspectives from the various actors involved and considering their roles in forming the intraverting human–AI relations.

HC-based CS has thus far not been extensively analyzed in cultural and digital anthropological or STS investigations. However, important studies exist in these and related fields on crowdworking, CS, the relationship between play and work or science in the digital age. Such studies also extend to digital and media anthropology, the anthropology of technology or STS on human–technology relations and AI. I build upon and discuss these in Chapter 2.

---

16 I first discussed this idea at the conference "Breaking the Rules: Power, Participation, Transgression" of the International Society for Ethnology and Folklore in 2021 (Vepřek 2021b).
17 Like anthropologist Brian Larkin, I am interested in forms of everyday breakdown: "the small, ubiquitous experience of breakdown as a condition of technological existence" (2008, 234).
18 One of the few exceptions is the anthropologist Mary Gray and computer scientist Siddarth Suri's work on "ghost work" or crowdworking (2019), which I discuss in the related work section in Chapter 2.

In my research, I conducted inductive ethnographic fieldwork, drawing on praxiography (e.g., Knecht 2012) and grounded theory (Glaser and Strauss [1967] 1971; Charmaz 2000; 2014), over the course of three years (Oct. 2019 – Nov. 2022) in the US and Germany. To gain an in-depth understanding of how HC systems in the field of CS are developed and maintained, I joined the Human Computation Institute as an "intern" and worked together or co-laborated (Niewöhner 2016) with Michelucci and the institute's team. In the remainder of this chapter, I briefly introduce this field site for my research as well as the HC-based CS games I studied, and, finally, provide an outline of this book.

The Human Computation Institute refers to itself as an "innovation center" (Human Computation Institute, n.d.), and was initially founded as a Limited Liability Company in 2014 before reincorporating as a nonprofit in 2017. The first and most popular project of the institute, Stall Catchers forms the main example in my research, providing a suitable focus not only because it is a highly successful HC-based CS vis-à-vis participant engagement and "analytical throughput" (Michelucci, fieldnote Nov. 2, 2022), but also given the access to the institute I obtained. Stall Catchers was launched in October 2016, and currently (May 2024) has over 71,000 registered participants.[19] While most participants contribute during special events or only occasionally, the average number of monthly participants is 313;[20] the core, persistent and committed on a daily basis, consists of around 21 participants.[21] Stall Catchers can be accessed via a web browser or a mobile app and has been designed as a "casual" game that one can engage with for only a few minutes or over several hours. Contributions from Stall Catchers participants have also been recognized by referencing them as coauthors on scientific publications related to the project (Bracko et al. 2019; Ali et al. 2021).

I also draw from the analysis of two other HC-based CS games—Foldit and AR-Tigo—to contextualize Stall Catchers and its different perspectives more accurately and to gain a better understanding of elements that generally apply to HC-based CS. I briefly introduce these two comparative examples in what follows.

Foldit (Center for Game Science [University of Washington] et al., n.d.a) is one of the most long-term and successful CS projects utilizing HC. This online puzzle video game focuses on protein folding, in which participants are challenged to fold the structures of proteins as efficiently as possible.[22] Promising protein structures developed by partici-

---

19 I noted in my field diary on September 18, 2020, that the current number of registered participants was 29,314. By mid-2023, there were over 52,000 registered participants, meaning the number of participants had increased by nearly 80 percent over the course of my research.
20 I obtained the number of monthly participants from the Stall Catchers database using a SQL (structured query language) query. The query was run on March 27, 2023. This number reflects the search period, from October 1, 2016 to March 27, 2023.
21 I obtained the average number of daily participants from the Stall Catchers database using an SQL query, run on March 27, 2023. This number reflects the search period, from January 1, 2019 to December 31, 2022.
22 The 3D structure of proteins defines how they interact with other molecules and their biological function. Therefore, knowing how sequences of amino acids, the building blocks of proteins, fold into such a 3D structure is important in medicine (such as in developing drugs), biotechnology, and other scientific fields. Protein structure prediction attempts to infer a protein's structure from its amino acid sequence, taking into account the various forces determining a structure, such as hydrogen bonds. However, due to the many possible amino acid arrangements in space, predic-

pants together with algorithmic tools and automated scripts (and, more recently, with the assistance of the AI program AlphaFold) are tested in a wet lab. On their website, the Foldit team emphasizes the enormous contribution of human participants to research on diseases such as COVID-19, influenza, and even cancer and Alzheimer's (Center for Game Science [University of Washington] et al., n.d.c). Similar to the example of Stall Catchers, participants have been included as coauthors in scientific publications (Khatib et al. 2011).

Foldit is a collaborative project, specifically between the Center for Game Science and the Institute for Protein Design at the University of Washington in Seattle (USA) and other, mainly US-based research institutions.[23] Since its launch in 2008, more than 460,000 participants have contributed to Foldit, although the active player base consists of only a small fraction of these participants (Curtis 2015, 729). Foldit, as a "multiplayer online scientific discovery game" (Khatib et al. 2011, 18949) serves as an interesting example of HC-based CS, because it appeals to users in their creativity and spatial reasoning skills, as well as their skills in the development of 3D patterns. Thus, Foldit presents a rather different problem compared to Stall Catchers. Players—even those without prior biochemical knowledge—become highly qualified in the task by playing it extensively (*cf.* Khatib et al. 2011; Ponti et al. 2018), rendering Foldit particularly interesting for participants eager to learn and develop new skills. Compared to Stall Catchers, Foldit features a steep learning curve, causing some participants to drop out of the game early.

The field of protein structure prediction has significantly developed in recent years thanks to AI models such as DeepMind's AlphaFold and RoseTTAFold developed by the Institute for Protein Design at the University of Washington. Such developments have led to the declaration that the scientific problem of protein structure prediction is effectively solved (Moult in Callaway 2020). Foldit offers an intriguing example of the relations between AI and HC endeavors and how they influence each other, starting from the problem of protein structure prediction as a way to bridge the gap between what computers and AI can do and what humans alone can achieve. While first focusing on the problem of protein structure prediction, over time Foldit has shifted its focus to protein structure design, for which no fully automated solution currently exists. Just as Foldit's purpose has shifted over the years, the human–software and human–AI relations in the project are continuously intraverting, as I show in this book.

The third HC-based CS project I analyze is ARTigo (Ludwig-Maximilians-Universität n.d.a), a game platform developed at LMU Munich, Germany, resulting from a collaboration between the computer science and art history institutes. This project, similar to Stall Catchers, is also primarily related to AI problems within computer vision, but with a different twist: the tagging of digital representations of artworks. As an interdisciplinary

---

ting that structure remains a challenging task. Protein design, in contrast, describes the process of building new proteins that have specific functions or characteristics rather than modifying existing proteins. Both computational and experimental methods can be used.

23 The Cooper Lab at Northeastern University, the Khatib Lab at the University of Massachusetts–Dartmouth, the Siegel Lab at the University of California–Davis, the Meiler Lab at Vanderbilt University, and the Horowitz Lab at the University of Denver (Center for Game Science [University of Washington] et al., n.d.a).

project which began in 2007, ARTigo affords an interesting comparative example since it is both a long-term example of HC-based CS and a project *avant la lettre* in terms of its purpose and developments in the field of computer vision and deep neural networks. In 2017, ARTigo included a large database of more than 65,000 images of artworks (Bogner et al. 2017, 53). The integration of this large database into various GWAPs aimed to achieve two goals: first, to generate keywords for individual artworks facilitating a semantic search engine, and, second, to engage "lay art historians" (Kohle 2018, 1) by providing them with new learning opportunities while also potentially influencing established approaches to art history. By 2017, more than nine million annotations had been collected (Bry and Schefels 2016; Bogner et al. 2017, 53). However, after thriving in its initial years, ARTigo navigated a lean period, during which it was temporarily inaccessible or only accessible through the LMU network. This rendered the collection of empirical material difficult. Consequently, my analysis of ARTigo here is less detailed than that of Stall Catchers and Foldit, primarily featuring a subsection of Chapter 6 only. Nonetheless, the project regained momentum in November 2022,[24] when relaunching on a new platform, featuring new games and addressing new image recognition issues. ARTigo's relaunch and the evolution of its human–technology relations accompanying the new computational possibilities and advancements can also be understood through the concept of intraversions.

This book spans eight chapters of which Chapters 4 through 7 present the core analysis of my research. Chapters 2 and 3 lay the groundwork for the empirical chapters. More specifically, in Chapter 2, I discuss the related literature from cultural and digital anthropology, STS, and related fields, and the theoretical perspectives I build upon in my analysis, drawing specifically from assemblage theory and thinking, relational conceptualizations of technology and human–technology relations, moral anthropology, and the ethics of technology. Moreover, building upon cognitive anthropologist Edwin Hutchins (1995b; 1995a) and Barad (1996), I develop and discuss the concept of intraversions. In Chapter 3, I present and discuss the methodological approach, based on co-laborative ethnographic fieldwork combining classic ethnographic methods such as participant observation, qualitative interviews, and media analysis with the more experimental analysis of code, the Stall Catchers in-game chat, and collaboration with the Human Computation Institute. I consider my own role in the development of HC-based CS, which I ultimately revisit in the conclusions to the book to discuss how cultural anthropology can contribute to the development of (hybrid)[25] sociotechnical AI systems.

The first empirical chapter (Chapter 4) then focuses on the visions, imaginaries, and designs of advocates, designers, and developers of such HC-based CS systems in the context of HC research. Here, I explore how they are imagined as counternarratives to strong AI, while sharing much in common, and what visions of human–technology relations in the future underlie the design of these ethical projects. The fact that they are designed

---

24   Information was kindly provided by the ARTigo team via an email exchange (Mar. 26, 2023).
25   In the following, "hybrid" can refer to both a term used in the fields of HC and HI, or one used by STS and feminist researchers and in the philosophy of technology. Wherever I refer to it or use the term, it should be clear from the context to which field or scientific tradition the term refers.

as open systems that can be understood as in-betweens on the way to developing HI is crucial to the possibilities of human–technology relations intraverting. Since imaginaries and visions must materialize to drive the development of HC, the second part of this chapter analyzes examples of infrastructuring (Bossen and Markussen 2010; Niewöhner 2015) performed at the Human Computation Institute.

This perspective, which focuses on design and initial implementation, is not only informed by directions and visions, but just as much by everyday negotiations. I turn to these in the next chapters. Chapter 5 discusses how these inscriptions of visions, values, and norms are continuously contested and negotiated in everyday life through various motivations, interests, and aims. Such inscriptions drive, for example, CS participants and the software's affordances (Gibson 1977; 1979; Bareither 2020a), materialities, and action potentials emerging from human–technology relations, all of which are situated in the entanglements of play and science. I focus here on discussing the multiple meanings of the case studies. To do so, I employ the example of how participants challenge the designer's image of them, adapt systems in their own ways, and ascribe different meanings to them than those intended by design. My analysis illustrates how these projects are included in the participants' everyday lives, which, for example, for some, are marked by a deadly disease. In the second part of Chapter 5, I turn to the interferences of play and science, which, on the one hand, productively create a seamless space, while, on the other hand, create tensions and frictions, together forming the assemblage. Within these entanglements, human–technology relations in HC-based CS unfold and continuously change.

While Chapters 4 and 5 provide a fundamental understanding of the formation of HC-based CS assemblages, in Chapter 6 I further build upon this analysis, investigating examples of human–technology relations in the HC-based CS games in detail and how they continuously intravert over time. Using the concept of intraversions, I demonstrate how tasks are redistributed, how practices change, and how role allocations shift alongside intraversions. First, the focus lies on participant–technology relations in Stall Catchers and Foldit and how they evolve in daily life and over time through the meshing of HC visions and everyday negotiations, the multiplicities of meaning of HC systems, and within the space created by the play/science entanglements. HC-based CS projects not only consist of sociomaterial participant–technology relations, despite the HC literature commonly only referencing participants when it talks about "humans in the loop." With the aim of developing a better understanding of the different actors and relations forming HC systems in CS as sociotechnical assemblages, I move to the human–technology relations developers and researchers of HC-based CS projects enter into, using the example of researcher–technology relations in Stall Catchers. Just as participant–technology relations intravert over time, researcher–technology relations are never fixed, but evolve alongside the introduction of new tools or automated steps in the data infrastructure connecting the biomedical engineering laboratory to the CS gaming platform. This infrastructure is never complete, requiring continuous work and improvements. My analysis of human–technology relations employing the concept of intraversions brings forth a pattern, which, by turning away from the microanalytical perspective of HC systems in their everyday situatedness, also allows for a better understanding of the nontrivial and dynamic relations between HC and AI research. Finally, at the end of this chapter,

I demonstrate how the concept of intraversions can be applied in a fruitful way to describe the evolution of HC-based CS in relation to AI advancements using the example of ARTigo.

In the continuous formation of the HC-based assemblages, intraversions destabilize established practices, requiring different processes of alignment. Chapter 7 then focuses on trust as an example of such alignment processes and how it is built in and with HC. Trust, as I understand it in this chapter, is not a mere cognitive phenomenon, but unfolds in sociomaterial practices and within human–technology relations. It plays a constitutive role in assemblages and needs to be adjusted and reestablished with intraverting relations. In this chapter, I, thus, analyze trust in HC-based CS collaborations, how it is programmed algorithmically, and the role it plays from the perspective of participants.

In this book's conclusions, I bring together the various perspectives and scales of analysis, arguing that cultural anthropological and STS research can help us understand HC systems and the hybrid modes of becoming of their human–technology relations. Inspired by STS anthropologist Lucy Suchman (2007b; 2021) and cultural and feminist anthropologist and early STS researcher Diana Forsythe's (2001f) fieldwork, my study also illustrates the value of ethnographic research in HC development. Here, I emphasize the need for HC development to consider all actors and human–technology relations involved. It should integrate various aspects of everyday life, as well as the relations' historical evolution, path dependencies, and existing relations remaining from the past, and understand and make explicit its underlying future imaginations and visions.

2 Approaching Human Computation-Based Citizen Science Analytically

To the best of my knowledge, HC-based CS has not yet been thoroughly analyzed from a cultural or digital anthropological perspective. A few prior investigations were undertaken, such as in the field of crowdworking (Gray and Suri 2019), the interplay of AI technologies and humans in CS (e.g., Ponti et al. 2021; Ponti and Seredko 2022), and the example of Foldit (Curtis 2015; Ponti et al. 2018). Nevertheless, my research builds on existing literature on related fields, such as crowdworking and CS in general, the relationships between play, work, and science in the digital age, and analyses of infrastructures, sociotechnical systems, and human–technology relations (with a focus on digital technologies), especially in the broader field of AI.[1] I mostly draw on research in cultural and digital anthropology, media anthropology, anthropology of technology, and STS. In the first part of this chapter, I give an overview of related work without claiming, or aiming for, completeness. Instead, I present and discuss selected work that is of particular interest to my research endeavor and helps to situate my research in both related fields and existing studies. For my analysis, I refer to and draw on theoretical concepts from the scientific fields mentioned above, in addition to philosophy of ethics and technology and moral anthropological theory. The theoretical concepts that form the overarching foundation upon which this work is built are assemblage theory, human–technology relations, sociotechnical imaginaries, ethical projects, and the care for our hybrid modes of being. In the second part of this chapter, I discuss these concepts in detail, followed by the introduction of the concept of intraversions. I apply this concept to analyze how human–technology relations unfold and continuously develop in HC-based CS.

---

[1] The related work discussed here refers to research on topics and fields that are of overarching importance for my own research. Where necessary throughout this work, I discuss additional literature on subtopics, such as research on trust in Chapter 7.

Crowdsourcing and Crowdworking

According to anthropologist Mary Gray and computer scientist Siddarth Suri, who describe HC as "people working in concert with AIs" (2019, x), "[t]his fusion of code and human smarts is growing fast" (2019, x). Gray and Suri refer here to the context of crowdworking. While I focus on CS as another application of HC in my research, HC generally intersects with phenomena such as crowdworking and crowdsourcing.[2] Most cultural anthropological and related investigations that directly or indirectly address HC are concerned specifically with the latter. I would like to briefly discuss crowdsourcing and crowdworking and selected work in these fields due to the overlap of practices observed in HC-based CS and the concurrency of the phenomena in neoliberal economies and "ludic capitalism" (Dippel 2018, 125).[3]

Tasks in crowdsourcing are outsourced from "traditional human workers" to "members of the public" and do not always rely on humans taking over computational tasks (Quinn and Bederson 2011, 1405). In contrast to crowdsourcing, Quinn and Bederson argue that "human computation replaces computers with humans" (2011, 1405). While HC-based crowdsourcing applications such as reCAPTCHA can be considered a form of "unpaid labor practices" (Scholz 2013, 2), in crowdworking, understood as a form of crowdsourcing, people are paid for their contribution to HC systems. Media and communication scholar Ayhan Aytes argues in his analysis of the crowdworking platform Amazon Mechanical Turk that "[c]rowdsourcing is a hybrid concept that merges the neoliberal outsourcing paradigm with the crowds on the digital networks" (2012, 88). The development of crowdworking is closely linked to the rise of digital platforms (Srnicek 2017). As a "neoliberal reincarnation of the chess-playing automaton" (Aytes 2012, 81) of Wolfgang von Kempelen's eighteenth century "Mechanical Turk," the crowdworking platform's algorithms produce and discipline the workers "into a particular cognitive mode and problem solving that eventually determines the efficiency of their labor and thus their livelihood" (Aytes 2012, 94). In most crowdworking applications involving HC (such as Amazon Mechanical Turk), people perform so-called "microtasks"—including, for example, the categorization of products—that are largely standardized and automated (Felstiner 2011, 150). The repetitive and often only poorly remunerated tasks resemble Taylorist forms of work, which is why cultural studies and political science scholar Moritz Altenried describes crowdworking as "digital Taylorism" in his work on digital factories and their human labor (2020; 2022).

Sociologists Frank Kleemann, Günter G. Voß, and Kerstin Rieder analyzed crowdsourcing and the phenomenon of the "working consumer" (2008) from a theoretical perspective, and sociologist Elisabeth Vogl (2018) has studied organizational models of work

---

2   On the term crowdsourcing, see Howe (2006). For a broader understanding, see Felstiner (2011, 145) and Vogl (2018, 8). The term HC is sometimes even synonymous with crowdsourcing or crowdworking. The only annual conference specifically focusing on HC is the *AAAI Conference on Human Computation and Crowdsourcing* (HCOMP), and combines HC and crowdsourcing (Association for the Advancement of Artificial Intelligence n.d.).

3   I translated the direct quotes from non-English literature and sources.

on three different crowdsourcing platforms, the latter's impact on society, and the reorganization of work. Political scientist Doris Allhutter (2019) also discusses the important role of human microwork for building the foundations of a semantic infrastructure and related power structures based on ethnographic research on the infrastructuring practices for creating commonsense ontologies in the field of semantic computing. Crowdworking can also go beyond so-called microtasks and include a wide variety of practices, such as design or programming. Anna Oechslen (2020), who argues for a differentiation of crowdworking forms, analyzes crowdworking practices of (graphic) designers from a cultural anthropological perspective. In contrast to microtasks, the designers' practices include designing logos, for example, which can be described as "macro tasks."[4] Platforms often include elements of gamification (Detering et al. 2011; Rackwitz 2015), i.e., game design elements that are introduced into nongame contexts (Detering et al. 2011), to make crowdworking more attractive. One of the first and a remarkable contribution to the analysis of crowdworking was provided by Gray and Suri in *Ghost Work: How to Stop Silicon Valley from Building a New Global Underclass* (2019), in which they investigate the workers behind crowdworking platforms, what drives them to this kind of work, how they creatively try to earn a living with, and, ultimately, what this kind of work means to them (2019, xxvii). At the same time, the study analyzes the business models of such platforms and how ghost work is organized (Gray and Suri 2019, xxvii). Gray and Suri apply the term "ghost work" to show how the work conditions and work performed by humans to drive and enable AI are made invisible by the platforms. Their APIs reduce ghost workers to "a string of letters and numbers instead of a name and a face" (Gray and Suri 2019, 34).

Analyzing how humans are included in such sociotechnical and HC-based work or game platforms, such as the example of CS games, will probably only gain importance in the coming years. A total of 25 million "ghost work" opportunities existed by 2019 and Gray and Suri anticipate this number to increase further (2019, 169). The COVID-19 pandemic probably acted as an additional catalyst for such digital working practices. The field of HC-based CS, and especially projects that focused on research on the coronavirus, saw an increase in participation over the course of the pandemic (Vepřek 2020). Gray and Suri argue that the question to be addressed concerning the increase of crowdworking or ghost work opportunities should be: "If we imagine AI and humans augmenting and supplementing each other, the next issue is not whether humans are necessary. The real question will be: When are they in demand, and for what purpose?" (2019, 192). I aim to contribute with this research to addressing this question regarding HC systems in the field of CS. Unlike crowdworking, however, where human labor is usually exploited for financial compensation, participants in CS are included in scientific knowledge production.

---

[4] On related work on crowdworking in general, see, for example, contributions in Altenried, Dück, and Wallis (2021) and on macro tasks specifically, see Oechslen (2020).

Citizen Science (Games) and the Entanglements of Play, Work, and Science

While crowdworkers are financially compensated for their contribution to AI, HC-based CS systems rely on the voluntary contribution of participants, raising issues such as acknowledgments and forms of engagement, and complex negotiations involving not only designers and developers of such systems but also participants and scientists. Computer scientists and ARTigo creators François Bry and colleagues understand CS as a subfield of HC (Bry, Schefels, and Wieser 2018, 1). The goal of the development of HC-based CS is often finding a solution to a specific scientific problem which cannot be solved with today's computational technologies. While these problems are not always defined as AI problems by the designers, they, nevertheless, are at the edge of AI research and contribute to the advancement of AI in general. Because HC-based CS projects rely on voluntary engagement, they need to be carefully designed and be engaging, entertaining, and rewarding to attract and retain participants.

However, CS generally goes beyond HC, and, according to philosopher of science and cultural ecologist Peter Finke, describes science beyond science (2014, 14), questioning the understanding of science as an expert phenomenon and creating new forms of scientific collaboration. Citizen science generally refers to actively involving members of the public in various scientific research activities (Vohland et al. 2021, 1) which can happen across all stages of the scientific process.[5] CS projects can, for example, take "bottom-up" or "top-down" approaches. While, in the first case, projects are launched by "citizen scientists" themselves or at least built upon the co-construction of research agendas (Hecker et al. 2018, 234), in the latter case, professional scientists invite participants to their research projects. In most cases, volunteer participants contribute to data collection and analysis (Land-Zandstra, Agnello, and Gültekin 2021, 244). The CS projects often arise from professional scientists needing more computing power or human assistance in data analysis steps or data collection (Vepřek 2022b, 31). The goal is to speed up the research progress by out- and crowdsourcing a time-consuming and laborious task that cannot be solved computationally, as in the case of Stall Catchers. In other examples, such as Foldit and ARTigo, CS is based on the assumption that new scientific knowledge can be obtained through novel approaches to scientific problems due the creativity of nonprofessionally trained scientists, their potential for "out of the box thinking," and "wisdom based on life experiences of great crowds" (Görsdorf 2007, 8).

There is a rich body of literature on CS, ranging from the field of ecology (e.g., Irwin 1995; or, as recent publications, Lepczyk, Boyle, and Vargo 2020; Fraisl et al. 2022) and astronomy (e.g., Westphal et al. 2005; Marshall, Lintott, and Fletcher 2015; Lintott 2019), to the history of science and STS research (e.g., Kimura and Kinchy 2016; Burri 2018; Strasser et al. 2018), analyzing, *inter alia*, the motivations of CS participants (e.g., Geoghegan et al. 2016; Land-Zandstra et al. 2016; Larson et al. 2020) and CS games (e.g., Curtis 2015;

---

5   The term CS, as we know it today, emerged at the end of the twentieth century, and has seen a huge upswing with the spread of the Internet. Nevertheless, it can be considered a phenomenon that has been practiced for several centuries (*cf.* Finke 2014; Hecker et al. 2018). The role of "lay persons" for and even before the constitution of the German-speaking cultural anthropology *Volkskunde* is discussed in Cantauw et al. (2017).

Tinati et al. 2016; 2017). I will discuss participants' motivations as identified in the literature and how these findings align with my empirical observations in Chapter 5. Citizen science has become a subject of cross-national associations, such as the European Citizen Science Association (ECSA) (Verein der Europäischen Bürgerwissenschaften – ECSA e.V, n.d.) or the Association for Advancing Participatory Sciences (Association for Advancing Participatory Sciences, n.d.). As much as it is of interest to scholarly research, there is debate and discussion around issues such as meaningful engagement of participants and power hierarchies and their reproduction (e.g., Cooper, Rasmussen, and Jones 2021),[6] the role of neoliberalization (Kimura and Kinchy 2016), ethical issues in CS, and the need for ethical frameworks (e.g., Resnik, Elliott, and Miller 2015; Rasmussen and Cooper 2019; Cooper, Rasmussen, and Jones 2021; Vohland et al. 2021). However, there still seems to be a gap of research around the question of how CS impacts professional science in concrete ways (*cf.* Wynn 2017). James Wynn (2017) addresses this gap through the lens of rhetoric, focusing on the results of CS projects, on the one hand, and on the interactions between nonprofessional and professional scientists and policymakers, on the other. In my research, I address this gap by focusing on how HC-based CS games create scientific data through human–technology relations with the examples of Stall Catchers, Foldit, and ARTigo, and discuss how the introduction of CS influences and changes the working practices of biomedical researchers in the case of Stall Catchers.

Not only have possibilities to participate in research multiplied with the Internet, but digital technologies, software, and code have also entered the domain of CS, creating new modalities for scientific knowledge production and new human–technology relations. An example is the emergence of "voluntary distributed computing" (VDC), which describes CS in which volunteer participants lend the computational power of their computers to scientific research (Holohan 2013). In fact, as I discuss in Chapter 6, Foldit emerged from such a VDC project. However, while volunteers in VDC simply let their computers contribute to help solve a scientific problem (such as the search for prime numbers), in HC-based CS, they actively perform a specific task themselves. Sociologist Anne Holohan's findings on the transformation of the scientific field, the motivations of VDC volunteers, and observations on "altruistic game-playing," nevertheless, form important references for my work (2013, 27, 71–75).

To the best of my knowledge, there is currently not much research focusing on the interplay of AI and humans in HC-based CS. Two of the few exceptions should be mentioned here. The first is the paper by information scientist Marisa Ponti and colleagues (2021) summarizing the discussion panel "Citizen Scientists Interacting with Algorithms: The Good and the Bad," that was organized by the authors at the *3rd ECSA Conference* in 2020. The aim of this panel was to discuss the collaboration of CS participants with ML algorithms. Using the HC-based CS project and platform Zooniverse as an example, the paper argues that the human–machine combinations can increase the efficiency of data classification.[7] Yet, it also discusses open issues that need to be considered in human–AI collaborations in CS, such as transparency and data ownership

---

6 On CS as "controversial space," see Jung (2015) and Starzmann (2015).
7 Previous research by AI researcher David Watson and philosopher Luciano Floridi has used the same example of Zooniverse to demonstrate how knowledge on the CS platform is produced in

(Ponti et al. 2021). The second work is Ponti and Alena Seredko's literature review on task distribution between AI technologies and humans in CS (2022). With the aim to close the research gap existing in this field, the authors emphasize that the tasks or "cognitive work," as they call it, "between humans and computational technologies will be shifting, challenging the ontological boundaries between them" (Ponti and Seredko 2022, 11). Therefore, they argue that it is important not to "essentialize the qualities of humans and machines, both of which are constantly evolving, and whose lists of what each is 'good at' (whether relative or absolute) are constantly changing" (Ponti and Seredko 2022, 11). My study aims at contributing to this line of thought by presenting an analysis of how the relations, including the task distribution, between humans and (AI) technologies continuously change in HC-based CS projects.

Projects in the field of HC-based CS are often designed as GWAPs,[8] computer games which, in addition to their core purpose as a game, have the purpose of solving a particular (scientific) problem. The HC-based CS games are also described as "dualpurpose Human Computation systems" (Bogner et al. 2017) and (often) rely on the assumption that people spend much time and energy playing (computer) games, which can be directed incidentally to solving computational or scientific problems as well as training AI algorithms (Von Ahn and Dabbish 2008, 60).[9] Citizen science games have been the subject of different studies, such as interdisciplinary ones in the field of human–computer interaction (HCI; e.g., Iacovides et al. 2013; Tinati et al. 2016; 2017; Díaz et al. 2020), which focus mainly on the participants' perspectives and their motivations. From the examples I study in my research, Foldit has been analyzed regarding, for example, participant motivation (Curtis 2015) or how participants engage with automated scripts (so-called "recipes") in the game and develop a "professional vision" by playing Foldit (Ponti et al. 2018). CS games were also studied by Dippel and Fizek (Fizek 2016; Dippel 2017; Dippel and Fizek 2017a; 2019).

Dippel's investigations of the relations between science and play and CS games as part of the ludic or playful aspects of knowledge production form an informative starting point for my analysis. Based on her ethnographic research in high-energy physics at the European Organization for Nuclear Research (CERN), Dippel analyzes different dimensions of playful aspects of knowledge production in her article "The Big Data game"

---

the sociotechnical interplay between human actors and technological components (Watson and Floridi 2018).

8   GWAPs and CS games are sometimes also referred to as "serious games," which are broadly defined as games that are designed "to be more than entertainment" (Ritterfeld, Cody, and Vorderer 2009a, 6), most often with a focus on education (cf. Abt 1987; Ritterfeld, Cody, and Vorderer 2009b; Dörner et al. 2016; Söbke et al. 2022). I refer to this term in the following because GWAPs focus more specifically on HC-based projects.

9   Participation in GWAPs might be considered a form of "free labor" (Terranova 2000; 2012), which, according to digital media cultures theorist Tiziana Terranova, is important for the constitution of the Internet and digital economy. However, I focus in my research on how participants perceive and describe their contribution to HC-based SC projects (see Chapter 5). It should, nevertheless, be noted that even if the games studied in my research are developed by nonprofit research institutes and collaborations, they are, nonetheless, influenced by the logic and the "agonal principle of free-market competition [which] has now gained a perfect system of rules" (Dippel 2018, 125).

(2017). Dippel writes that everyone wins in CS games (2017, 511), which are part of the third dimension:

> [W]hile the individual motivation of players participating in *Citizen Science* is characterized by idealistic traits and the desire to participate in a game that has another goal outside of itself, the experts in the *challenges* and the cernies themselves benefit from Big Data because the game in this case brings about a concrete improvement in artificial intelligences. [...] And last but not least, research on general artificial intelligence benefits from this deal in the long run. (Dippel 2017, 511, emphasis in the original [i.o.])

Together with Fizek, Dippel explores CS games as "laborious playgrounds" producing new forms of work-play relations (Dippel and Fizek 2019) and "playbouring cyborgs" (Dippel and Fizek 2017b) that consist of humans and algorithms. They use the term "ludification of culture" to describe "a societal phenomenon that points to an ever-increasing importance of games in everyday life. The concept goes far beyond the use of specific game mechanics that are used to control people's behavior" (Dippel and Fizek 2017a, 368).

Along the same lines, Dippel and Fizek introduce the term "interferences" (Dippel and Fizek 2017a, 377) to describe the mutual overlaps between the different spheres of everyday life. I adopt their term in my research to study how HC-based CS assemblages are formed between science and play and how human–technology relations, or "playbouring cyborgs" evolve and change continuously. Ideas and concepts from play theory that discuss the ambiguity of play (Sutton-Smith 2001; 2008) and how play is permeated with seriousness (Turner 1995) also inform my analysis, which will be discussed in Chapter 5 (Dippel and Fizek 2017a; 2019; Abend et al. 2020; Dippel 2020).[10] Studies from the extensive body of cultural and digital anthropological and cultural and media studies research on games focus on the actors behind games and situate them in broader power and regulatory structures (e.g., Malaby 2012; Cassar 2013).

Other relevant research examines the developer side of software, and particularly of computer games, focusing on their (working) practices and everyday lifeworlds (e.g., Coleman 2013; O'Donnell 2014; Amrute 2016; Bachmann 2018; Plontke 2018; Tischberger 2020). In his research on the Dynamic Medium Group, a San Francisco Bay Area-based research collective, European ethnologist Götz Bachmann investigates the work of engineers on a new digital medium (2018). He shows the importance of including the imaginations of engineers who guide the development of new digital media systems—or, as in my research, HC systems— in order to understand systems in their ongoing emergence and interaction with people and the environment (Bachmann 2018). Social psychologist

---

10   *Cf.* McGonigal (2012) on the entanglements of work and play. Cultural historian Johan Huizinga's *Homo Ludens* ([1938] 2016) continues to form influential theories of play. In the fields of media and cultural studies as well as philosophy, Valerie Frissen et al. (2015) update Huizinga's theory to apply to digital technologies, which Huizinga had still understood as opposites of play. Emotional practices in (video) games have been studied by cultural and digital anthropologist Christoph Bareither (2020b). On theoretical concepts of play in general see, *inter alia*, Adamowsky (2018), for anthropological approaches to play in general see, e.g., Malaby (2009) and Dippel (2020).

and social anthropologist Sandra Plontke examines design methods and programming practices in game development from the perspective of STS and ANT, focusing on the representation of the player in the code (2018). I begin my analysis of HC-based CS from the perspective of developers and designers of HC-based CS, which has, so far, received less attention in CS literature (Miller et al. 2023). I do this to understand their aims, their imaginations of desirable futures, and of the humans in the loop, and how they materialize these in their everyday development, maintenance, and infrastructuring practices. The humans in the loop imagined in HC-based CS refer to the users or participants.

Studies in STS have focused on how users are imagined and represented in technology and how they, at the same time, shape technology (e.g., Woolgar 1991; Akrich 1995; Grint and Woolgar 1997). The edited volume *How Users Matter* (Oudshoorn and Pinch 2005) discusses how users and technologies are co-constructed and, thus, points to the active role of users in the formation of technology. Emphasizing these forms of co-construction, Malaby suggests defining games as processes: "Games can change as they are played, and this passage points to how this can be done intentionally [...]. But games can also change through the unintended consequences of practice, such as when talented individuals or teams find new ways to play the game" (Malaby 2007, 102). How users appropriate games in situated practice in their own ways has also been explored by Internet studies scholar Michele Willson and cultural anthropologist Katharina Kinder-Kurlanda using the concept of "tactics" developed by theologian, historian, cultural theorist, and psychoanalyst Michel de Certeau ([1980] 2013). While Willson and Kinder-Kurlanda focus on tactics that users employ to make themselves less visible to the game platform, I use de Certeau's concept of tactics to analyze participant–technology practices that go beyond the play practices intended by design and enhance, adapt, or work against the game mechanisms. Games, and HC-based CS games in particular in my research, are, thus, "grounded in (and constituted by) human practice and are therefore always in the process of becoming" (Malaby 2007, 103). Therefore, I also include the perspective of participants who play a fundamental role in the everyday becoming of HC-based CS assemblages by bringing their own motivations and creative practices to the projects which do not always align with the developers and researchers' aims. I will return to this perspective when discussing European ethnologist Stefan Beck's relational concept of technology below.

Sociotechnical Systems and the Study of Algorithms, Computer Code, and Artificial Intelligence

While crowdworking, CS, games, or play/work interferences constitute research fields related to HC-based CS, research on sociotechnical systems, algorithms, and AI in general provide helpful insights that I build upon to study sociotechnical HC-based CS assemblages and their human–technology relations.[11]

---

11   Due to the rapid technological developments in AI and its increasing use in various areas of everyday life, which have attracted increased attention in the humanities and social sciences in re-

In cultural anthropological discourse, digital anthropology has emerged as a new field of research since the 1990s (Fleischhack 2019, 197).[12] Today, it describes an interdisciplinary field that focuses particularly on human–technology relations (Bareither 2022, 29). While different foci can be observed in the research agendas of individual contributing disciplines, the boundaries are blurry, and I concentrate here on research that centers on sociotechnical systems and human–technology relations in everyday life. In German cultural and European anthropology, technology is conceived as a "cross-cutting phenomenon" (Schönberger 2007)[13] and has been analyzed in terms of "culture(s) of technology and the technology(ies) of culture(s)" (Hengartner 2012, 119). Human experience, practice, and constructions of meaning are understood as technologically shaped and mediated (Hengartner 2012, 119). Beck's "complex situational analysis" (1997) of technology in practice is fundamental to my research, specifically from an analytical perspective and its methodological consequences. Beck emphasizes that human experience and practice is not only formed by technology, but vice versa, the social must be situated in technological arrangements and considered a fundamental part of them (2019, 12).

The anthropology of futures and emerging technologies is another productive research area that focuses on how anthropologists can study, engage with, and critically intervene in future-making practices in the field of emerging technologies (e.g., Salazar et al. 2017; Pink 2022; 2023; Lanzeni et al. 2023).

In this section I aim to discuss selected work from the field of digital anthropology, but also from sociology, STS, and related fields. My research here takes particular inspiration from early studies on human–computer relations and the field of AI, of which I would like to discuss five studies in particular.

Sociologist and STS scholar Sherry Turkle conducted a long-term ethnographic study in the late 1970s and 1980s on the computer as an "evocative object for thinking about human identity" (2005b, 3) when computers were mainly regarded as tools, the World Wide Web was not yet born, and mobile computational devices, such as smartphones and smart watches, were far away. In her groundbreaking research, Turkle analyzes how the computer influences human thinking and human nature itself. Her focus on early AI research as one example of computer culture is of particular interest for my research. Turkle aptly shows AI theorists' attempts to create a new philosophy in times when the aim of building AI with superhuman power is still out of reach (Turkle 2005a, 244). What these AI theorists share, Turkle argues, is "an emphasis on a new way of knowing. The new way of knowing asks that you think about everything, especially all aspects of the mind, in computational terms, in terms of program and information processing" (2005a, 225). Understanding this new way of thinking is crucial for Turkle because it shapes how AI theorists think about themselves and human life in general (2005a, 231–232). Turkle's

---

cent years, the following considerations are sure to be incomplete and refer mainly to selected research that has been published up to March 2023.

12   Cultural anthropologist Julia Fleischhack (2019) summarizes the main positions, methodological approaches, questions, and foci in these discussions. On digital anthropology in general, see, e.g., Boyd (2009); Boellstorff et al. (2012); Horst and Miller (2012); Koch (2015; 2017a); Pink et al. (2016).
13   Cultural and media anthropologist Manfred Faßler had already described information technology as "cross-cutting technology" in 1997 and an evolutionary project in his publication "media interaction" on human–computer interaction (1996, 17).

early analysis of the development of the field of AI is helpful for understanding how HC distances itself from other AI endeavors and is, at the same time, situated in these historical developments.

Furthermore, Suchman's work is instructive in understanding how HC-based CS systems are never fully defined by the designers' imaginations, but are instead *situated* in everyday practice. Around the same time as Turkle's publication of *The Second Self*, Suchman investigated human–machine interactions at Xerox Palo Alto Research Center (PARC) (Suchman 2007b). Suchman's research not only transformed the understanding of HCI in the field of computer science but is also still highly informative today for social and cultural analyses. One of the core contributions of her research is the elaboration of an understanding of "situated actions" that determine human–machine interaction. Against the technical understanding of plans as algorithmic specifications determining action that she observed in her fieldwork, Suchman argues that "given the contingencies of any actual occasion of action, every plan presupposes capacities of cognition and (inter)action that are not, and cannot ever be, fully specified" (2007b, 78). In the expanded 2007 edition, Suchmann, similar to Barad, understands plans as "socialmaterial be-ins" (Barad 1996, 188) that challenge the understanding of autonomously interacting actors and entities.

In reflecting on my own role in the field of HC as an anthropologist (see Chapter 3), Forsythe's extensive and pioneering research on medical informatics and AI is particularly helpful. Forsythe conducted ethnographic fieldwork at different knowledge-based system laboratories in academia and industry in the US in the 1980s and 1990s. Her research, which focuses on software design, presents fascinating insights into AI research culture, one of her central themes being an "attempt to unpack intelligent systems conceptually, from a cultural and disciplinary standpoint" (Forsythe [1996] 2001e, 94). Forsythe showed, for example, how assumptions of AI researchers were inscribed in *intelligent* computational systems that are always "cultural objects as well as technical ones" ([1996] 2001e, 94) and what role users, in her case patients, play.

Like Forsythe, sociologist and STS scholar Susan Leigh Star was one of the first STS scholars to actively engage in the field of AI. Star's work spans various topics, such as infrastructure (see below), classification and standardization of ideas, grounded theory, and distributed AI (Star 2008).[14] Based on the observation that AI relies on social and natural metaphors to fill the void between what computers can currently do and what advanced computer science systems are capable of, and to serve as attempts to make AI intelligible (Star [1988] 2015, 244), Star "argues that the development of distributed artificial intelligence should be based on a social metaphor rather than a psychological one" ([1988] 2015, 243). She suggests using the concept of "boundary objects" (Star and Griesemer 1989; Star 2008; Bowker et al. 2015) as a data structure for the field of distributed AI: "Boundary objects are objects which are both plastic enough to adapt to local needs and

---

14 Distributed AI is a research subarea of AI. In a nutshell, it is concerned with the development of AI systems that build upon different forms of concurrency, such as parallel computer architecture or multi-agent systems. It can be considered a predecessor of multi-agent systems (Bond and Gasser 1988, 3).

the constraints of the several parties employing them, yet robust enough to maintain a common identity across sites" (Star and Griesemer 1989, 393). Ultimately, she argues:

> The more seriously one takes the ecological unit of analysis in such studies, the more central human problem-solving organization becomes to design—not simply at the traditional level of human-computer interface, but at the level of understanding the limits and possibilities of a form of artificial intelligence. (Star 1989; cited in 2015, 249–250)

Consequently, in order to understand human–technology relations and how they form HC-based CS systems, it is necessary to analyze them from multiple perspectives and viewpoints, including the material infrastructures supporting them.

Finally, in the context of early research in German-speaking cultural anthropology, Gertraud Koch's groundbreaking work on the technological processes of becoming of AI (2005) is to be mentioned. Koch investigates practices, policies, and knowledge cultures contributing to the cultural production of technology with the aim of exploring the link between culture and technology. Based on written scientific sources and qualitative interviews, and following Pfaffenberger's concept of the "technological drama" (1992), Koch analyzes the (mainly) German-speaking AI discourse of the late 1970s until 1990s that accompanied the emergence of AI as technology in Germany. She shows how different AI advocates positioned themselves and performed "boundary work" (Gieryn 1983) to legitimize their viewpoints on AI, which is helpful for my analysis of the boundary work performed by HC advocates.

Together, the examples of early ethnographic research on HCI and AI discussed form a fruitful starting point for my research to understand how HC is imagined as being a counter-imaginary to AGI, while, at the same time, sharing common understandings, and how HC-based CS systems are situated in the everyday practice of various actors and human–technology relations. I aim to contribute to this rich ethnographic knowledge by studying a subfield of AI research that has, so far, not been extensively analyzed by ethnographers. My study, furthermore, aims to provide a contribution that shows how HC as a specific branch of research emerged from the fields studied by Turkle, Forsythe, and others, and how the AI research culture has, thus, changed since the research presented.

My research also takes into account previous work that focuses on computer code and algorithms,[15] particularly in the field of AI. I follow the understandings established in cultural anthropological and STS research that algorithms are part of sociocultural networks or entanglements (e.g., Haraway 1991; Mathar 2012, 178) and, thus, on the one hand, have agency (e.g., Kunzelmann 2015; Amelang and Bauer 2019), and, on the other hand, have meanings, values, norms, and (in)equalities inscribed in them. "[A]lgorithm

---

15   Algorithms are formalized instructions for solving a specific problem that can be expressed differently, for example, as verbal ideas, or implemented using programming languages. When implemented, they can then be executed in computer programs and become part of computer code, which includes all instructions and steps that are machine executable and can, thus, include several (nested) algorithms. On the importance of defining algorithms precisely and as distinct to computer code and programs, see Dourish (2016).

cultures" (Seyfert and Roberge 2017, 18) are always multiple, performative, and never neutral (Plontke 2018). This also corresponds to the understanding that the digital is, in a certain sense, already "coded culture" (Koch 2017a, 11).

Research fields, such as software studies (e.g., Fuller 2008) and critical code studies (CCS; e.g., Marino 2016; 2020), which focus specifically on the analysis of source code have emerged over roughly the last 20 years and provide fundamental insights into the functioning, formative power, and societal meaning of code. Researchers have focused on the agency of code and algorithms, and how they "make the world they work in hang together" (Mackenzie 2005, 13; for selected studies see, e.g., 2006; Kitchin and Dodge 2011; Gillespie 2014; Kitchin 2016; Seaver 2017).[16] In order to understand the agency of algorithms and code and how they operate, they have to be studied in practice, which is where they unfold (Amelang 2017, 359). Cultural anthropologist Katrin Amelang, for example, shows how algorithms are "sensually known and experienced" (2017, 358). She and sociologist Susanne Bauer follow a risk-predicting epidemiological algorithm in its multiple trajectories from development to its validation and use in practice and the infrastructures involved (2019). By adapting the approach of "following the actor" to "following the algorithm," they show how the risk score, predicted by the algorithm, is integrated into accountability practices related to health care and public health (Amelang and Bauer 2019, 495). I take this as inspiration to "follow the data" in Stall Catchers, from *in vivo* microscopic images of mice brains to analyzable video clips in the HC system and beyond (see Chapter 6).

Furthermore, code and algorithms do not simply do things *by themselves* but unfold in dynamic human–machine relations (Lange, Lenglet, and Seyfert 2019) and are embedded in and part of heterogeneous sociomaterial assemblages (Ananny 2017). In fact, code can itself be considered a sociomaterial assemblage (Carlson et al. 2021), though, in this work, I consider it one element of HC-based CS assemblages. Code and algorithms are interwoven with other elements from developers and users (and their practices) to infrastructures, such as servers, APIs, databases, other software libraries, and, ultimately, rely on physical circuits and logic gates that evaluate Boolean functions. Digital media scholar Wendy Hui Kyong Chun argues that *"source code [...] only becomes source after the fact"* (2008, 307, emphasis i.o.).

Due to their embeddedness in these assemblages, code and algorithms must be analyzed as part of these assemblages in practice (Introna 2016, 20). Furthermore, researcher of technology and ethics Lucas Introna writes, following Barad's (2007b) understanding of intra-relating actions, "[w]hat we see is that the action, the doing, of the code has a *temporal flow*. Every particular 'doing' happening in the present already assumes some *inheritance* from antecedent 'prior-to' actions, and it already anticipates, or *imparts* to, the

---

16   Algorithms and code, and how they influence, change, and are part of governance, and the exercise of power have been studied in various fields. In addition to the fields already discussed, such as crowdworking and games, research exists on, for example, predictive policing (e.g., Brayne 2017; Egbert 2017; Bennett Moses and Chan 2018; Singelnstein 2018; Egbert and Krasmann 2019), surveillance (Introna and Wood 2002, Introna 2016, Zuboff 2019), and finance (Muniesa 2011), and how they change political protest (Kunzelmann 2021). This list is by no means complete but rather aims at emphasizing the multitude and variety of studies (*cf.* Vepřek et al. 2023).

subsequent 'in-order-to' actions" (Introna 2016, 21, emphasis i.o.). To understand such temporal flows of action (Mousavi Baygi, Introna, and Hultin 2021), I analyze not only the source code of Stall Catchers, or more specifically, the flow of human–code intraactions in it (see Chapter 3), but also its human–technology relations and both their instantaneous and gradual temporal unfolding (see the section on "Intraversions of Human–Technology Relations" in this chapter). I return to the discussion of code in Chapter 3 when discussing the analysis of computer code in my ethnographic approach.

In this work, I consider AI to be the broad field of research that seeks to build "intelligent" machines. Machine learning, then, is a subset or technique of AI which builds on data to create a decision or prediction model without directly programmed instructions. Deep learning and artificial neural networks, subsequently, are subsets of ML. Other subsets of AI include NLP, computer vision, robotics, and expert systems.

In recent years, more and more research in cultural and digital anthropology, STS, and related fields has discussed and analyzed AI. These include Turkle (2005a), Nilsson (2010), and Engemann and Sudmann (2018), who discuss the historical developments of AI, or media studies researcher Anja Bechman and STS researcher Geoffrey Bowker's analysis of the knowledge production of AI using the framework of classification theory (Bechman and Bowker 2019). Another example is the exploration of the relation between magic and AI by social anthropologist Simon Larsson and cognitive science, psychology and philosophy scholar Martin Viktorelius (2022). Based on the observation that AI advocates use the imagination of AI "as working like magic and glossing over the limitations of technological systems," (Elish and boyd 2018, 74) cultural anthropologist Madeleine Clare Elish and technology and social media scholar danah boyd argue that this hype can lead to poorly constructed models that are understood to be infallible, undermining their power and potential. This also specifically results in "limited space for interrogating how cultural logics get baked into the very practice of machine learning" (Elish and boyd 2018, 74). Therefore, they call for grounding both the rhetoric and practices of AI. History of media, technology, and society scholar Alexander Campolo and media, technology, and AI scholar Kate Crawford use the term "enchanted determinism" to describe how AI discourse is often characterized by references to magic and the inability to create a complete understanding about the generation of results (Campolo and Crawford 2020).[17] In STS research, I would like to highlight the contribution of feminist scholarship, or feminist STS, to the analysis of "sciences of the artificial," as Suchman calls the subfields of science

---

17 In addition to the literature discussed here, it should be noted that attempts to establish a disciplinary research agenda of AI have also been undertaken in other fields, such as sociology, with a focus on how inequalities are (re)produced by AI technology (Joyce et al. 2021). Furthermore, AI's societal implications are also increasingly becoming the subject of computer and information science literature. To provide one example, the question of how AI affects and changes work has been part of AI discourse from its early days. A recent publication on this topic is the special issue of the *Journal of the Association for Information Science and Technology*, which discusses the "mutual transformations" (Jarrahi et al. 2023, 303) of AI and work and organization. The editors of the special issue argue for the need for practice-focused studies that analyze "the technology at work not in isolation but in conjunction with organizational policies and routines" (Jarrahi et al. 2023, 304).

and technology that include cognitive science, AI, robotics, and related fields in critical adoption of political scientist Herbert Simon (Suchman 2007a).[18]

Sociologist Adrian Mackenzie's "auto-archeological" (2017, xi) study of ML practices and machine learners, which he conceives as referring to both humans and machines, is particularly insightful for this work. Mackenzie aptly describes the diagrammatic practices that form machine learners to show how ML is both a strategy of power and a form of knowledge production (2017, 9). I am particularly interested in his focus on subject positions and the distribution of agency between machines and humans. Drawing on Foucault (1972), Mackenzie asks: "Who is the machine learner subject?" (Mackenzie 2017, 179).

> Oscillating between cognition and infrastructures, between people and machines, neural nets suggest a way of thinking not only about how "long-term knowledge" takes shape today but about subject positions associated with machine learning. As infrastructural reorganization takes place around learning, and around the production of statements by machine learners, both human and nonhuman machine learners are assigned new positions. These positions are sometimes hierarchical and sometimes dispersed. The machine learner subject position is mobile rather than a single localized form of expertise (as we might find in a clinical oncologist, biostatistician, or geologist). Because machine learners vectorize, optimize, probabilize, differentiate, and refer, what counts as agency, skill, action, experience, and learning shifts constantly. (Mackenzie 2017, 186)

Although my focus is more on participant–software and researcher–technology relations in HC-based CS systems and less on programmer-machine relations, Mackenzie's analysis of the "mobile" subject positions and the shifts in, for example, agency and skill, nevertheless, provides rare and valuable points of reference on how subject positions are redistributed and continuously changing in ML processes (2017, 186). Similar to Mackenzie's approach to the machine learner subject, political geographer Louise Amoore, in her book *Cloud Ethics. Algorithms and the Attributes of Ourselves and Others* (2020), defines the "we" of ML as "a composite figure in which humans learn collaboratively with algorithms, and algorithms with other algorithms, so that no meaningful outside to the algorithm, no meaningfully unified locus of control, can be found" (2020, 58). Amoore, therefore, calls for a relational understanding and analysis of the ethicopolitics of ML (e.g., 2020, 7).

Artificial intelligence has also been analyzed from media theoretical perspectives. Andreas Sudmann, for example, whose work is concerned specifically with ML and DL,

---

18 Among the guiding and common questions driving feminist STS is the "ongoing project of unsettling binary oppositions, through philosophical critique and through historical reconstruction of the practices through which particular divisions emerged as foundational to modern technoscientific definitions of the real" (Suchman 2007a, 140). Even though addressing feminist STS distinctively in the broader field of STS can be important boundary work, I here consider STS to include feminist scholarship. As will become apparent in this chapter and specifically in the second part, where I discuss theoretical conceptualizations, I draw from research and scholars who aim to move beyond dichotomies and binary oppositions.

investigates the role of documentary practices in DL technologies (2015), or media-political dimensions of DL and the company OpenAI's agenda of "democratizing AI" (2018). The volume edited by Christoph Engemann and Andreas Sudmann (2018) provides insights into the development, media, infrastructures, and technologies of AI from the perspective of cultural and media studies and the history of science. However, in these works, HC or HI have received little attention so far.

The development of DL and how it became successful is further studied by Rainer Mühlhoff (2020) from a media-philosophical point of view and social-theoretical critique. He traces the success back not only to the advances in computing power—as is commonly argued—but to "a fundamental structural change in media culture and human–computer interaction (HCI) at societal scale" (Mühlhoff 2020, 1869). He uses the term "Human-Aided AI" to describe a media-cultural dispositive in which different forms of human contributions to DL systems take place and which is based on "socio-economic conditions, technological standards, political discourses, and specific habits, subjectivities and embodiments in the digital world" (Mühlhoff 2020, 1881). Mühlhoff identifies five types of human involvement in hybrid human–AI systems and their corresponding power relations, including gamification as introduced with von Ahn's GWAPs. Together with these forms of human engagement, Mühlhoff argues, comes a shift that leads to a new understanding of intelligence, where human cognitive abilities are integrated into "machine networks" (2020, 1870), changing the role of humans from being simulated by machines and replaced to active embedded cognitive resources. For this understanding of intelligence as relational and distributed across humans and AI, Mühlhoff introduces the term "cybernetic AI" (2020, 1880). His article is an important contribution to the analysis of hybrid human–AI systems from a philosophical and social scientific perspective. While the forms of human participation in AI described by Mühlhoff are helpful for identifying human–AI relations in concrete examples, my research focuses on *how* these relations unfold. The analysis will show that, even though the human–AI (power) relations are indeed initially defined by the creators of such systems, they are, nevertheless, distributed across different actors, who also shape the relations according to their own needs.

In the interdisciplinary volume *The Democratization of Artificial Intelligence. Net Politics in the Era of Learning Algorithms* (Sudmann 2019a),[19] Dippel's contribution *Metaphors We Live By* is especially noteworthy. Building on her study of scientists working with ML and evolutionary algorithms at CERN, Dippel argues for the "paramount importance [... of] investigat[ing] artificial intelligence not only from a specifically technical angle, but in a broader socio-cultural and political context" (2019a, 39). Artificial Intelligence, according to Dippel, should be considered as a "technological alien" (2019a, 39) to be able to think about a different future of the relation between AI and humans than positivist and neoliberal imaginaries propose.

---

19   The focus here lies on the political dimensions of AI, specifically on discourses and understandings of the *democratization* of AI technologies. Democratization is defined as "the realization of an ethic, aiming at political information, a willingness to critique, social responsibility and activity, as well as of a political culture that is critical of authority, participative, and inclusive in its general orientation" (Sudmann 2019b, 11).

Other work has studied the imaginaries and narratives of (mostly strong) AI (e.g., Cave and Dihal 2019; Cave, Dihal, and Dillon 2020; Fjelland 2020; Bareis and Katzenbach 2022) that fundamentally influence public discourses and AI development. Since the imaginary of HC builds on these narratives to form a counter-imaginary, I will return to this work and discuss it in more detail in Chapter 4.

Finally, sociologist of science, technology and computing Florian Jaton conducted a laboratory study on the constitution of algorithms, regarding how algorithms come into being, at a computer science laboratory for digital image processing (2021). Such constitution, as Jaton shows, is always "open-ended and amendable" (2021, 289). He pays particular attention to the three activities of "ground-truthing, programming, and formulating" (Jaton 2021, 17) that shape algorithms. By following an "enactive conception of cognition" (Ward and Stapleton 2012) which frames cognition "as a local attempt to engage *with* the world" (Jaton 2021, 130, emphasis i.o.) through actions, he situates programming in experience. Jaton's work, therefore, is an insightful contribution to studying "algorithms from within the places in which they are concretely shaped" (2021, 286). In my research and with the Human Computation Institute as an example, I investigate HC-based CS systems, in a similar way, from within the places in which they are formed.

Infrastructures and Infrastructuring

In order to analyze the sociotechnical assemblages and their human–AI or –technology relations in general, it is crucial to include the infrastructures enabling and forming such relations in the first place. As will become clear in the following elaborations, I consider infrastructure both as a subject of study and as an analytical lens. Without neglecting that most of the literature cited above includes an infrastructure perspective, I here want to explicitly point to research on infrastructures and the field of "information infrastructure studies" (Bowker et al. 2009).[20] Anthropologist Brian Larkin defines infrastructures as follows:

> Infrastructures are built networks that facilitate the flow of goods, people, or ideas and allow for their exchange over space. As physical forms they shape the nature of a network, the speed and direction of its movement, its temporalities, and its vulnerability to breakdown. They comprise the architecture for circulation, literally providing the undergirding of modern societies, and they generate the ambient environment of everyday life. (Larkin 2013, 328)

Infrastructures are essential for all areas of social organization of which they form the "backstage" (Koch 2017b, 117). They "mediate between scales, connecting local practices with global systems" (Star and Ruhleder 1996, 114; cited in Hallinan and Gilmore 2021, 6). Despite their importance, they often remain invisible both physically and in discourse

---

20   Due to the growing interest of social science and humanities researchers in infrastructure in, *inter alia*, Internet studies, media, or urban studies, the trend is referred to as the "infrastructural turn" (e.g., Hesmondhalgh 2021).

(Bowker and Star 2008; Niewöhner 2015). This invisibility is no coincidence but part of the inner workings of power. Bowker and Star showed how working infrastructures go hand in hand with classification and standardization systems and what roles the latter play (2008). Infrastructures in the words of communication scholars Blake Hallinan and James Gilmore, are "agents of power" (2021, 2), and their disappearance contributes to the fiction that they are objective as well as acultural and asocial, and, thus, reliable. If they work smoothly and as intended, the data infrastructures underlying Stall Catchers are not the focus of the researchers' work or even the subject of laboratory discussions. However, infrastructure becomes visible upon breakdown (Star and Ruhleder 1996, 113; *cf.* Star 1999, 381–382) and,[21] as in the example studied in this work, when they are changed and new aspects are introduced to them. The geographer and scholar of urbanism and the sociology of technology Stephen Graham writes that these moments of disruption, when infrastructures in the background stop working as they should, are the "most powerful way of really penetrating and problematizing those very normalities of flow and circulation to an extent where they can be subjected to critical scrutiny" (Graham 2009, 2).

Susan Leigh Star and computer and information scientist Karen Ruhleder argue, following computer scientist Tom Jewett and social informatics scholar Rob Kling (1991), that the concept of infrastructure is "a fundamentally relational concept, becoming real infrastructure in relation to organized practices" (Star and Ruhleder 1996, 113). They define nine properties of infrastructure, including transparency and embeddedness (Star and Ruhleder 1996, 113; Star 1999). Based on such a relational understanding, social anthropologist Jörg Niewöhner considers infrastructures "as transient embodiments of social, technical, political, economic, and ethical choices that are building up incrementally over time" (2015, 2). With respect to my research field, data infrastructures are, thus, a co-constitutive part of HC-based CS games as sociotechnical assemblages (Niewöhner 2015, 6–7). An analysis of the data infrastructures behind Stall Catchers, therefore, requires not starting with the imaging processes in the biomedical laboratory and stopping at the stage of analyzable data, but rather including the practices of researchers and developers, the game's code infrastructure, databases, and the social organization and values that flow into and shape the infrastructures, as Bowker and colleagues argue for infrastructures in general (2009, 99). The latter authors also argue that when infrastructure is considered as a concept, it "consists of both static and dynamic elements, each equally important to ensure a functional system" (Bowker et al. 2009, 99).[22] As I will show in this work, infrastructure and their related researcher–technology relations often resist the attempts to be stabilized and cleaned up. They are constantly in the making.

---

21   Here, breakdown is not considered to be an exceptional state of infrastructure but "a condition of technological existence" (Larkin 2008, 234), as shown by Larkin in his ethnography of media in Nigeria.
22   Bowker argues for performing "infrastructural inversion" (1994; *cf.* Bowker and Star 2008, 34) to analyze infrastructures: "Infrastructural inversion means recognizing the depths of interdependence of technical networks and standards, on the one hand, and the real work of politics and knowledge production on the other. It foregrounds these normally invisible Lilliputian threads and furthermore gives them causal prominence in many areas usually attributed to heroic actors, social movements, or cultural mores" (Bowker and Star 2008, 34).

What constitutes infrastructure, then, depends on perspective. Larkin writes that the "act of defining an infrastructure is a categorizing moment" (2013, 330) that, as I would add, is more an ongoing process than a concrete moment.

This understanding of infrastructures as being always in the making, moreover, shifts the focus to "infrastructuring as a material-semiotic practice" (Niewöhner 2015, 5). Information scientist and STS scholar Claus Bossen and historian and STS researcher Randi Markussen use infrastructuring as a verb to point to, among other things, "the efforts required for their integration, and the ongoing work required to maintain it" (2010, 618; *cf.* Jackson 2014). In my study I consider infrastructuring practices at both the Human Computation Institute and the biomedical laboratory whose Alzheimer's disease research data is analyzed on the Stall Catchers platform. These infrastructuring practices are part of the HC-based CS assemblages and present informative examples of human–technology relations in these sociotechnical systems.

The aim of my research is to understand how HC-based CS assemblages come into being in the interplay of different human and nonhuman actors, and how their human–technology relations change over time and in everyday life. Despite this focus, my research approach is also inspired by the analytical genre of laboratory studies common in STS (Jaton 2021, 19–20). These studies focus on the analysis of how scientific knowledge is produced (famous examples are Lynch 1985; Latour and Woolgar [1979] 1986; Traweek 1992; Knorr-Cetina 1999), or recently, how algorithms are constituted (Jaton 2021). While I mainly conducted participant observation at the Human Computation Institute, my observations at the Schaffer–Nishimura Lab, and the focus on the participant's perspective also played a crucial role in my research (see Chapter 3).

A Theoretical Framework for Analyzing Emerging Hybrid Systems

Alongside the related work I have discussed, my research further builds upon key theoretical concepts which I combine to form a theoretical framework to address the scientific questions central to this study. These theoretical lines, which I discuss separately, consist of, first, the assemblage concept, which is particularly well-suited to adhering to HC-based CS projects' dynamic, procedural, and complex nature and directs the focus onto relations between the humans and nonhumans forming the assemblages. Second, I discuss a relational understanding of technology. Here, I draw specifically from Beck's "complex situational analysis" (1997) and a postphenomenological understanding of human–technology relations, which views human experience as always mediated by technology (e.g., Ihde 1990; Verbeek 2001; 2005; Rosenberger and Verbeek 2015c; Dorrestijn 2017). Third, I discuss moral anthropological and ethics of technology approaches, which form a crosscutting perspective in my research and help analyze how HC-based CS is imagined by designers and developers. This extends to how, for example, participants relate to and cope with their engagement in such systems. From these conceptualizations, I develop the concept of intraversions to capture how human–technology relations in HC-based CS intravert along the dimensions of instantaneity and gradual temporal development.

From Assemblages to Assemblage Thinking

Human computation systems are not static, distinct objects, but continuously becoming in the interplay of and relations between different human and nonhuman actors, themselves shaped and co-constituted as part of the process of becoming. While human–technology relations and, thus, HC systems themselves stabilize over a certain timeframe, they also carry the potential for change. In order to analyze HC-based CS projects, the concept of assemblages offers a helpful theoretical approach to capture such projects in both their temporal consistency and volatility (Welz 2021a, 161).[23] Simply stated, assemblages can be understood as compositions of various heterogeneous elements, including human and nonhuman actors and their relations, which temporally come together in specific configurations (Welz 2021a, 162). The resulting sociotechnical assemblages do not merely constitute the sum of their individual elements; instead, something new emerges, rendering HC systems in their multiplicity unique. As I show in this work, only a highly specific interplay of various sociomaterial relations allows HC-based CS projects to meaningfully contribute to the scientific analysis of a problem.[24]

Assemblage theory recently received considerable attention from across the social sciences (Hansen and Koch 2022, 3). It has also been increasingly well-received in (cultural) anthropology since the turn of the millennium (Welz 2021a, 161). In recent years, European ethnology and digital anthropology employed assemblage theory to analyze "nonlinear processes, unstable states, and unexpected effects" (Welz 2021a, 168).[25] Originally, assemblage theory was developed by Deleuze and Guattari, specifically in their book *A Thousand Plateaus: Capitalism and Schizophrenia* (1980). Here, an assemblage is formed in a multiplicity of heterogeneous elements and "necessarily changes in nature as it expands its connections" (Deleuze and Guattari 2013, 7). Moreover, assemblages act "on semiotic flows, material flows, and social flows simultaneously" (Deleuze and Guattari 2013, 24). In dialogue with the French journalist Claire Parnet, Gilles Deleuze describes assemblage as follows:

> It is a multiplicity which is made up of many heterogeneous terms and which establishes liaisons, relations between them, across ages, sexes and reigns—different natures. Thus, the assemblage's only unity is that of a co-functioning: it is a symbiosis, a 'sympathy'. It is never filiations which are important, but alliances, alloys; these are not successions, lines of descent, but contagions, epidemics, the wind. (Deleuze and Parnet 2007, 69)

---

23   Citizen science systems have already been analyzed as assemblages by virtuality design scholar Nathan Prestopnik and information science scholar Kevin Crowston (2012). However, the authors did not draw from assemblage theory as discussed in this research and considered CS system assemblages as "a collection of interrelated functional components and social activities" (Prestopnik and Crowston 2012, 1).

24   In fact, in solving a certain problem, new problems emerge (or existing problem definitions change) to be addressed with HC. I will elaborate on this, specifically in Chapter 6.

25   See, for example, the application of assemblage theory to the analysis of AI in museums (Bareither 2023).

These connections between different elements are, in a certain way, random and non-linear. According to Deleuze and Guattari, they are rhizomatic in that they "connect [...] any point to any other point" (2013, 21). The rhizome, in their thinking, lies in opposition to arborescence, a tree-like hierarchical structure that has a beginning and an end. Rhizomes are nonhierarchical, undirected, and heterogeneous, always multiple and can never be broken.[26]

The attempt to fully characterize Deleuze and Guattari's concept of "assemblage" fails due to the various and often inconsistent definitions they offered. Moreover, any attempt to define the concept in English already falls short given the term itself since the English translation "assemblage" does not capture both meanings of the original French "*agencement*" (Phillips 2006, emphasis i.o.). The translation of the term originates from the Canadian philosopher and social theorist Brian Massumi, who first translated *A Thousand Plateaus* and introduced the term "assemblage," adopted by other translators and recipients in subsequent years (Brenner, Madden, and Wachsmuth 2011, 227; cited in Welz 2021a, 163). In its French meaning, the term refers to both the "action of matching or fitting together a set of components (*agencer*) [...], as well as to the result of such an action: an ensemble of parts that mesh together well" (DeLanda 2016, 1, emphasis i.o.). According to sociologist John Law, the English translation does not reflect the uncertainty related to the process (2004, 41).

For Deleuze and Guattari, assemblage was a "provisional analytical tool rather than a system of ideas geared towards an explanation that would make it a theory" (Müller 2015, 28). Nevertheless, various scholars have attempted to explain Deleuze and Guattari's concept or further define it without strictly adhering to their thinking—attempts that others have also criticized, including philosopher and critical and cultural theorist Ian Buchanan (e.g., 2015). My aim in this section, however, is to discuss assemblages in a way that is conducive to my analysis of HC-based CS, admittedly leaving the discussion of different interpretations incomplete. Extensive discussions of the concept of assemblage and its various receptions can be found in, for example, DeLanda (2006; 2016), Müller (2015), Welz (2021a), and Hansen and Koch (2022).

Theorist, artist, and philosopher Manuel DeLanda's work on Deleuze and on Deleuze and Guattari's assemblage concept offers a detailed account of their thinking. DeLanda even advanced Deleuze and Guattari's formulation to a "neo-assemblage theory" or "assemblage theory 2.0" (DeLanda 2006, 4), more tailored towards use as an analytical tool. DeLanda, in fact, introduces a new approach to social ontology based on Deleuze's assemblage theory in *A New Philosophy of Society: Assemblage Theory and Social Complexity* (DeLanda 2006). Assemblages, he argues,

> being wholes whose properties emerge from the interactions between parts can be used to model any of these intermediate entities: interpersonal networks and institutional organizations are assemblages of people; social justice movements are assemblages of several networked communities; central governments are assemblages

---

[26] The latter means that rhizomes do not cease to exist upon rupture but instead change along a different line. Deleuze and Guattari list six defining characteristics of rhizomes in the introductory chapter "Rhizome" (2013). Here, I focus on the assemblage concept.

of several organizations; cities are assemblages of people, networks, [...]. (DeLanda 2006, 5)

All these intermediate entities can be modeled and analyzed using the concept of assemblages and historical processes. This approach, according to DeLanda (2006, 4), presents a "realist social ontology" that does not need to rely on essentialism to explain the identities of organic, inorganic, or social assemblages by focusing on the processes of production and the maintenance of assemblages "instead of the list of properties characterizing the finished product" (DeLanda 2006, 39). Maintenance, in the form of territorializing processes, is important because deterritorializing (see below) processes continuously destabilize assemblages (DeLanda 2006, 39). Following this thinking, then, DeLanda defines the ontological status of assemblages as individuals and singular (2006, 40). Taking assemblages as starting points helps to follow the processes that form them and those that destabilize them simultaneously.

Assemblage theory is particularly useful for analyses across scales because it understands phenomena as always consisting of various interwoven scales. This allows one to analyze how assemblages come into existence through the interaction of their elements and how they, in turn, influence these individual parts (DeLanda 2016, 34).

Following Deleuze and Guattari, and DeLanda's interpretation, geographer Martin Müller (2015) summarizes five characteristics of assemblages. First, they are "relational," meaning that assemblages come into being through the relations between different elements (Müller 2015, 28): "In a multiplicity, what counts are not the terms or the elements, but what there is 'between', the between, a set of relations which are not separable from each other" (Deleuze and Parnet 2007, viii). These relations are not fixed but temporal (Deleuze and Guattari 2013, 98), and, as DeLanda argues, they are *"relations of exteriority*. These relations imply [...] that a component part of an assemblage may be detached from it and plugged into a different assemblage in which its interactions are different" (DeLanda 2006, 10, emphasis i.o.). This understanding of relations also means that it is not the properties of such components that describe the relations of the assemblage, because the realization of their capacities relies on relations, on references to "the properties of other interacting entities" (DeLanda 2006, 11).

Second, Müller (2015, 29) argues that assemblages create new actors and actions, relations, expressions, and territorial organizations and are, thus, *productive*. As I will show in this work, HC-based CS projects and their intraverting human–technology relations generate new subjectivities, tasks, and purposes. In fact, the purposes of the projects themselves sometimes change.

Third, "[a]ssemblages are *heterogeneous*" (Müller 2015, 29) and, as such, always sociomaterial. "There are no assumptions as to what can be related—humans, animal, things and ideas—nor what is the dominant entity in an assemblage" (Müller 2015, 29, emphasis i.o.).

Moreover, assemblages are formed through the constant processes of deterritorialization and reterritorialization: "the assemblage has both *territorial sides*, or reterritorialized sides, which stabilize it, and *cutting edges of deterritorialization*, which carry it away" (Deleuze and Guattari 2013, 103, emphasis i.o.). Deterritorialization and reterritorialization can be understood as processes that act upon assemblages by disembedding and destabilizing them (deterritorialization), and by restructuring and stabilizing them (reterritorialization). As assemblages, HC-based CS projects are constantly marked by processes of reterritorialization that bring together, align, and stabilize heterogeneous relations and actors, while those of deterritorialization tear them apart, destabilize the assemblage, and increase gaps and frictions between different relations. While such processes play a role in all of the subsequent empirical chapters, Chapter 7 explicitly focuses on the example of building trust as a (re)territorialization process. Despite the continuous work of various processes on the assemblages, they are, nevertheless, not randomly changing and not everything is in motion (Müller 2015, 36). Similar to Beck's observation of the "use potentials" of technology (1997, 223), assemblages are multiple but not arbitrary (see below for more detail).

Finally, and importantly to Deleuze and Guattari, desire is fundamental to fusing the elements of and forming them into an assemblage (Müller 2015, 36). In Deleuze and Guattari's words, "The rationality, the efficiency, of an assemblage does not exist without the passions the assemblage brings into play, without the desires that constitute it as much as it constitutes them" (2013, 465). It follows, then, that assemblages as collections of heterogeneous elements are never neutral but always driven by desire, which can take many different forms of "passion," such as power, pity, cruelty (Deleuze and Guattari 2013, 466), or ethical principles. In this work, I analyze the various passions, the "effectuations of desire" (Deleuze and Guattari 2013, 466) that create HC-based CS assemblages and are created by them.

While DeLanda's work made assemblage theory accessible to the social sciences in general (Welz 2021a, 162), other modes of interpretation have been developed which can be summarized by the term "assemblage thinking" (Anderson et al. 2012; Welz 2021a). These often moved away from the poststructuralist philosophy of Deleuze (Welz 2021a, 164). Here, assemblage thinking encompasses not only ontological but also methodological or empirical approaches (Brenner, Madden, and Wachsmuth 2011, 230), serving as a "descriptive emphasis of how different elements come together" (McFarlane 2011b, 652). The anthropologists Aihwa Ong and Stephen Collier's (2005) edited volume *Global Assemblages. Technology, Politics, and Ethics as Anthropological Problems*, for instance, influenced the cultural anthropological reception of the assemblage concept (Welz 2021a, 165; Hansen and Koch 2022, 5). "Global Assemblages," as the title suggests, focused on questions of globalization at the beginning of the twenty-first century. Here, the term unites the tension in the terms "global", implying a "broadly encompassing, seamless, and mobile" (Ong and Collier 2005, 12) perspective, and "assemblage" as "heterogenous, contingent, unstable, partial, and situated" (Ong and Collier 2005, 12). Contrary to DeLanda, Ong and Collier (2005) do not aim to define the concept further but to make it accessible to empirical research, presenting "assemblage thinking as a heuristic for emerging globalization research in cultural anthropology" (Welz 2021a, 166). They describe assemblages as "ensembles of heterogeneous elements" (Ong and Collier 2005, 4), which are "the product of multiple determinations that are not reducible to a single logic. The temporality of an assemblage is emergent. It does not always involve new forms but forms that are shifting, in formation, or at stake" (Ong and Collier 2005, 12). The latter is especially essential to my research, since the relations within sociomaterial assemblages in HC-based CS intravert such that they, for example, change without necessarily involving new elements or actors. According to cultural anthropologists Lara Hansen and Gertraud Koch, the focus in the current empirical cultural studies of assemblages lies on

> the emergence and unfolding of socio-material fields and the tracing of just such inherent processes of change by the various human and non-human actors. In particular, the randomness of connections and the recognition of the ambiguity of social realities in which the disruption of existing structural categories and dichotomies such as social-material, animate-inanimate, nature-culture, human-non-human, object-subject, micro-macro, or structural-practices is prevalent. The unmasking of these dichotomies as specific, often anthropocentric world views offers fruitful starting points for ethnographic research in the fields of political, educational, environmental or medical anthropology. (Hansen and Koch 2022, 4–5)

I argue for adding digital anthropology and the anthropology of technology to this list of fields of ethnographic research that can benefit from an assemblage approach. Moreover, assemblage thinking and assemblage theory have been of particular interest in urban studies (Farías and Bender 2010; Brenner, Madden, and Wachsmuth 2011; McFarlane 2011a; 2011b; Färber 2014),[27] where the concepts have proven quite useful when employed in combination with ANT (e.g., Farías and Bender 2010; Färber 2014). While my research does not lie within the field of urban studies, the interpretation of assemblage theory from the perspective of ANT, the concepts and perspectives of which serve as important starting points for my research, represents an important trajectory which I will discuss in what follows along with a brief overview of ANT itself.

Actor–network theory, specifically going back to philosopher, anthropologist, and sociologist Bruno Latour ([1988] 1993), sociologist Michel Callon (1984), and John Law (1984; *cf.* Latour 2005, 10), focuses on the analysis of actor networks and their properties (Latour 1996, 369), thereby following a flat ontology (Latour 2005, 16). The focus of ANT lies on overcoming binaries and "does not limit itself to human individual actors, but extends the word actor—or actant—to *non-human, non-individual* entities" (Latour 1996, 369, emphasis i.o.),[28] conceived as "circulating objects" (Latour 1996, 374). These objects are formed through and in actions with other actants and associations. The agenda of ANT, according to Latour, is

> [t]he attribution of human, unhuman, non-human, inhuman characteristics; the distribution of properties among these entities; the connections established between them; the circulation entailed by these attributions, distributions and connections;

---

[27] The concept of assemblage highlights the fundamental "human—non-human multiplicity of relations" (McFarlane 2011b, 651), attending to "why and how multiple bits-and-pieces accrete and align over time to enable particular forms of urbanism over others in ways that cut across these domains, and which can be subject to disassembly and reassembly through unequal relations of power and resource" (McFarlane 2011b, 652). Moreover, it allows researchers to ethnographically analyze cities as interconnected elements in everyday practice without having to determine the nature of their connections *a priori* (Färber 2014, 98).

[28] The term "actant" refers to whatever or whoever "acts or to which activity is granted by others. It implies *no* special motivation of *human individual* actors, not of humans in general" (Latour 1996, 373, emphasis i.o.).

the transformation of those attributions, distributions and connections of the many elements that circulate, and of the few ways through which they are sent. (1996, 373)

According to Star, ANT "opened up a whole new way of analyzing technology" ([1991] 2015, 276) which has much in common with assemblage theory. In fact, Latour, in defining ANT, referred to assemblages and Deleuze's term "rhizome" (e.g., in Latour 1996, 370). Therefore, Müller described ANT as "an empirical sister-in-arms of the more philosophical assemblage thinking" (Müller 2015, 30) developed by Deleuze and Guattari. In contrast to DeLanda, for whom relations of exteriority characterize assemblages, ANT follows a relationalist ontology, according to which relations of interiority form assemblages. For the latter, relations "define the very identity of the terms they relate" (Ball 2018, 242) to. With assemblages, ANT focuses on the becoming of their components (Schwertl 2013, 118),[29] which are conceived as processes of "recursive self-assembling" (Law 2004, 41). Thus, here, the components are co-constructed and shaped in the entanglement with each other and are not preexisting (Law 2004, 42).

Consequently, ANT decenters subjects and artifacts by placing associations, networks, and translations at the center of concern, focusing specifically on processes, changes, and stabilizations (Schwertl 2013, 113). The productive power of relations—especially human–technology relations—plays an important role in understanding how intraversions form.

Additionally, a helpful approach to agency building upon the ANT conception of assemblage is political scientist and philosopher Jane Bennett's notion of "distributive agency" (2010).[30] This understanding decenters agency from being a capacity solely ascribed to humans and sees it, instead, as "distributed across an ontologically heterogeneous field" (Bennett 2010, 23). Following this understanding, assemblages, then, are also not defined and conducted by individual agents but possess their own agency:

> [N]o one materiality or type of material has sufficient competence to determine consistently the trajectory or impact of the group. The effects generated by an assemblage are, rather, emergent properties, emergent in that their ability to make something happen (a newly inflected materialism, a blackout, a hurricane, a war on terror) is distinct from the sum of the vital force of each materiality considered alone. Each member and proto-member of the assemblage has a certain vital force, but there is also an effectivity proper to the grouping as such: an agency *of* the assemblage. And precisely because each member-actant maintains an energetic pulse slightly "off" from that of the assemblage, an assemblage is never a stolid block but an open-ended collective. (Bennett 2010, 24, emphasis i.o.)

---

29   Within the example of studying science, the concept, according to Law, allows one to "recognise and treat with the fluidities, leakages and entanglements that make up the hinterland of research" (2004, 41).

30   Welz (2021a, 172) also acknowledges the important contribution of feminist STS to assemblage theory, which considers agency as distributed in assemblages across human and nonhumans.

Importantly, while understanding agency as distributed does not directly link agency to a moral subject, it also does not neglect intentionality, albeit consigning it "as less definitive of outcomes" (Bennett 2010, 32; *cf.* Hansen and Koch 2022, 9–10). Thus:

> Agency is, I believe, distributed across a mosaic, but it is also possible to say something about the kind of striving that may be exercised by a human within the assemblage. This exertion is perhaps best understood on the model of riding a bicycle on a gravel road. One can throw one's weight this way or that, inflect the bike in one direction or toward one trajectory of motion. But the rider is but one actant operative in the moving whole. (Bennett 2010, 38)[31]

Despite the parallels between assemblage thinking and ANT, they are also different in several ways. Cultural anthropologist Maria Schwertl (2013), for instance, pointed to the diverging forms of analyzing networks. While ANT is primarily interested in how networks emerge and are stabilized, assemblage theory focuses on its continuous recompositions (Schwertl 2013, 117).[32] Furthermore, ANT focuses on situational development, while assemblage concentrates on pervasive structures and logics (Schwertl 2013, 118).[33]

Focusing on irreversibilities within networks and how they are introduced can be important for analyzing power in the field of sciences (Star [1991] 2015, 275). However, ANT has been criticized for focusing too heavily on the analysis of specific associations and alignments between heterogeneous elements of networks (Beck 1997, 288).[34] According to Beck, this results in an (at least temporarily) stabilized network and the "assignment of specific, stable roles for all human and nonhuman actors" (1997, 288). Additionally, although ANT's contribution to overcoming binaries and boundaries was significant, Star writes:

> [O]ne of the features of the intermingling that occurs may be that of exclusion (technology as barrier) or violence, as well as of extension and empowerment. I think it is both more analytically interesting and more politically just to begin with the question, *cui bono?* than to begin with a celebration of the fact of human/non-human mingling. (Star [1991] 2015, 276–277, emphasis i.o.)

Despite the differences between ANT and assemblage theory discussed here, McFarlane concludes that they "nonetheless exist [...] in similar conceptual terrain attempting to confront the complexity of sociomaterial relationality" (2011b, 655).[35]

---

31 This reminds one of Suchman's (2007b) insightful canoeing example, with which she emphasizes the contingencies of action and the necessary incompleteness of plans.
32 Deleuze and Guattari, therefore, call the analysis "nomadology" (2013, 409ff.).
33 For further differences between ANT and assemblage theory, see Müller (2015).
34 It should be noted that writing about "ANT" in fixed terms simplifies its ideas and approaches. Of course, ANT has evolved since its introduction and adaptation (e.g., Law and Hassard 1999; Gad and Jensen 2010). One example is the work of ethnographer and philosopher Annemarie Mol, who demonstrated how phenomena are enacted in multiple versions and through different networks (Mol 2002a; 2002b; *cf.* Gad and Jensen 2010).
35 There are also similarities between the assemblage concept and Foucault's concept of the dispositive (Schwertl 2013, 118). However, since the dynamic nature of assemblages—which does not play

In contrast to ANT's rather unidirectional approach to the stabilization of associations, my research focuses on the continuous changes in role allocations, power distributions, and responsibilities that occur in human–technology relations.

Having discussed assemblage theory or thinking with a focus on DeLanda's and ANT's conceptualizations,[36] I summarize five main points related to how the assemblage concept—which I apply both empirically and analytically (Brenner, Madden, and Wachsmuth 2011, 231), as object and orientation (McFarlane 2011b, 653)—provides a useful theoretical approach to the analysis of HC-based CS projects. First, it allows me to remain open to the complexity of HC-based CS systems (Dietzsch 2022). Indeed, following DeLanda's (2016, 3) interpretation, assemblages are always assemblages of assemblages. To provide an example from my field, biomedical engineering in the laboratory itself can be understood as an assemblage that is part of the Stall Catchers assemblage.[37] I demonstrate in Chapter 6 how it can also be helpful to think of the scientific process in the example of Stall Catchers as an assemblage in and of itself.

Second, the concept of assemblage also directs the focus onto the relations between different human and nonhuman actors, which together form assemblages (Welz 2021a, 164). As a concept, it "highlights the dynamic, the procedural and the inconsistent dimensions of social orders [and sociomaterial practices] rather than their structural dimensions in and beyond societies" (Hansen and Koch 2022, 4). Using the concept of assemblage, then, it is possible to analyze the processes of stabilization (reterritorialization) and destabilization (deterritorialization) simultaneously acting upon Stall Catchers, both forming and changing it.

Third, and related to this latter point, assemblage thinking conceptualizes an assemblage's agency itself in processual terms (*cf.* Hansen and Koch 2022, 9). ANT, as well as its

---

an important role in dispositives—is particularly useful for my analysis of HC-based CS, I do not pursue this line of thought further here.

36   I briefly mention Buchanan's critique of both lines of thought, although I focus on how assemblage thinking provides a fruitful starting point for my analysis of HC-based CS systems—or assemblages. Buchanan criticizes ANT's and DeLanda's interpretations of assemblage theory for "cloud[ing] our understanding of Deleuze and Guattari" (Buchanan 2015, 383). Among his greatest points of criticism are the following: On the one hand, and even though ANT's focus on the agency of materialities and nonhuman entities is in line with Deleuze and Guattari's thinking, Buchanan argues it should not form the core of the analysis, and ANT, therefore, misses "what is central to the assemblage" (Buchanan 2015, 385). Furthermore, ANT, according to Buchanan, understands assemblages as indetermined collections and not as purposeful as "the deliberate realization of a distinctive plan (abstract machine)" (2015, 385). On the other hand, DeLanda, from Buchanan's point of view, focuses too much on the becoming of assemblages (2015, 382), and, in that sense, on the "how:" "Worrying about *how* a particular authority structure actually changes forgets that the real question here, at least insofar as assemblage theory is concerned, is *what* is that structure of authority? How is it constituted?" (Buchanan 2015, 388, emphasis i.o.). Deleuze and Guattari, by contrast, were interested more in questions related to the idea of the state itself (Buchanan 2015, 389). As I have stated previously, I do not aim to strictly follow Deleuze and Guattari's formulation of assemblages but, instead, focus on the assemblage concept in a way that is helpful to my analysis, and, thus, pursue a different goal from Buchanan with my discussion of assemblages.

37   It is, thus, important to reveal the cuts I place (Barad 1996, 170–171), the boundaries I draw, and specific regions of the assemblage I focus on in my analysis of HC-based CS projects as assemblages (see Chapter 6).

conceptualization of assemblages, has been criticized for its symmetric relational approach to agency, making it impossible to ascribe accountability to actors.[38] My perspective, therefore, departs from this approach. As I have already briefly mentioned in the introduction, I, instead, follow an asymmetric understanding of human and nonhuman agency, similar to Barad, who, following physicist Niels Bohr, argues as follows:

> (i) [N]ature has agency, but it does not speak itself to the patient, unobtrusive observer listening for its cries—there is an important asymmetry concerning agency: we do the representing, and yet (ii) nature is not a passive blank slate awaiting our inscriptions, and (iii) to privilege the material or the discursive is to forget the inseparability that characterizes phenomena. (Barad 1996, 181)

This understanding is also reflected in the relations, temporalities, and spatialities that form the assemblage, as some actors and processes have a greater influence on structuring, narrating, and forming them than others (McFarlane 2011b, 655). From this, it follows that assemblages are "structured, hierarchised, and narrativised through profoundly unequal relations of power, resource, and knowledge" (McFarlane 2011b, 655).

Fourth, thinking about HC-based CS systems as (sociotechnical) assemblages directs the focus to the different sociotechnical relations in which different human and nonhumans are linked and mutually form each other. Different human–technology relations are interwoven and influence each other productively. Simultaneously, they pull and push against each other, creating frictions while continuously forming and reforming, or reterritorializing and deterritorializing, the assemblage.

Finally, these relations are continuously changing alongside the assemblages, which, as Deleuze and Guattari (2013, 7) have pointed out, change themselves as they extend. I focus on selected human–technology relations and perspectives, elements, and dimensions of HC-based CS assemblages to answer the questions guiding my research.

Using the assemblage concept as a theoretical starting point, I now turn to selected theoretical conceptualizations of human–technology relations, and how these can be analyzed.

Human–Technology Relations

Beck writes that, at least since the era of modern industrial societies, the "relation between self and world is comprehensively technologically mediated and moderated" (1997, 248). Given the dense "texture of the 'technosphere' within which we undertake our daily affairs" (Ihde 1975, 271), technology forms an omnipresent condition of everyday life,[39] which it shapes and, at the same time, is shaped by (Beck 1997, 10). Therefore, the attempt to grasp and understand technology and its role in human life has occupied researchers in various scientific fields. I focus on relational conceptualizations of technology, specifically following Beck's complex situational understanding of technology and

---

38 As described above, Bennett (2004), therefore, emphasizes intentionality within distributed agency.
39 For these reasons, Hengartner (2012, 120) argues, technology can be understood as culture.

postphenomenological approaches to human–technology relations. Following relational conceptualizations of technology, the focus always lies on the relation between humans and nonhumans, such as humans and technology,[40] and considering both simultaneously. Humans and technology mutually constitute each other within and through their relations. While Beck writes about *user*–technology relations, I refer here to humans in order not to exclude diverse related actors, such as developers, providing specificity whenever necessary. From there, I develop the concept of intraversions, which adds a particular processual and temporal focus to the analysis of the evolution of and continuous changes within human–technology relations along instantaneous and gradual temporal developments in the rapidly advancing fields of HC and HI. I first turn to Beck's work on the use of and engagement with technology (*Umgang mit Technik*).

Beck's *Umgang mit Technik* (1997) has been highly influential in the German-speaking fields of cultural and digital anthropology but to date has not been published in English. In this book, Beck developed the analytical framework of a "complex situational analysis of the use of everyday artefacts," building upon a praxiological perspective on technology in order to take into account the materialities and use of technology in addition to the dimension of meaning, which, thus far, had been the focus in German "*Sachforschung*" (1997, 18, 20). He builds upon concepts from German cultural anthropology and European ethnology and its predecessor *Volkskunde*, as well on philosophical, ethnological, and sociological concepts of technology. Beck develops this framework to analyze the *usage of* what he calls "classical" *technology*, which he contrasts with information and communication technologies, such as the computer (1997, 232–233). Despite this demarcation, Beck's framework constitutes a landmark in the multidimensional and multiperspective sophisticated analysis of technology. Therefore, here I summarize the main elements of Beck's approach relevant to my research.

Beck describes the relationship between users and technology as an interactive process, a "feedback structure" (Zimmerli 1990, 252), in which technology has a formative influence on its possible uses. At the same time, users can render technical artifacts usable for their purposes in creative ways. The relational concept of technology stresses the analysis of *the use* of technology. It, therefore, establishes a "*situational understanding of technology*, which ultimately also opens up the perspective on the *cultural and social contexts of the use act*" (Beck 1997, 224, emphasis i.o.).

In analyzing technology in use and the relations between humans and technology, Beck combines a perspective which focuses on materiality ("*sachtheoretische Perspektive*"), taking technology as a starting point, through a perspective focusing on practice ("*praxistheoretische Perspektive*"), taking actors as its starting point (Bareither 2013, 32).

Focusing on technology itself, Beck builds on a differentiated understanding of context, which he divides into "hard, material con-texts," describing technological object potentials (dimension of practice) and "soft, discursive co-texts" (dimension of meaning), that together lead to the configuration of a user and orient their practice (Beck 1997, 294). Following literary scholar and philosopher Mikhail Bakhtin (1981), Beck stresses the

---

40   The focus in my study lies on human–technology relations, although these generally also include animals, such as mice, and other nonhuman entities, such as plants, trees, microorganisms, and other formations.

polyphony of co-texts, which can be simultaneously present in one situation and "juxtaposed to one another, mutually supplement one another, contradict one another, and be interrelated dialogically" (Bakhtin 1981, 291–292; cited in Beck 1997, 343). In these situations, different discourses and meanings are negotiated and define acceptable or appropriate practices around technology (and, thus, around HC-based CS projects as well). In Chapter 5, I specifically focus on the multiple meanings present in HC-based CS systems given their situatedness in various fields, such as play and science, and, in the case of Stall Catchers, the powerlessness experienced by some participants toward Alzheimer's disease.

For Beck, con-text refers to the situational practice in which technology's affordances are realized (1997, 342).[41] He reinterprets and expands upon the concept of affordance introduced by psychologist James J. Gibson (1977; 1979),[42] who introduced the term to describe how the physical properties of an object, independent of the user's perception, afford certain behaviors and interactions. Beck criticized Gibson's conceptualization for excluding the social and cultural conditions of technology use (1997, 244), the co-texts in Beck's terms, which restrict these affordances (1997, 304).[43] Furthermore, Beck explains that Gibson's definition did not consider the social context in which user–technology interactions occur or relations unfold. He adds that the same object can provide different opportunities to people depending upon their abilities, intentions, and social contexts. Nevertheless, Beck argues, the concept, if expanded, can be useful for the analysis of the everyday use of technology in two ways:

> First, [Gibson's] hint that perception is directly bound to bodily movement, to the actors' ability to act in space—their kinesthetics; and second, that the relation of user and object is to be regarded as the decisive "unit of analysis", characterized by affordances, by manifold and hardly clearly determinable object potentials. (Beck 1997, 244)

The concept of affordances, thus, focuses on the situated relations between humans and technologies. Beck's emphasis on the manifold and indeterminate object potentials opens up the analysis to human counteractions and creative practices,[44] which de Certeau described as "tactics" ([1980] 2013). These challenge discursive regulations as well as designed and programmed ways of use (Beck 1997, 244–245). The framework of a complex situational analysis, therefore, pays particular attention to resistance and creative

---

41   Beck further differentiates between *manifest* and *latent* con-texts. While technology offers various latent affordances, these are always constrained by specific co- and con-texts. Therefore, from many possible ways of acting, only a few are considered acceptable and favored (Beck 1997, 348).
42   Gibson's affordance concept has since also been used for and adapted to ethnographic and praxeological research by various authors (e.g., Boyd 2010; Costa 2018; Bareither 2019; 2020a).
43   Beck (1997, 244), for example, shows that the realization of the affordances of a letterbox depend on the knowledge of how a letterbox is embedded in the sociotechnical system of sending a letter.
44   This also relates to sociologist Ian Hutchby's interpretation of the affordance concept. Hutchby added a relational focus to Gibson's functional affordance concept that "draws our attention to the way that the affordances of an object may be different for one species than for another" (2001, 448).

practices and, thus, to the agency of users actively shaping their relations with technology. Beck also suggests moving beyond Gibson's affordance concept, which neglects social and intersubjectivity by including "social configurations" (Elias [1970] 2012) and their influence on how affordances are realized in social interactions (Beck 1997, 246). In accordance with, *inter alia*, psychologist William Noble (1981) and philosopher of science and ecological psychologist Edward Reed (1991), Beck refers to this dependence on social configurations as "socially mediated affordances" (1997, 246). Ultimately, however, the complex relationship should be considered a *"con-figuration* of technological artifacts and society: technological development can, thus, be understood as the result and condition of (not only) modern societies, in that new objects provide new affordances, which in turn enable different kinds of social figurations, from which in turn new objects emerge" (1997, 246, emphasis i.o.).

From a practice-theoretical perspective, it is crucial to focus on how affordances and usage instructions are activated and differently realized in everyday use. In this context, Beck conceives practice as both processual and reflexive, embodied and situated (1997, 298–299), which describes an "active and recognizing mode of being in the everyday life world" (1997, 298–299). He argues that as users realize the co- and con-texts in everyday use, technology as a social and cultural construct is then transformed into a "Tat-Sache" (Beck 1997, 295), a matter (or "thing") of practice. The analysis of technology or human–technology relations, therefore, always requires the analysis of both technology as a "use complex and use configuration" ("*Nutzungskomplex* und *-figuration*") (Beck 1997, 294, emphasis i.o.). This perspective emphasizes users' formative role in creatively constructing these relations and their ability to seize the situated contingency. Building upon the sociological theory of "contingency," Beck then conceptualizes technology as a "materialized *form* of contingency management" (1997, 223, emphasis i.o.). As such, "technology allows *diverse but not arbitrary* use possibilities" (Beck 1997, 223, emphasis i.o.). Beck writes that,

> through the design and functional specification of technical artifacts, options are provided within a spectrum that—in a creative process—must be realized by the users. Additionally, it must be emphasized that, in addition to the materiality of the artifacts, use instructions also discursively stabilize the possible uses. Culturally and socially bound technology can thus be conceptualized as *material and immaterial constraints on the contingency* of technical action—with technology and users being embedded in a sociomaterial feedback structure. (1997, 223, emphasis i.o.)[45]

Considering these different, but not arbitrary, possible uses in the analysis of HC-based CS projects is important for understanding how, for example, user–technology relations change and intravert over time as participants realize the options offered by the design of the platform and software in different ways. In Chapter 5, for instance, I discuss participants' practices, which by design go beyond the intended task contribution, and in Chapter 6 I show how participant–technology relations intravert.

---

45   Beck differentiates between, "*Gebrauchs- und Nutzungsweisen*" and "*Technik und Technologie,*" which are difficult to adequately translate into English.

Taken together, the multiperspective approach to the different dimensions (the co- and con-texts) and the distinction between technology as a "use and orientation complex" allows for the analysis of the "complex interwoven relations of artifact, culture, and user" (Beck 1997, 247). This analysis takes into account both the perspective of actors and their biographies[46] as well as spatial circumstances and temporal processes that go beyond the actual situation (Beck 1997, 344, 347–348).

Focusing on temporalities to study how human–technology relations unfold and change over time is specifically important to my research. Beck cites anthropologist Arjun Appadurai here, who argues that understanding the meanings of objects, which "are inscribed in their forms, their uses, their trajectories" (Appadurai 1986, 5), requires studying "things in motion" (Appadurai 1986, 5), that is, the "*total* trajectory from production, through exchange/distribution, to consumption" (Appadurai 1986, 13, emphasis i.o.). Different studies in cultural and digital anthropology and STS have already employed such an approach by tracing the different stages of the "biography" (Beck 1997, 291) of technology or objects (e.g., Bijker and Pinch 1984; Bijker and Law 1992; Löfgren 1994). Indeed, I adopt such an approach to things in motion, focusing specifically on selected human–technology relations within HC-based CS assemblages. This focus allows for a microperspective analysis, which considers particular human–technology relations while simultaneously considering how different relations are entangled with and shape each other, together creating assemblages.

Instead of demonstrating how researchers and developers strive to build HC-based CS as *black boxes*, I show how HC-based CS systems specifically rely on staying open to future changes, remaining at the edge of AI and scientific research. Crucially, not only does technology change over time but, along with it, so do the distribution of agency and the subject and object positions of the actors (Beck 1997, 292). I return to this point when introducing the concept of intraversions below.

The question of subjectivities and self-experience regarding human actors in relation to the world mediated by technology, which, according to Beck, must be considered a part of technology as a "use complex" (Beck 1997, 353–354), has been the focus of the philosophical branch of phenomenology. Phenomenology, which, since Edmund Husserl, has focuses on analyzing the world in the everyday (Verbeek 2001, 145),

> seeks to overcome the classical, Cartesian dichotomy between subject and object. Against this dualistic notion, phenomenology holds that subject and object [...] cannot be thought independently of each other, but only as always already related. Humans cannot be conceived apart from their relations to the world, and the world cannot be conceived apart from people's relations to it. (Verbeek 2001, 120)

The postphenomenological approach, developed in the philosophy of technology and connected to a broader change in the 1980s in the form of an empirical turn (Achterhuis 2001b), is helpful for my endeavor given its practice-oriented perspective and focus on

---

46 For example, as becomes clear in Chapter 5, for some participants, their background and relation to Alzheimer's disease play essential roles in their engagement with Stall Catchers and Foldit.

the embeddedness of technology in the everyday.[47] Dorrestijn summarizes this turn with the new theme, "for better or worse, humans have become hybrids with technology" (2017, 316). This understanding of technology as an intrinsic part of culture and human praxis (Ihde 1990, 20) is similar to digital anthropology and the anthropology of technology as described above. Haraway's "cyborg" (1985) and Latour's "hybrids" (1993), or philosopher of science and technology Don Ihde's "human-technology relations" (1975; 1990), serve as examples of this approach to technology (cf. Dorrestijn 2017, 316). Ihde's (1975) work on human–technology relations was particularly important for postphenomenology. Specifically, he developed a philosophy of technology, in critical engagement with Heidegger, focusing on technologies or technological artifacts "to reflect [upon] technology as it is concretely present in our daily experience" (Verbeek 2001, 122). Ihde, therefore, places the relation of humans to technologies at the center of his analysis and defines four types of human–technology relations and the world. Among these, the first three— *embodiment*, *hermeneutic*, and *alterity* relations—are "focal" human–technology relations, and the fourth are *background* relations (Ihde 1990, 98).

Embodiment relations ("[Human-technology] → World") describe contexts in which "I take the technologies *into* my experiencing in a particular way by way of perceiving *through* such technologies and through the reflexive transformation of my perceptual and body sense" (Ihde 1990, 72, emphasis i.o.). One of the most apparent examples of such relations are eyeglasses, which mediate the eyeglass wearer's relation to the world in an embodied and perceptual way. The second focal human–technology relations are *hermeneutic* ("Human → [technology-World]"). By referring to the hermeneutics in philosophy as an interpretation, these relations are characterized by "a special interpretive action within the technological context" (Ihde 1990, 80). Users of, for example, a wristwatch, experience the world differently through reading and interpreting the watch's display, explain philosophers Robert Rosenberger and Peter-Paul Verbeek (2015a, 17). In these two relations of mediation (Verbeek 2001, 124), technology forms a "means through which something else is made present" (Ihde 1990, 94). By contrast, in the third relation, the *alterity* relation ("Human → technology-[-World]"), humans have a "relation to technologies as relations *to* or with technologies, to technology-as-other" (Ihde 1990, 98, emphasis i.o.). The AI bots introduced in Stall Catchers represent an example of the latter: participants interact *with* the AI bots, which "emerge as the foreground and focal quasi-other" (Ihde 1990, 107). In contrast to algorithms operating in the background, the AI bots present artificial fellow (or competitive) participants to the human participants.

Finally, the fourth *background* human–technology relation refers to "technologies which ordinarily occupy background or field positions" (Ihde 1990, 108) *designed* and meant to function in or form the environmental context of humans (Rosenberger and

---

47 While "classical" philosophers of technology, such as Martin Heidegger, focused more on uncovering the essence of technology as an autonomous force determining society (Poel 2020, 500), empirical philosophers of technology understood technology as "fundamentally intertwined with" (Dorrestijn 2017, 316) the human condition. For an overview on the empirical turn, see Achterhuis (2001a) and Dorrestijn (2017, 316), and on different philosophical perspectives on the relation between society and technology, see Van de Poel (2020). For the differences between postphenomenology and "classical" philosophies of technology, see Dorrestijn (2012a).

Verbeek 2015a, 19). Examples mentioned by Ihde include heating or cooling systems but could also refer to infrastructures in general. Despite remaining (if functioning) backstage, "[b]ackground technologies, no less than focal ones, transform the gestalts of human experience and, precisely because they are absent presences, may exert more subtle indirect effects upon the way a world is experienced" (Ihde 1990, 112). Importantly, these different relations form a continuum and sometimes overlap (Ihde 1990, 93, 107) and "stand within the very core of praxis" (Ihde 1990, 108), since, following this postphenomenological perspective, "[t]here is no 'thing-in-itself'" (Ihde 1990, 69).

From Ihde's pioneering work and conceptual framework on human–technology relations, postphenomenology developed into an empirical, philosophical perspective (Rosenberger and Verbeek 2015a, 30). This perspective emerged in various fields, such as STS and the philosophy of technology, to analyze relations between humans and technology to which researchers from diverse disciplines, *inter alia*, sociology, anthropology, and philosophy, contribute (Rosenberger and Verbeek 2015b, 1). Postphenomenology combines the two philosophical traditions: phenomenology and American pragmatism. It does so by critically connecting the understandings of human–world relations from phenomenological scholars such as philosophers Maurice Merleau-Ponty and Martin Heidegger and their approach to *describing* phenomena and human experience "from a closer engagement" (Merleau-Ponty 1962; cited in Rosenberger and Verbeek 2015a, 11), instead of *analyzing* them from afar. At the same time, contrary to analyzing, for example, the alienation of the human experience from the world and from itself by technology, as in the work of Heidegger (1996; Rosenberger and Verbeek 2015a, 10), postphenomenology investigates how technologies "help to shape our relations to the world, rather than merely distancing us from it" (Rosenberger and Verbeek 2015a, 11); thus, how it mediates human experience. As such, following such a theory of technical mediation, "human existence is always, and inescapably, marked and influenced by technology" (Dorrestijn 2017, 312), a perspective that goes beyond both utopian and dystopian perspectives on human–technology relations. Building upon American pragmatism, it follows that the analysis of such relations focuses on the practices of engaging with technology and materiality (Rosenberger and Verbeek 2015a, 12): "It is in practices of interacting with technologies where the phenomenon of technological mediation occurs and can be studied. Human–world relations are practically 'enacted via technologies'" (Rosenberger and Verbeek 2015a, 12). This turn was also particularly important, making it possible for a cultural or digital anthropological analysis to connect to such a postphenomenological understanding of human–technology relations.

In line with ANT, postphenomenology seeks to move beyond the divide between subjects and objects and, following phenomenology's criticism of modernism used to ascribe the source of knowledge to objective facts *or* to subjective ideas, also concentrates on how subjects and objects rely on their interrelations (Rosenberger and Verbeek 2015a, 11). At the same time, postphenomenology focuses on the "fundamentally mediated character" (Rosenberger and Verbeek 2015a, 12) of these relations with technologies as mediators between subject and object. It, thus, does not consider subjects and objects *a priori* but as constituting each other (Verbeek 2005, 112); human subjectivity and objectivity are always shaped in mediation. In contrast to ANT, however, postphenomenology does not adhere to a strict symmetry (Ihde 2015, xv) between humans and objects or nonhuman actors. Instead, it is concerned with the mutual constitution between them and with overcoming their "separation" (Verbeek 2005, 166–168; cited in Rosenberger and Verbeek 2015a, 20):

> In order to see these processes of mutual constitution, and to do justice to human experiences of being subjectively "in" a world, it remains very relevant to make a distinction between humans and things. When we give up this distinction, we also give up the phenomenological possibility to articulate (technologically mediated) experiences "from within." (Rosenberger and Verbeek 2015a, 20)

In this research, I acknowledge how human actors and technology co-constitute each other in their interrelations and how these relations describe more than a connection between two independent entities (which they are not). However, to analyze how the role distributions, responsibilities, and tasks shift over time, I sometimes specifically consider the perspectives of humans and technologies independently and move the focus from one to the other.[48]

Here, Ihde's relations offer helpful starting points to think analytically about the relations between participants and software in HC-based CS projects and how Stall Catchers' contributors experience the world through and with the computational elements. However, in contrast to Ihde and technical mediation, my focus moves between relations. As suggested by Beck, I attempt to consider technology as both a use and orientation complex. Instead of applying Ihde's four types of human–technology relations to HC-based CS, I develop the concept of intraversions to study specifically how these relations change over time and along different temporalities. My focus, therefore, does not remain on the human experience of the world through and with technologies but moves to how the relations of humans and technology unfold and the potential that emerges from them. Nevertheless, a postphenomenological and practice-oriented approach to analyzing technology and human–technology relations in the everyday and concrete contexts of production and use (Beck, Niewöhner, and Sørensen 2012, 41) forms the starting point of my research. Following Beck, Niewöhner, and anthropologist of knowledge, STS, and data Estrid Sørensen (2012), agency and creativity are then studied "at the level of concrete practice and thus as distributed across human and non-human actors" (2012, 41).

Additionally, Beck raises a critical limitation of theoretical approaches building on phenomenology, which must be considered. According to Beck (1997, 312), phenomenology does not sufficiently consider social and intersubjective contexts, and, I would add, the broader embeddedness of human–technology relations in assemblages. Moreover, even if phenomenology does not actively neglect these contexts, it carries an "individualistic bias" (Beck 1997, 312). Beck, therefore, recommends focusing on approaches suggested by American pragmatism and Marxism. While postphenomenology combines phenomenology with American pragmatism to overcome this problem, I, nevertheless, additionally draw from the assemblage concept described above to place human–technology relations within sociotechnical assemblages. I, therefore, pay attention to how

---

[48] For a comparison of the "more complementary than combative" (Ihde 2015, xvi) styles of analyzing technology, see, among others, Ihde (2015).

they are interwoven with other relations, which they form and are formed by, also showing how other human and nonhuman actors influence these relations. Moreover, combining approaches to human–technology relations with assemblage thinking allows me to focus on the continuous process of their formation (Cassirer 1985, 43).

Suchman's (2007b) concept of "situated actions" in human–machine configurations also proves helpful here, since it not only turns away from the idea of human intention as the driving force of action but also, to some extent, questions the understanding of causal sequences of action in linear time. This opens up the perspective to both temporalities and contingencies, also stressed by Beck (see above).

Combining assemblage thinking with human–technology relations as analytical perspectives, thus, helps one to focus on the various nonlinear processes that form, stabilize, and destabilize HC-based CS systems and their human–technology relations. This combination also connects to a "becoming ontology" (Hultin 2019), such as agential realism (Barad 2007a), with the understanding that humans and nonhumans are entangled and continuously formed as intrarelations (Hultin 2019, 92). Sociomaterial studies following a becoming ontology pay particular attention to how "agency emerges, transforms, and enacts as a temporal and performative flow of practices," summarizes economist Lotta Hultin (2019, 93). Specifically, in my research I focus on the temporal becoming of human–technology relations both along instantaneous everyday life and gradual temporal developments, what digital innovation and information systems scholar Reza Mousavi Baygi and colleagues have described as a "flow-oriented genealogical" analysis (Mousavi Baygi, Introna, and Hultin 2021). Hansen and Koch, following anthropologists Paul Rabinow and George Marcus' concept of an "anthropology of the close future" (Rabinow et al. 2008), have pointed to the potential of examining temporality in addition to space in assemblage thinking, bringing together "past, present and anticipated actions" (Hansen and Koch 2022, 6). I hope to contribute a tool for this endeavor with the concept of intraversions. As I argue later in this work, my position is that, here, in bringing together past, present, and the anticipated future, the potential of ethnographic research lies not only in the deconstruction and criticism of studied phenomena, but also in the contribution to shaping HC-based CS assemblages in ways that acknowledge the different interests and perspectives of actors involved and which embraces the contingencies and multiplicities of everyday life.

Before turning to the concept of intraversions, I discuss one final line of theoretical thinking important for my research, one which is concerned with ethics. Such an analytical focus is necessary given that HC-based CS projects (and specifically in the example of the Human Computation Institute's projects) are often imagined and legitimized as the *good* and *right* way to solve specific computational and scientific problems. This naturally leads to the ethical framing of the phenomenon studied. Following anthropologist Michael Lambek, I consider ethics as always representing an "intrinsic dimension of human activity and human lifeworlds" (2015b, 18) and, therefore, investigated in and with the everyday (Lambek et al. 2015, 3; Fassin 2015). Such a moral anthropological perspective directs the focus in ethnographic field research to the question of what is understood and problematized as "good." I discuss in the following section the concept of "ethical projects" (Ege and Moser 2021a), "sociotechnical imaginaries" (Jasanoff and Kim

2009; 2015), and Dorrestijn's "subjectivation and technical mediation" (2012a) in order to do this.

Ethical Projects, Imaginaries, and the Care for Our Hybrid Modes of Being

> Ethics of technology does not entail defending what is genuinely human, but caring for the quality of one's hybrid mode of being. (Dorrestijn 2012b, 234)

The field of moral anthropology has garnered much interest in the last few years (e.g., Faubion 2011; Fassin and Lézé 2014; Fassin 2014; 2015; Lambek et al. 2015; Dürr et al. 2020; Ege and Moser 2021b).[49] In my research, moral anthropology and the ethics of technology form a crosscutting analytical perspective which cannot be clearly separated from other analytical perspectives. For the sake of comprehensibility and to provide an overview, however, I will discuss moral anthropology and the ethics of technology separately in this section.

From a moral anthropological perspective, I aim, first of all, to analyze on an empirical level what is understood and problematized as "good" in the design and development of HC-based CS projects. Specifically, I focus on the Human Computation Institute (see Chapter 4) and why participants contribute to such sociotechnical systems (see Chapter 5). The question is as follows: "[A]ccording to which values, by which means, to what ends, and with what deviations or lapses do people try, and in fact, *do*, make their way?" (Lambek 2015, 9, emphasis i.o.). Here, the analytical concept of "ethical projects" (Ege and Moser 2021a), developed by cultural anthropologists Moritz Ege and Johannes Moser in the context of the interdisciplinary research group "Urban Ethics" (Ludwig-Maximilians-Universität München, n.d.b), which focuses on questions of urban life and its ethical dimensions, is helpful to my analysis. How human actors relate to HC applications alongside how they shape them and are shaped by them can be discussed with reference to

---

49 It should be noted that a long debate exists on the question of the difference between ethics and morality. In fact, there not only exist different understandings of the terms, but sometimes, they are used interchangeably (Fassin 2012, 6). Foucault, for example, distinguishes between three different forms of morality (1988, 25–26), which Fassin summarizes as "moral code", "moral behavior," and "ethical conduct" (2012, 7). The last form refers to "the manner in which one ought to form oneself as an ethical subject acting in reference to the prescriptive elements that make up the [moral] code" (Foucault 1988, 26). It is this form which Foucault deals with in his work and that has influenced one branch of anthropological approaches to ethics and morality (*cf.* Fassin 2012, 7) as well as Dorrestijn's (2012a; 2012b) ethics of technology, which I discuss later. However, as Fassin emphasized, despite the importance that the distinction between morality and ethics might have from philosophical and conceptual perspectives, in the empirical situations examined by anthropologists, these demarcations blur (2012, 8). Here, I follow Ege and Moser's (2021a) pragmatic approach to this distinction and their understanding of ethics "as the ways in which individuals engage with and relate to moral codes, as socially legitimated and, in that sense, normative 'good' behavior and 'proper' (or 'right') conduct of life. Ethical practice is a form of subjectification or subjectivation, of becoming a type of subject. It is also a form of subjection that relates individuals and groups' regimes of living to broader configurations of power and rule" (Ege and Moser 2021a, 13).

Foucault's analysis of the subject and subjectification (e.g., 1983; 1988), in addition to his reflections on technologies and power. Dorrestijn (2012a), following Foucault, developed a theoretical framework of "subjectivation and technical mediation," which I discuss below. This framework helps in the analysis of how individuals relate to technology and technologies, and how these, in turn, affect actors and predefine certain forms of action.

Considering the perspective of designers and HC developers who strive for "better" human–AI systems with "unprecedented capabilities," HC-based CS projects can be understood as "ethical projects" as defined by Ege and Moser (2021a). In the context of urban anthropology, these are:

> future-oriented undertakings with a certain amount of pre-planning, self-awareness and intentional communication that promise better or more just cities and a better urban life through assemblages of policy, technology, buildings, aesthetics and institutions, and also a ethico-moral sense of "something better." (Ege and Moser 2021a, 7–8)

The aim of such projects, therefore, is not only to improve the quality of urban life but "the ethical character and the ethical valence of urban life" (Ege and Moser 2021a, 7). Applying this concept to the development of HC as new sociotechnical systems allows us to analyze what is understood as ethical or moral by designers and developers of such systems rather than normatively evaluating *if* something is moral or ethical (Lambek 2010; cited in Fassin 2012, 6). Thus, ethical projects are formed by the imagination of a "good life."

The STS scholars Sheila Jasanoff and Sang-Hyun Kim introduced the concept "sociotechnical imaginary" to investigate the relationships between technologies, imagined futures, and society by focusing specifically on how these imaginations are normativized and on the materialities which are part of sociotechnical networks (Jasanoff 2015a, 19). In their first definition, they described sociotechnical imaginaries as "collectively imagined forms of social life and social order reflected in the design and fulfillment of nation-specific scientific and/or technological projects" (Jasanoff and Kim 2009, 120). While this understanding focuses on the analysis of national imaginaries, which they developed in their work analyzing nuclear power and South Korean and US responses to it in their later book *Dreamscapes of Modernity* (Jasanoff and Kim 2015), they also broaden the concept "to do justice to the myriad ways in which scientific and technological visions enter into the assemblages of materiality, meaning, and morality that constitute robust forms of social life" (Jasanoff 2015a, 4). The revised definition includes, *inter alia*, social movements, organizations, and professional associations that advocate for their sociotechnical imaginaries, which can also stem from individual visionaries and are often taken up and spread by powerful institutions, including the media or lawmakers (Jasanoff 2015a, 4). Sociotechnical imaginaries are, thus, shared visions of futures that are desirable, and which are both demonstrated in public and "institutionally stabilized" (Jasanoff 2015a, 4). Such imaginaries form through shared ideas of and the belief that progress and developments in science and technology can lead to and support desired forms of social organization and ways of life (Jasanoff 2015a, 4). By considering the "normativity of the imagination with the materiality of networks" (Jasanoff 2015a, 19), the concept forms a bridge for my research between an assemblage analysis focusing on the interplay of human and nonhuman actors and their relations, and a moral anthropological approach investigating the understandings of the "good" in and with HC-based CS.

However, following Jasanoff and Kim's thinking, imaginations in the field of HC might not currently have risen to the status of an imaginary but can be better described by the term "vanguard visions," introduced by STS scholar Stephen Hilgartner (2015). As a relatively new concept and phenomenon, HC is still highly flexible, partly unstable, and under constant development and change. That is, a coherent history has yet to be established. Even if the term used, applied, and referenced today by scientists with varied backgrounds in their work and a growing scientific community identifies itself as HC (or HI) researchers, the visions, shared concepts, and ideas can be traced back to a few individual "sociotechnical vanguards." I return to this concept in Chapter 4.

Bringing together the concepts of sociotechnical imaginaries and ethical projects, then, according to Ege and Moser, ethical practice is both a "form of subjection that relates individuals and groups' regimes of living to broader configurations of power and rule" (2021a, 13) and "a form of subjectification or subjectivation, of becoming a type of subject" (2021a, 13; see also Foucault 1983; 1988). The latter refers to the other important consideration in my research: how other human actors involved in HC-based CS relate to these defining and forming suggestions of ethical projects. As an example, how do participants engage in and relate to HC-based CS (see below)? Ethical projects and sociotechnical imaginaries do not occur in a vacuum. Instead, they are materialized and maintained in everyday life through, for example, the practices of infrastructuring (see Chapter 4) or trust (see Chapter 7). In this everyday enactment of human–technology relations and the becoming of HC-based CS assemblages, different actors contribute to the reterritorializing and deterritorializing processes that form the assemblages and, therefore, bring in their own motivations and goals, which do not always align with the ethical projects and imaginaries but potentially contest them.

How participants relate to the HC applications designed and how they engage with technology in HC-based CS systems, thus, how they shape and are shaped by them, can be researched with reference to Foucault's analysis of the subject and subjectification (e.g., 1983; 1988), as well as his considerations on technologies and power (e.g., 1995; 1998). The approaches of the anthropology of ethics (Faubion 2011) and the postphenomenological approach to the ethics of technology by Dorrestijn, which I build upon in my research, are inspired by Foucault. Dorrestijn (2012a; 2012b) brought together the theory of technical mediation described above with ethical subjectification and technology (or technologies in Foucault's thinking) for an ethics of technology. Dorrestijn's approach to "subjectivation and technical mediation" links the analysis of human–technology relations and the moral anthropological perspective outlined above. In what follows, I briefly outline Dorrestijn's approach, which connects back to the postphenomenological understanding of human–technology relations and technical mediation.

By considering humans as hybrids,[50] the theoretical approach to technical mediation, which Verbeek (e.g., 2005) in particular further developed, "undermines the ethical

---

50   Even though my focus lies on hybrids of humans and technology, humans are hybrids not only in relation to technology but in multiple ways (e.g., Haraway 1991; Star [1991] 2015).

stakes that inspired much of the philosophy of technology" (Dorrestijn 2012b, 226–227). The problem with this, according to Dorrestijn, is that, if there are no autonomous but instead only hybrid subjects entangled with technology mediating our experiences and relations to the world, the question arises: How then can ethics be considered?[51] To solve this problem, Dorrestijn's ethics of technical mediation follows Foucault's thinking, and approaches human–technology relations by focusing on *"caring for the quality of the interactions and fusions with technology"* (2017, 317, emphasis i.o.). This ethical approach to human–technology hybridization allows one to not commit to either a technology pessimistic/dystopian or optimistic/utopian perspective. Dorrestijn's perspective corresponds to moral anthropology in that it does not seek to normatively evaluate the morality of a given situation but, instead, to understand what values people follow, for what reasons, and how (Lambek 2015, 9), as well as how they "cop[e] with the technical conditions of [their] existence" (Dorrestijn 2017, 317).

Dorrestijn's "technical mediation and subjectivation" framework requires bringing together Foucault's early work on (disciplinary) power and later work on the subject:

> While Foucault's earlier work is rightly seen as a dramatic attack on the autonomous subject presupposed in modern ethics, his later work is concerned with developing an alternative ethical framework wherein "the subject" is not eliminated by revealing its external conditions. Foucault begins to understand ethics as the active engagement of people with governing and fashioning their own way of being in relation to conditioning circumstances. An extension of that framework to the problem of technical mediation opens up a new perspective for ethics in relation to technical mediation. (Dorrestijn 2012b, 227)

Through this interpretation of Foucault's work, "the ethics of technology means an ongoing 'problematization', or a 'critical ontology,' of our technically mediated existence. The aim is finding, or forcing, openings to possible transformations of our way of being" (Dorrestijn 2017, 319). Instead of studying subjectivation in regard to sexuality, as Foucault did, Dorrestijn investigates subjectivation in relation to technology:

> The question is then how people perceive and conceptualize the influence of technology on themselves (and others, human beings in general). [...] Articulations of the mediating effects of technology are simultaneously ethical problematizations of how one's own mode of existence is affected by technology. (Dorrestijn 2012b, 234)

Dorrestijn's work, therefore, is of particular importance in analyzing how human–technology relations come into being and how intraversions of human–technology relations "contribute to the coming about of new forms of subjectivity" (2017, 318). Ultimately, this understanding allows us to analyze the practices of actors within such relations in order to cope with changing roles and power distributions. Dorrestijn used design and engineering pilots and usability tests as examples regarding where the approach to the ethics of technical mediation can provide interesting insights:

---

51   Latour's (1992) answer to the question of morality regarding sociotechnical or hybrid human–nonhuman entanglements is that morality is (partly) delegated to nonhumans.

Tests are normally performed to examine the technical functioning of new products. These moments also offer a privileged possibility to observe technologies in use for the first time. However, from the perspective of subjectivation, it should be stressed that testing must not be seen as a last check moment, which marks the transfer of a product from its design phase to its use phase. Instead, pilots and tests offer the possibility to see how the accommodation of technology by users takes place, in an experimental setting, and with the possibility of making adjustments to the technology. (Dorrestijn 2012b, 237)

In these testing environments, it can be observed how users relate to the technology under examination, how they are "conditioned by their environment [, and] how people transform themselves, [to] become subjects in an environment" (Dorrestijn 2012a, 119). According to Foucault's and Dorrestijn's understandings of ethics, "freedom," then, is precisely such a "'practice' of conducting oneself by actively coping with external power" (Dorrestijn 2012b, 238).

In Chapter 6, I discuss how Stall Catchers participants relate to the introduction of AI bots as new participants on the platform as an example from my research. As I will show, from the perspectives of participants—and not specifically in line with how human–AI bot teams were envisioned by the Stall Catchers team—AI bots become the other (Ihde 1990, 98). In relation to technical mediation, the human is only "knowable and recognizable" (Rhee 2018, 3) in technology. By linking these elements, we can ultimately ask how actors perceive the functioning of technologies and their own role within them. This allows us to examine the values motivating participants, developers, and researchers, to which actors orient themselves in relation to technologies (Dorrestijn 2012b, 221–222). Dorrestijn argues that this can lead to the design of an "ethics of technology," whereby one cares "for the quality of one's hybrid mode of being" (2012b, 234).

In this chapter, I have discussed three different theoretical approaches: assemblage thinking, human–technology relations, and moral anthropological and ethical concepts, which together form the analytical foundation of my research and support the development of the *intraversions* concept. In what follows, I introduce intraversions as a concept, which attends to the evolution of and continuous changes occurring within human–technology relations in HC-based CS systems.

Intraversions in Human–Technology Relations

Once HC-based CS assemblages are formed—despite continuous processes of deterritorialization and reterritorialization changing the assemblage—they also (re)configure their elements and formative relations in everyday life. Human–technology relations in HC-based CS, therefore, also do not converge to any specific pattern that is stable and remains the same. Actors and materialities do not always engage with relations and sociotechnical systems in the same way. For instance, materialities and infrastructures, such as servers, break down from time to time requiring human interventions, the life cycles of mice can diverge from research agendas, and human actors can intentionally act differently by finding new ways to engage with a game interface and software.

In addition, the *raison d'être* of HC systems, as I will show in the chapter on HC's imaginaries (Chapter 4), requires that they remain at the edge of technological/AI capabilities and—in the field of CS—of scientific research. This edge is continually moving given the last mile of automation, which Gray and Suri call the *"gap between what a person can do and what a computer can do"* (2019, xxii, emphasis i.o.).

Therefore, HC system's human–technology relations must remain open for future tweaking and change. This results in additional forces (primarily, but not entirely, coming from HC developers) acting on and changing human–technology relations, specifically the distribution of agency and the role assignments of subjects and objects, tasks, and responsibilities within these relations over time.

Using the concept of *intraversions*, I attempt to describe processual forward movements and shifts within relations between humans and technology. These movements result from the introduction of new computational capabilities or through new potentials arising from existing relations, which form directly from human actors' practices or algorithmic and material affordances. Along these processual forward movements, various forms of reconfigurations occur. These include 1) shifts in the role assignments of subjects and objects—or, more precisely, in the distributed agency across the different actors—, which can never be fully attributed to one side or the other; or 2) redistributions of tasks or practices, which result in reconfigurations of power dynamics. Intraversions can, thus, be understood as oscillations or weighted shifts appearing within these relations and which I chart along both everyday instantaneity and gradual development. I combine these two temporal dimensions in the analysis of evolving human–technology relations in HC-based CS in order to analyze their intraactions beyond specific moments.

Merely focusing on singular moments and situations has been criticized by European ethnologist Jens Wietschorke, who argues against ANT's emergence-theoretical approach:

> Social actors constantly negotiate meanings, and they do so within conflictual networks of relationships. They negotiate their affairs also within the framework of networks involving nonhuman actors and actants, forming chains of action that converge in concrete situations. But that is only one side. The social is not always newly constituted in the moment, but is embedded in discursive formations and structures of history and society that remain indispensable heuristic categories for social and cultural analysis. (Wietschorke 2021, 64)

Following Wietschorke, if the moment forms the sole focus of analysis, it fails to include historical developments and discursive formations forming and acting upon relations in those specific moments (such as through path dependencies leading to certain configurations of humans and nonhumans) even if these forces remain intangible in the moment itself (2021, 57).

As an analytical tool, the concept of intraversions aims to contribute to the digital anthropological and STS analysis of human–technology relations in HC-based CS. Evolving from my analysis of human–technology relations in this field, the concept of intraversions serves as a magnifier of the forward circular movements within relations between humans and technology, combining the theoretical concept of assemblages, and relational and processual approaches to technology described above, alongside the concept of distributed cognition put forth by Hutchins (1995a) discussed below.

Various terms, such as "transformation" or "change" in general, describe movement and development. However, the concept of intraversions differs from these terms in important ways. While the concept of transformation in general describes an unspecific, not necessarily goal-oriented reshaping or transformation of something, intraversions are more specific, capturing forward-pushing transformative processes that take place within relations. I derive the term intraversion primarily from inversion, which describes the reversal of what existed before, creating the exact opposite or turning something upside down. The Latin prefix "intra" (meaning "within") is invoked in reference to how changes and shifts in human–technology relations do not necessarily lead to the exact opposite of what previously existed but instead identifies movements *inside* and *across* relations in which humans and technology become interwoven. Thus, intraversions focus on inner changes, the partly unpredictable emergent oscillations appearing in various dimensions of human–technology relations unfolding within HC systems. The term also draws upon Karen Barad's term "intra-actions" (1996), since it concentrates on relations between human and nonhuman actors, who and which dissolve and are formed through the relations in which they engage. Humans and technology are, therefore, not independent of each other but form within their specific intraactions with one another. Instead of referring to interactions between fixed and independent entities, Barad uses the notion of intraactions to move beyond such dichotomies (1996, 179). Thus, the concept of intraversions also connects to Karen Barad's described asymmetry regarding human and technological agency.

My definition of intraversions can be further specified using five characteristics which I discuss below. These characteristics cannot be completely separated from each other but are discussed individually here for better comprehensibility.

First, intraversions are not static or fixed entities but processes. In these processes, human–technology relations temporarily stabilize. I take inspiration here from Latour and sociologist Steven Woolgar's discussions in *Laboratory Life: The Construction of Scientific Facts* ([1979] 1986). Specifically in their book, Latour and Woolgar focus on how statements stabilize, analyzing how scientific facts are constructed in the everyday activities of scientists in a laboratory. These processes lead to temporarily stabilized relations (such as algorithmic tools assisting humans in folding protein structures or humans assisting computational models in analyzing data). Yet, at the same time, human–technology relations always remain open for future intraversions. Various actors, such as developers and participants in HC-based CS projects, also contribute to this by intentionally pushing for new developments or action potentials. In my own work, I discuss, for example, how participants in Foldit's predecessor Rosetta@Home demanded the possibility to intervene or how Stall Catchers participants explored new ways of engaging with the platform (Chapters 5 and 6). In this sense, intraversions continuously evolve as modifications of existing relations.

Second, intraverting relations reconfigure actors' tasks, practices, forms of engagement, and even subjectivities within sociotechnical systems. This characteristic builds upon Hutchins' observation of the development of distributed cognition,[52] which he made when studying naval navigation (Hutchins 1995a) and airplane cockpits (Hutchins 1995b). Hutchins' work contributes to understanding cognition as well as human–technology relations.[53] In *Cognition in the Wild*, he aims to contextualize cognitive activity within a sociocultural and material world, "where context is not a fixed set of surrounding conditions but a wider dynamic process of which the cognition of an individual is only a part" (Hutchins 1995a, xiii). Distributed cognition resembles assemblage thinking in that it shows how "various elements [are brought] into coordination with each other" (Hutchins 1995a, 123). Hutchins speaks about a "cognitive ecology," where different tools and humans relate to and mutually support each other in different tasks (Hutchins 1995a, 114). Tools here are also understood as mediating technologies (Hutchins 1995a, 154). Hutchins vividly demonstrates, through the example of solving distance–rate–time problems in naval navigation, how a task that needs solving changes depending upon how a problem is presented as well as how the introduction of new tools changes the task for human problem-solvers (1995a, 147–55).[54] This example demonstrates that the

---

52 Hutchins' goal is to replace both mentalist and behaviorist approaches to cognition using distributed cognition (1995a, 129). In this way, the concept of distributed cognition can also be understood as a critique of the cognitive approach to AI and HC systems, which considers intelligence as information processing. I return to this aspect when discussing HC imaginaries in Chapter 4. Hutchins moves from a "classical cognitive science approach" (1995b, 266), whose research object is an individual human, to an analysis of sociotechnical systems. It should be noted that Hutchins referred to cognitive science at the end of the twentieth century, which might not reflect the state of the art in cognitive science today. Taking cognitive science's guiding metaphor of cognition as computation as the starting point but applying it to sociotechnical systems, he analyzes the representations internal to sociotechnical systems (Hutchins 1995b, 266). "For our purposes, 'computation' will be taken, in a broad sense, to refer to the propagation of representational state across representational media. This definition encompasses what we think of as prototypical computations (such as arithmetic operations), as well as a range of other phenomena which I contend are fundamentally computational but which are not covered by a narrow view of computation" (Hutchins 1995a, 118). While the internal processes of humans remain hidden in observational studies (Hutchins 1995a, 49), based on Hutchins' approach, "internal" can refer not only to tools and technologies but also to human action, for example. To Hutchins' definition, I would add transformation, not only propagation, "of representational state across representational media." Through this understanding of computation, cognition can then be analyzed in the various interactions between humans and nonhumans.

53 This could also be included in the discussion of human–technology relations in the previous section, but here I discuss it separately to directly connect a specific element of Hutchins' theory to intraversions.

54 He depicts different ways of solving such distance–rate–time problems. A task performer, for example, could rely on their algebraic knowledge and pencil and paper or, instead of paper and pencil, a pocket calculator, or they could have either a three-scale nomogram or a nautical slide rule as tools at their disposal. With algebraic knowledge, solving individual arithmetic operations is no problem. However, the difficult part in the first two scenarios is the coordination of these operations with each other. While the calculator makes it easier for the task performers to solve the individual arithmetic operations, coordinating them remains tricky. In the third scenario, the coordination of the operations is already built into the tools and algebraic knowledge is not required: "The nomogram and the slide rule transform the task from one of computational planning (figuring out what to divide by what) to one simple manipulation of external devices" (Hutchins 1995a,

task or problem that a task performer has to solve is presented to them differently, depending on the individual tools, requiring "a different set of cognitive abilities or a different organization of the same set of abilities" (Hutchins 1995a, 154). Through the introduction of new technologies or tools to a cognitive ecology—and solving problems such as the distance–rate–time problem should never be considered separately but always within cognitive ecologies—the tasks themselves change. Based on this understanding, I demonstrate how in HC-based CS, tasks, practices, and subject–object position assignments change within the human–technology relations, such as through the introduction of new automated tools in Foldit or the introduction of AI bots in Stall Catchers. Regarding the relations studied in HC-based CS, these changes result from the introduction of new computational capabilities and through new potentials arising from relations themselves when, for example, participants actively use the timing of a particular moment to their advantage (Mousavi Baygi, Introna, and Hultin 2021), as well as through algorithmic and material affordances. I then demonstrate how, in relation to the data pipeline, participants in HC-based CS games and researchers also adapt to changes in their subject positions (Beck 1997, 292) and how they reposition themselves and relate to technologies (Dorrestijn 2012a) because of intraversions. In fact, the tasks and purposes of the sociotechnical assemblages themselves continue shifting. This understanding connects back to Deleuze and Guattari's understanding of assemblages as "chang[ing] in nature" (2013, 7) with new and spreading connections, as I specifically show through the examples of Foldit and ARTigo in this study.[55]

Third, intraversions refers to how power dynamics, responsibilities, and agency are redistributed across relations through their reconfigurations. Power is not understood here as belonging to one party or actor. Instead, I adopt Foucault's concept of power as something that is distributed. "Power is everywhere," Foucault wrote (1998, 93), in the first volume of *The History of Sexuality*. Power, then, "must be understood in the first instance as the multiplicity of force relations immanent in the sphere in which they operate and which constitute their own organization" (Foucault 1998, 92). Understood as force relations, power is dynamic, not static. Key to this dynamic understanding of power is that it includes resistance (Foucault 1998, 95). Importantly, Foucault describes these power relations as "both intentional and nonsubjective" (1998, 94). Thus, individuals are not mere passive objects of power, but also exercise power.[56] Power relations "are imbued, through and through, with calculation: there is no power that is exercised without a series of aims or objectives" (Foucault 1998, 95). While power is intentional, the outcome of these intentions remains uncertain and depends upon interactions with other

---

150). The tools, here, both "constrain the organization of action of the task performer" (Hutchins 1995a, 151) and "they are representational media in which the computation is achieved by the propagation of representational state" (Hutchins 1995a, 154). The tools, then, transform the task for the task performer.

55 If assemblages change as a whole and are flexible, uncertain structures are debated and questioned by, for example, Buchanan as a misinterpretation of Deleuze and Guattari (Hansen and Koch 2022, 8). Since I do not strictly follow one interpretation of the assemblage theory but refer to the concept in ways that help me better understand and analyze HC-based CS and human–technology relations, I acknowledge these discussions but do not go into more detail here.

56 See Heller (1996) for more on power and intentionality in Foucault's work.

power relations. While the developer's intention to create teams of humans and AI bots by introducing bots into Stall Catchers without including the participants in all development steps leads to new user–AI bot relations, these can sometimes unfold in unintended ways. In Chapter 6, I demonstrate how participants do not necessarily engage with AI bots as team partners but also as competitors. Along similar lines, agency in intraverting human–technology relations is understood as distributed and continuously redistributed. As discussed above, Bennett (2010) defines agency in way consistent with Foucault's understanding of power as distributed. However, this does not exclude intentionality or "striving that may be exercised by a human within the assemblage" (Bennett 2010, 38). Following these definitions of power and agency, such dynamic changes within human–technology relations can be grasped through the concept of intraversions. Anthropologist Tim Ingold's notions of the weaving and knotting of relations (2007) also come close to delineating these changes, although intraversions describe the circular forward movements and their related oscillations within relations. They are nearly circular because they evolve and are pushed into new, imagined-to-be better relations between humans and technology in HC-based CS systems.

Fourth, and moving on to the next characteristic, intraversions are not only imagined but are also material, situational, and contingent. Referring to the third characteristic, it follows that intraversions are always situated and materialized in specific configurations of humans and technology (Suchman 2007b). At the same time, they are, to a certain degree, contingent and unpredictable due to the multiplicity of relations interacting with one another and due to different actors' intentions. This also extends to "serendipitous discover[ies]" (Schaffer in Human Computation Institute 2018, 00:33) and instant breakdowns in some entities, such as servers.[57] Therefore, new action potentials can emerge from existing human–technology relations. In addition, however, intraversions do not merely occur via the relations' own momentum but are also differently imagined by various actors. In Chapter 4, I focus on how HC advocates and developers imagine HC systems and their user–technology relations in developing HI, and how these imaginations themselves are always emergent and rendered material.[58] Intraversions are, thus, led by both constant attempts at structuring by different actors according to their imaginaries alongside situational and incidental failings—that is, their contingency.

Finally, intraverting relations are multiples. Anthropologist and philosopher Annemarie Mol (2002b) studies how atherosclerosis disease is enacted and practiced in multiple ways and forms in her book *The Body Multiple: Ontology in Medical Practice*. Instead of defining atherosclerosis disease as a singular object, she studies it through multiplicities of practices. Human–technology relations in HC-based CS are similarly enacted through different assembling practices. User–technology relations in Stall Catchers, for example, can be enacted along the lines of the game and HC system design. However, they can also be enacted through new participant engagements, which circumvent the patterns designed. At the same time, such relations can carry various meanings for different actors and from different perspectives (see Chapters 5 and 6).

---

57 This holds not only for HC-based CS systems but for sociotechnical systems in general.
58 I will also discuss how intraversions of researcher–technology relations are, for example, imagined by researchers with the aim of someday fully automating the data pipeline.

Although intraverting relations are multiples (Mol 2002b), despite their diversity—as Beck pointed out for use possibilities of technology—, they are not arbitrary (Beck 1997, 223). Similarly, intraversions are not arbitrary, and, in analyzing them, attention must be paid to the processes, discursive elements, specific materialities, and contingencies forming them.

To summarize, the concept of intraversions allows us to investigate from within, showing how human–technology relations continually evolve. As a concept and tool, it attends to the continuous development of these relations both along everyday instantaneity and gradual development. This concept facilitates analyses regarding how previous relations and power dynamics shape sociotechnical assemblages and how they form and influence current and future relations. In this way, intraversions as a concept analyzes sociotechnical systems and their human–technology relations across the past, present, and anticipated future. Notably, given the nature of human–technology relations in HC-based CS as always unfinished and open, their intraversions cannot be explored in their entirety—that is, they are always excerpts of continuous motion.

Following the theoretical approaches and conceptualizations of assemblages and human–technology relations discussed in this chapter, I investigate the continuous formation and reformation of HC-based CS assemblages through the intraactions of their various elements. My investigation includes a focus on intraverting human–technology relations guided by the imaginations of how hybrid systems should be built in the future. Bringing these different lines of theoretical thinking together, my discussion and the argument brought to the fore in this chapter demonstrate the usefulness of such a "flexible" (Heimerdinger and Tauschek 2020, 16) engagement with theoretical concepts for analyzing and better understanding HC-based CS systems.

3 Methodology: Encountering Human Computation Ethnographically

The anthropological tradition of extended field research stems in part from the recognition that in human affairs, the relation between beliefs and practice is invariably complex. In any given situation, what people believe that they should do, what they report that they do, and what they can be seen to do by an outside observer may all differ somewhat. However, in the absence of systematic participant observation, such disparities are difficult to detect. If we base our study of science solely on scientists' self-reports, we may fail to realize what the reported actions or tools actually consist of; if we look only at observed practice, we may miss what particular objects or actions mean to the scientists involved; and if we limit ourselves to introspection about a particular problem addressed by scientists, we may learn little or nothing about how the scientists themselves approach the problem. (Forsythe [1993] 2001h, 36)

In my investigation of how HC-based CS are formed in the interplay of different human and nonhuman actors and into the intraversions of their human–technology relations, I followed the tradition of cultural and digital anthropological inductive ethnographic fieldwork based on (constructivist) grounded theory (Glaser and Strauss [1967] 1971; Charmaz 2000; 2014), inspired by Forsythe's (2001f.) pioneering research on medical informatics and AI in the 1980s and 90s and her reflections on ethnographers in these fields. Owing to her work and that of other ethnographers discussed in the related work section, my research builds on their achievements in STS research, ethnographic reflection, and new collaborative approaches.[1] I conducted field research over the course of nearly three years in Germany and the US. My ethnographic approach, which draws on

---

[1] However, since Forsythe's research, a lot has also changed, such as the field of AI research itself. Human computation had not yet been born as a sub-research field of AI. Another difference in my field work experience lies in the openness of my research partners toward other epistemologies.

praxiography,[2] combines multiple perspectives and methods and considers assemblages and their human–technology relations across different scales and dimensions.[3]

I started from the perspective of designers and developers of HC-based CS but also included the perspectives of participants and researchers. Since HC-based CS assemblages are always sociomaterial and formed by human–technology relations, I also included nonhuman actors in my ethnographic analysis, including code, infrastructures, data, and how it circulates. This required navigating between different perspectives, diverse practices, and processes through which the relations between these elements became observable, as well as the application of experimental methods to include the relevant materials (e.g., source code and large digital textual chat data) in my analysis to supplement the more well-established methods, such as participant observation and qualitative interviews, I applied.

In this chapter, I first discuss methodological considerations that include co-laborative ethnography inspired by praxiography, and the methodology of (constructivist) grounded theory. This is followed by its operationalization in my research process, which includes reflections on my role in the field. Finally, I individually introduce each of the main methods I worked with.

Praxiographically Inspired Co-Laborative Ethnography

I followed a co-laborative (Niewöhner 2014) ethnographic approach to analyze HC-based CS assemblages in their complexity and their formation in the interplay of different human and nonhuman actors. I, first, discuss ethnography inspired by praxiography before turning to co-laboration.

Ethnography is particularly well-suited for investigating my research questions regarding sociotechnical phenomena as its intensive participation and observation enables the tracing of situations, interactions, and conflicts in their dynamics (Knecht 2013, 86). It is characterized by the collection of various complementary empirical material allowing for the inclusion of multiple perspectives (Amann and Hirschauer 1997, 16; cited in Knecht 2013, 86). Furthermore, due to its processuality and openness, the ethnographic approach can adapt dynamically in the course of research to shifting research questions and developments in the field (Pink et al. 2016, 11). Finally, ethnography allows for the encounter of aspects that had not been initially anticipated in the research design due to its participating, immersive, and open approach to the phenomena studied (Knecht 2013, 86). For these reasons, ethnography as an approach is always multi-methodical and

---

2   Following ethnologist Michi Knecht, I conceive praxiography as a specific position within ethnography (Knecht 2012, 249) I discuss this relation between praxiography and ethnography in more detail when discussing participant observation below.

3   My approach could, therefore, also be considered "assemblage ethnography," as described by anthropologist and sociologist Ayo Wahlberg, which studies configurations forming assemblages "across scales, sites, and practices" (2022).

-perspectival. I combined virtual digital ethnography with on-site in-person ethnography, also referred as "hybrid ethnography," (Przybylski 2021) in my research.[4]

My ethnographic approach is inspired by praxiography, which emerged as a new methodical approach in STS research and ANT, and aimed at moving beyond binary ontologies (such as the distinction between nature and culture) (Knecht 2012, 249; *cf.* Niewöhner 2019a), which fail to grasp how humans and nonhumans together shape sociotechnical phenomena. Praxiography describes the *"observation and description of actors and objects in action and interaction"* (Mol 2002b, 53–55; Hirschauer 2004, 73; Law 2004, 59; cited in Knecht 2013, 98, emphasis i.o.). As such, similar to ethnography, it focuses on processual and praxeological perspectives (Knecht 2013, 96), and is, therefore, particularly valuable for the analysis of sociotechnical assemblages and relations between humans and technology. In this sense, praxiography and ethnography have much in common (Knecht 2012), with both forms of knowledge production leading to a research practice that continuously reflects and critically reassembles its concepts and practices in the encounter with the field (Knecht 2012, 262).[5] However, unlike ethnography, praxiography places less emphasis on reflection the social dimensions of researchers' participation in the field. This approach can result in overlooking social dimensions (Knecht 2013, 97),[6] as well as aspects of meaning-making and perception. Therefore, while building on central perspectives and approaches from praxiography, with participant observation as one important method, this work does not pursue a radical praxiographical approach. An additional pragmatic reason for this is that HC-based CS projects come into being through the practices of various actors and materialities distributed across the world, in different time zones, which made it impossible to directly observe all practices through participant observation.

The understanding of praxis in this work follows Beck, Niewöhner, and Sørensen (2012), who define praxis as an analytical perspective regarding "human coexistence as manifoldly situated" (2012, 33). According to this understanding, ethnographic practice focuses on what may be experienced directly but simultaneously does not neglect their material, historical, and cultural conditions (Beck, Niewöhner, and Sørensen 2012, 33). "Praxis thus includes different time horizons, different spaces and the material-objective contexts in the analysis of the concrete *how* of human coexistence" (Beck, Niewöhner, and Sørensen 2012, 33, emphasis i.o.). Building on Beck's complex situational analysis of the use of technology, as discussed in the last chapter, I focused in my processual ethnographic approach on the dynamics of HC-based CS and analyzed both the sociotechnical systems and their surrounding practices (Beck 1997; *cf.* Bareither a.o. 2013, 32).

---

4   On digital ethnographic approaches and the combination of virtual and on-site ethnography, see, e.g., Boellstorff et al. (2012); Miller and Horst (2012); Pink et al. (2016); Fleischhack (2019); Przybylski (2021).

5   Different perspectives and understandings of praxiography exist (e.g., Mol 2002b; Niewöhner 2017). I aim in this section to describe how praxiography influences my ethnographic approach in a pragmatic way without going into details about different praxiographic approaches in general. For a historical overview of the emergence and different developments of practice theory see, e.g., Beck (2019) and Sørensen and Schank (2017).

6   Praxiography in its radical understanding focuses only on the practices, their constellations and structures, and not on actors or social collectives (Knecht 2012, 249).

In order to conduct research *with* rather than *on* my research partners (e.g., Ingold 2020, 210), I followed a collaborative approach with the Human Computation Institute. My approach in this was inspired by Forsythe's research on AI and Suchman's fieldwork at Xerox's Palo Alto Research Center and her engagement with some projects there, which she described as an "involved but also constructive engagement" (Suchman 2021, 70–71). I aimed at constructively but also critically contributing to the development of HC-based CS projects.

It has become common knowledge in cultural anthropology that we, as researchers, construct the phenomena we study. By asking certain questions, approaching the phenomena in specific ways, and defining boundaries—setting cuts, as Barad has described it (2007a)[7]—we not only describe but always also create our research fields and realities (Law 2004, 5). Design and futures anthropologist Sarah Pink wrote that "[w]hat was conventionally called 'the ethnographic field' is ongoingly made and remade through our active participation as ethnographers in collaboration with research participants, other stakeholders in research and future readers and viewers" (Pink 2018, 201). Ethnographic knowledge must, therefore, also always be considered both partial and situated (Haraway 1988). However, this understanding of the ethnographer's construction of the research field also means we (ethnographers) cannot sneak away once our research has been published. It has "real material consequences" (Barad 1996, 183) and, therefore, to stay with Barad, "[w]e are responsible for the world of which we are a part, not because it is an arbitrary construction of our choosing but because reality is sedimented out of particular practices that we have a role in shaping and through which we are shaped" (Barad 2007b, 390).

Working together with the Human Computation Institute, therefore, played a major role in my research. Even though I take a multi-perspectival approach, including both the participants' and researchers' perspectives, as well as the other two case studies of Foldit and ARTigo, the institute is a core focus and starting point of this work. My collaborative approach was inspired by that of co-laboration developed by Niewöhner (2014), which builds a complement to the ethnographic approach with the goal of strengthening and expanding anthropological reflexivity and the production of new knowledge (Niewöhner 2019b, 26–27; Bieler, Bister, and Schmid 2021, 88). Niewöhner defines co-laboration as "temporary, non-teleological, joint epistemic work aimed at producing disciplinary reflexivities not interdisciplinary shared outcomes" (Niewöhner 2016, 2; 2019b, 27). In this way, it differs from forms of collaboration that follow shared work objectives. Instead, it builds on shared epistemic work and on the "exchange between different epistemic cultures" (Klausner and Niewöhner 2020, 162; *cf.* Niewöhner 2016).[8] Niewöhner specifies that "practices of co-laborating help to diversify existing notions of reflexivity and critique, thereby broadening the analytical spectrum and adding interpretative degrees of freedom" (2016, 2). It is particularly useful in research fields characterized by high degrees of reflection (Niewöhner 2016, 41).

I engaged in a field wherein scientific methods already guided the development of HC systems, reflexive analysis, and publications, even if not ethnographically. The ques-

---

7   "Different agential cuts produce different phenomena" (Barad 2007a, 175).
8   "People do different things through the same process" (Niewöhner 2019b, 32).

tions that brought me to the field, regarding, for example, the subjectivities in HC-based CS, how actors involved strived to contribute to something greater, and the societal implications overlapped with questions already guiding the institute's work. This presented opportunities to think about them together from different perspectives and disciplinary viewpoints.

Finally, regarding co-laboration, I would like to highlight the aspect of moving between roles in the field. Through co-laboration, Niewöhner writes, clear roles, power relations, and relationships continuously change or break down between the parties involved from which "[a]mbiguous and searching moments emerge" (2019b, 39–40). Despite the advantages resulting from my collaborative engagement, such as gaining a deeper understanding of the phenomenon, this ambiguity and the uncertainties also presented challenges during my research process. Questions of how I could meaningfully and constructively but, at the same time, critically contribute to the development of new hybrid human–AI systems accompanied my research process, and, at times, proved to be more difficult than expected. These questions are not new to anthropological STS research in the field of computer science and AI, having already been discussed, for example, by Forsythe when "studying up" was still uncommon in ethnographic research (2001f). Forsythe mentions how this (at that time) new fieldwork context of studying up does not only "create new kinds of vulnerability, but the risks to both anthropologist and informants may extend far beyond the fieldwork itself" ([1999] 2001d, 125). It also creates new "dilemmas for the anthropologist as conflicting loyalties pull her in opposite directions and the collapsed roles of participant, observer, critic, employee, and colleague collide with one another" (Forsythe [1999] 2001d, 125).[9] Such dilemmas sometimes further complicated my own role within the field beyond the common challenge of closeness and critical distance to the field in ethnographic research. I sometimes, for instance, felt like I missed the chance to contribute my perspective based on my ethnographic observations simply because I was too slow in deconstructing and reflecting on current developments. Here, open discussion and reflection of my role both with team members of the Human Computation Institute and with colleagues in STS and cultural anthropology proved to be important to learn from these dilemmas experienced and reflect on the divergent different temporalities of HC development and cultural anthropological knowledge production.

Constructivist Grounded Theory

My co-laborative ethnographic approach builds on the methodology of (constructivist) grounded theory. How ethnographers approach the phenomenon they study, what they "bring" with them, and their position—as I have discussed above—in some ways, constructs the research object in the first place. Hultin aptly describes how her "position as a researcher [must be] understood as a genealogical line (rather than a dot) entangled with and conditioning what 'data' can become" (Hultin 2019, 100). This entanglement frames

---

9   Similar concerns and challenges have been described by European ethnologist and sociologist Rolf Lindner for ethnographic research in general (1981).

what ethnographers learn, even if it does not necessarily determine the knowledge to be gained from the field (Charmaz 2009, 48). According to Kathy Charmaz, this holds equivalently for theorization: "how we theorize reflects our interactions before we begin and those occurring within and beyond the field" (Charmaz 2009, 48). She, therefore, concludes that "[t]heorizing arises through analytic thinking about our field experiences, not merely recording and synthesizing them" (Charmaz 2009, 48). The experiences we (ethnographers) make in the research process, the methods we apply, and our knowledge and emotions also feed into the analysis, not only the (in my case) transcribed, digitized empirical data that I gathered and analyzed. Not all information contributing to theorization can be digitized in the first place, instead, it unfolds in the interaction with the field site and its actors. Grounded theory as a systematic but flexible method (Charmaz 2014, 1) allows not only for incorporating these diverse materials, experiences, and reflections but particularly encourages circular movements between data collection and analysis. It is, therefore, particularly well-suited for studying processes.

Grounded theory goes back to sociologists Barney G. Glaser and Anselm L. Strauss, who first developed it in their publication *The Discovery of Grounded Theory: Strategies for Qualitative Research* ([1967] 1971). Since then, grounded theory has been developed further by not only Glaser and Strauss themselves but by other researchers who, at times, also distanced themselves from the positivism that was underlying Glaser and Strauss' understandings of the method (Charmaz 2014, 12). In my research, I follow a *constructivist* grounded theory approach, first mentioned in Charmaz (2000). While embracing important aspects of Glaser and Strauss' approach, such as inductivity, comparison, emergence, open-endedness, and iterativity (Charmaz 2014, 12–13), it—in contrast to Glaser and Strauss' method (or "conventional" [Charmaz 2009, 52] grounded theory)—"highlights the flexibility of the method and resists mechanical applications of it" (Charmaz 2014, 13). It aligns with the understanding of situated and partial knowledge and construction of the field described above (Charmaz 2014, 13).

Starting from inductive research, grounded theory is a "comparative, iterative, and *interactive* method. The emphasis in grounded theory is on analysis of data; however, early data analysis informs data collection" (Charmaz 2012, 2, emphasis i.o.). It jointly considers data collection and analysis and, therefore, allows for the construction of middle-range theories that are "'grounded' in their data" (Charmaz 2012, 2).

Data analysis was part of my research process from the early collection of empirical data. Even though, as stated above, analysis cannot be reduced to coding digital data, such as interview transcripts, fieldnotes, media articles, and chats, I gathered most of the data, as well as my fieldnotes, in the qualitative data analysis software MAXQDA (VERBI – Software. Consult. Sozialforschung. GmbH, n.d.). My analysis of this data followed the two main phases described by Charmaz as the "initial phase," which included line-by-line coding of practices, processes, meanings, perceptions, and imaginations, and a "focused, selective phase that uses the most significant or frequent initial codes to sort, synthesize, integrate, and organize large amounts of data" (Charmaz 2014, 113). This approach is also well established in ethnographic coding (Breidenstein et al. 2020). Working with codes, which Star described as "transitional objects" ([2007] 2015, 130) using psychoanalyst Donald Woods Winnicott's (1965) term, helps one to understand the phenomenon studied better, while, at the same time, abstracting from it and creating something new

(Star [2007] 2015, 130). Notably, but not only, in the second phase, comparing codes with each other or with other data allowed me to further abstract and connect my codes to existing research (Charmaz 2012, 4). By then going back into the field and collecting more data with the already emerging categories in mind, these codes densified further. The second phase of my field research in Ithaca, NY, in 2022, was particularly constructive for my engagement with the data. This process of condensing and further abstracting codes is referred to as theoretical sampling in grounded theory terminology.

> Theoretical sampling stretches the codes, forcing other sorts of knowledge of the object. The theory that develops repeats the attachment-separation cycle [that already characterizes codes], but in this sense taking a code and moving it through the data. In so doing, it fractures both code and data. (Star [2007] 2015, 130)

This process of oscillating between analysis and data collection was complemented by writing notes, or memos, in addition to fieldnotes, in which, for example, I captured connections between data or helpful theoretical concepts and how they related to my data.

Following the discussion of methodological considerations guiding my research, the next section describes its operationalization in the research process.

Doing Research On, With, and Among Researchers and Developers

I first learned about HC by way of a computer science class by Bry at LMU Munich. I came across its description in early 2019 while searching for classes to attend in the following term: "Human Computation is a novel and interesting branch of Computer Science that intends to incorporate human intelligence to solve problems computers alone typically have problems to deal with." (Bogner n.d.) I chose to attend this class intrigued by this brief description of HC and the topics to be covered, including GWAPs, "Social Behavior Analysis," and "Participation and Ethics," (Bogner n.d.) which piqued my interest regarding questions about *how* humans and computers are combined in HC, what the image of humans and their intelligence is, as well as what the role of games was in this. From this first encounter with HC, my initial research interest focused on the human–technology relations and the role of the human in the sociotechnical systems that emerge in the "interferences" (Dippel and Fizek 2017a; 2019) of play and science.

I started exploring HC-based CS games by participating in various examples, including Stall Catchers, Foldit, and ARTigo, which I then selected as examples to study due to their visibility in online CS, and their differences in the scientific research area and modes of engaging participants. I gained my first ethnographic experiences with the ARTigo project, developed at LMU Munich by Bry and his colleagues in cooperation with the Institute of Art History at LMU Munich. Around the end of 2019 and beginning of 2020, I interviewed team members of the project from both the department of computer science and the art history institute. While getting into contact with the ARTigo team was straightforward, gaining access to ARTigo participants proved difficult. When I began my research, the project had already been ongoing for over a decade. Although the initial years were prosperous, the team's activities were reduced to platform maintenance after the funding period ended. The software had become partially outdated by that time, and there was no longer much active research being conducted on it, as team member Emilia explained in our interview (Nov. 8, 2019). ARTigo could be played without registration,[10] and there was no communication feature for contacting other participants. I describe my attempts to acquire interview participants from ARTigo's player base below.

At around the same time, in the fall of 2019, Bry introduced me to Michelucci, the director of the Human Computation Institute. As the only institute solely focused on the development of HC systems (to the best of my knowledge, and at the time), the institute presented a good starting point for learning how GWAPs or HC-based CS systems were being developed and maintained and how human–AI relations were envisioned by scientists and developers. After an initial call and the exchange of several emails to identify possible paths of working together, I joined the institute at the end of 2019 as an intern—specifically with the title of "ethics intern," a title that was suggested by the institute. In one of my first email exchanges with Michelucci, in which we tried to determine how we could best collaborate, I had described my research interest, which, at that time, was still broad and not yet clearly defined, to lie (among other aspects) in questions of digital ethics in the sense of what dignified contributions to HC systems were and could be and how participants could gain recognition through their participation. Despite the effect that my title would at times predefine or bias the questions and topics I would be actively included in, my research partners allowed me to participate in the institute's different working contexts and to get to know all the different perspectives and tasks related to developing and maintaining HC-based CS projects.

As an intern at the Human Computation Institute, I was listed on the institute's website and my intern role was, thus, visible to participants and the general public. Later, when I became involved in community management for Stall Catchers, my initial username, which I had chosen at the beginning of my engagement with HC-based CS, was changed to my name plus the suffix "+HCI" to identify me to participants as an institute contact. After a brief introduction in the chat at the beginning of my fieldwork, a more thorough introduction, supported by the institute, followed in August 2021 in the form of a blog post describing my work with the institute and ethnographic research in general (Vepřek 2021a). The institute also sporadically reported on my research via blog posts and tweets (e.g., Santander 2022).

My contributions to the institute's projects included a variety of activities. Initially, my primary role was to assume the position of community manager for the new soon-to-be-relaunched Dream Catchers[11] project (Ramanauskaite 2020). However, my responsibilities soon evolved into a diverse set of tasks ranging from assisting Ethical Review Board (or Institutional Review Board) applications, community management of Stall Catchers, for example, by helping organize Catchathons, to contributing to load and performance testing and optimization efforts of the platform's software and infrastructure. In addition, members of the biomedical engineering laboratory identified me, at least in part, with the institute during my fieldwork, as became clear during laboratory

---

10   Unfortunately, no public data on user and platform statistics exist.
11   At the time of writing this work, Dream Catchers has yet to be relaunched for various reasons, including prioritization of other projects and insufficient resources.

meetings, in which open questions for the institute would sometimes be directed toward me.¹²

My work with the institute also included collaborative work on the question of how to improve or adapt ethical review to online (HC-based) CS projects. We organized, for example, a workshop at the ECSA conference 2020 that brought together researchers, practitioners, and participants to discuss this topic. Furthermore, we designed a new ethical review workflow together with Institutional Review Board operations and administrator Patricia Seymour (Vepřek, Seymour, and Michelucci 2020). I then further developed and designed this workflow in my computer science master's thesis as a *Collaborative and Adaptive Ethical Review* platform (Vepřek 2022b). Even though this collaborative research also informed my understanding of the field of HC-based CS and is an important example of how I contributed to shaping the field through my engagement, this work only plays a marginal role in the ethnographic research to which I return in the following.

Engaging in forms of experimental "worldings" (Tsing 2011; Niewöhner et al. 2016) helped me reflect on my epistemological assumptions and thinking constraints brought to the field, and opened up new insights into and possibilities for joint reflection of the entanglements of relations within HC systems. This joint reflection with members of the Human Computation Institute and the researchers happened throughout the research process. However, it also built the explicit focus of my second research visit to Ithaca in 2022, while the first visit in 2021 focused on participant observation and collecting data. In the following, I describe the two research visits and their process in detail and include my approach to the Foldit and ARTigo examples.

From the beginning of my engagement with the institute, two research visits of about three months each to Ithaca were planned. Due to the COVID-19 pandemic, which began shortly after the start of my fieldwork, plans to join the institute in Ithaca had to be canceled and remade several times. Therefore, I initially began the collaboration and fieldwork remotely, conducting most of my ethnographic research online.¹³ Because of the very digital and web-based nature of large parts of the institute and the projects being studied, this did not (only) have disadvantages, as for most members of the institute, this form of engagement was also similar to their everyday experience with the Human Computation Institute. Except for annual in-person meetups, the team had been working remotely, communicating exclusively through web-based instant messaging services, email, and virtual meetings. Over the course of my research, the institute consisted of a core team of around eight members including the roles of developers, CS coordinators and strategic advisors, and a larger group of affiliated researchers. Since

---

12   My experience sometimes also resembled social and cultural anthropologist and migration scholar Maria Schiller's, who did research as a "research trainee" in municipal organizations of three European cities (2018, 67).
13   On remote ethnography, see, e.g., Postill (2017), and for a recent account on remote ethnography taking into account ethnographic fieldwork in times of the COVID-19 pandemic, see Podjed and Muršič (2021). Anthropologist Dan Podjed and cultural anthropologist Rajko Muršič point to the advantages remote ethnography can have, for example, by allowing interview participants to remain in their own environment, creating a more relaxed interview situation via digital communication technologies (2021, 45).

the institute is a nonprofit research organization, it is strongly dependent on funding. Accordingly, the team's size varied throughout my engagement.

During this first remote part of my fieldwork, I also conducted qualitative interviews with Stall Catchers participants and analyzed readily available data, such as the in-game chat and Stall Catchers' source code. These methods will be explained in more detail below. In addition, as the second comparative HC-based CS example, I conducted interviews with the US–based Foldit team in early 2020. For the latter, I used video conference services and followed Foldit's developments by playing the game occasionally and reading its newsletter and regular scientific and game updates shared on the website.

While remote engagement and digital methods helped me to gain valuable insights for my research, this approach also made some aspects of Stall Catchers and the Human Computation Institute's working practices invisible. A considerable part of the Stall Catchers project remained in the dark, most notably the biomedical research conducted at the Schaffer–Nishimura Lab. The Alzheimer's disease research underlying Stall Catchers, including the scientists' practices, the wet lab, experiments with mice, microscopes, lasers, dyes, medication, as well as the connected work on Stall Catchers' data pipeline, remained invisible from my screen in Germany. Moreover, and especially before the COVID-19 pandemic times, the Human Computation Institute organized various in-person events where people would, for example, come together to participate in Stall Catchers during the special Catchathon events, another aspect of the project I viewed as important for my research that required physical presence.

Therefore, as soon as was possible given the COVID-19 pandemic restrictions, my remote digital ethnographic research was complemented by two research visits to Ithaca. The first three-month field research period took place from August to October 2021, its foci being the collaboration with the institute and ethnographic data collection through participant observation and interviews.

In some ways, my collaboration with the Human Computation Institute during these three months continued as before with weekly Zoom meetings with team members but with the added benefit of in-person meetings and conversations with Michelucci several times a week, which often included or evolved into hours of informal conversations. I also had the opportunity to accompany Michelucci to meet researchers interested in collaborating to develop a new HC-based CS game to support their research, which proved insightful to better understand how HC-based CS projects are started and imagined before they are actually built.

During this three-month period, I participated in meetings and events of the biomedical engineering laboratory at Cornell University and observed the working practices of Alzheimer's disease research in the laboratory. The focus was on practices related to Stall Catchers, such as work on the data pipeline, including testing new software tools for preparing data or the manual analysis of research images. Researchers also walked me through the individual steps and related code structures. In addition to practices related to the data pipeline, I also observed and participated in onboarding meetings for new laboratory members, and observed scientific practices in Alzheimer's disease research, such as craniotomies,[14] drug and treatment injections at night for behavioral experiments with the mice the next day, imaging sessions, and, most challenging for me personally, mouse euthanasia and tissue collection. Even though my analysis focuses on human–technology relations and the assemblage starting from the imaging sessions and excludes the preceding research practices in the laboratory, insights from observing practices beyond this focus, such as the animal research practices, were important to situate these relations and the assemblage in their broader context of biomedical Alzheimer's disease research. These participant observations were complemented by qualitative interviews with laboratory members working with Stall Catchers. Taken together, the time spent in person at the Human Computation Institute and Schaffer–Nishimura Lab were invaluable to my research, especially thanks to the patience my research collaborators showed me.

Following this first research visit, my focus shifted to transcribing and analyzing the empirical material collected. During this time, I reduced my active involvement with the institute and the frequency of conducting participant observations both at the institute and on the Stall Catchers platform, although Michelucci and I continued a biweekly meeting schedule, and I occasionally attended larger team meetings and events. One exception to this reduced engagement was during the *Danish Science Festival* and the *Engaging Citizen Science Conference* in April 2022 which took place in Aarhus, Denmark. Eleven team members of the institute came together there to participate in the festival and conference and connect (some meeting during this event in person for the first time).

My second research visit to Ithaca, NY, lasted from October to November 2022. This time, I concentrated on discussing my observations and preliminary findings with my research partners. To avoid overly formal interview situations and, instead, concentrate on exchanging ideas, I decided to omit audio recording sessions and relied on taking field notes, which I later copied into my diary. The second field phase also allowed me to learn about further developments in the laboratory on the data pipeline since my last visit in 2021. During the twice-weekly meetings with Michelucci, in addition to discussing a variety of topics related to HC, Stall Catchers, the institute, and the insights I had gained over the course of my fieldwork, we also discussed early iterations of ideas, themes, and theses that are now included in this work.[15] I also met individually with all researchers from the laboratory who had participated in my research during my first field phase to discuss my observations. These meetings were of great value for refining my insights and were essential in giving them the opportunity to share their perspectives on my preliminary findings. I draw on these conversations and reflections extensively in later chapters of this work. Toward the end of my visit, I finally gave a presentation on my work at the laboratory's weekly meeting, which was also attended by Human Computation Institute team members. The presentation of my work and subsequent discussion presented a

---

14   Craniotomies are surgeries in which a part of the skull is removed from the head of a mouse and windows are installed to allow subsequent imaging of blood flow in the brain.
15   In one session, for example, we discussed play/science entanglements in HC-based CS, in another, my ideas on intraverting human–technology relations, and in yet another, we reflected on our collaboration and what ethnographers could bring to the field of HC.

helpful learning process in my ethnographic work (Klausner 2015, 49) and provided valuable insights into how my research partners perceived it. It also included joint reflection on how projects such as Stall Catchers could or should be run. During these conversations and meetings with research partners, the roles shifted between me as the ethnographer, learning from the institute's members and the researchers, and my research partners approaching me with questions about how to improve Stall Catchers and the collaboration between all different actors.

In the following section, I discuss the main individual ethnographic methods I applied in the course of this study.

A Toolkit of Methods for Emerging Hybrid Systems

Participant Observation

Participant observation[16] is one of the core research methods in ethnographic fieldwork and formed the basis of my analysis to examine everyday practices[17]—both use and development practices—of HC-based CS assemblages and their human–technology relations. Participant observation made it possible to analyze the practices' tacit and embodied knowledge as well as the processes, situations, and conflicts related to them. In this way, the innovation potentials and "tactics" (Certeau [1980] 2013) of routinized practices also became analyzable (Beck 1997, 346).

Participant observation at the Human Computation Institute was crucial for gaining insights into what it means to build "sustainable participatory systems" (Human Computation Institute, n.d.). Building such systems involves a wide range of tasks and practices beyond engineering, such as conducting meetings, team communication, writing papers, giving talks, software maintenance, management (of people, processes, and systems), and fundraising. These activities are similar to the tasks performed by members of the expert system laboratories studied by Forsythe in the 1980s and 1990s ([1993] 2001h, 23–24) and are generally representative of the work conducted in (university) research laboratories. Additionally, work at the Human Computation Institute also included community outreach and translating biomedical knowledge into content for the broader public in the form of blog posts and other formats, in other words, science communication. Due to my collaboration with the institute, I was able to closely follow developments, such as the first experiments with AI bots in Stall Catchers, and to do so from the different perspectives of both institute's team members and the project's participants. I followed participants and researchers in their engagement with and the

---

16   Participant observation can be traced back to anthropologist Bronislaw Malinowski ([1922] 2013) and has since been further developed into one of the core research methods in ethnographic fieldwork (cf. DeWalt and DeWalt 2011).

17   Bareither described this focus on practices as one of the strengths of cultural anthropology of technology because of its "sensitivity for everyday, routinized, culturally encoded and in social negotiated processes integrated *practice* of actors with and in relation to technology" (2013, 31, emphasis i.o.).

relations they enter into within the HC-based CS system, as well as the data moving through and along these relations and infrastructures.

While my engagement involved participation in the institute's everyday practices, focusing on the development and maintenance practices at the institute and the different game platforms, my involvement in the laboratory was typically limited to observation, as the focus of my research was not on the wet laboratory practices but rather on human–*technology* relations directly connected to the HC-based CS project. In studying the CS platforms of the studied examples from the users' perspective, I actively participated in all projects as a participant. I recorded my experiences in a field diary to better understand the participants' experiences and relations with the platforms. Here, participating included playing the games, engaging in the project's communication channels, and reading blog and forum posts as well as other updates provided by team members and participants on the project's websites. By being "digitally co-present" (Hamm 2011), I could move within the same digital space as and together with the projects' participants.

While participating in ARTigo and Stall Catchers was relatively easy, I experienced a difficulty that many Foldit beginners face, namely the steep learning curve of the game. My participation in Foldit can be described as experimental, since after completing the tutorials, I mostly tested different approaches to folding proteins by following tips other participants had shared with me in our conversations, experimenting with different recipes, and using other algorithmic tools provided by Foldit.

Because Stall Catchers formed my primary research example, I here also contributed to all Stall Catchers events, such as Catchathons. By engaging not only as a Human Computation Institute team member and observing ethnographer, but also as a participant, I had the chance to experience the team spirit that emerges during such events, especially in the final hours of a challenge when the competition between different game teams heats up.

Participant observation was also my primary approach to including materialities, technologies, nonhuman actors, and entities, such as data, in my analysis, and specifically to focus on the relations between them and human actors. I here want to briefly point to some selected instances highlighting the importance of this approach for my investigation. Researcher–technology relations, for example, only became fully comprehensible after observing how researchers use and work on the infrastructures, software, and microscopes, how imaging data were manually analyzed, or how data were transformed and prepared via the data pipeline before being sent to Stall Catchers. Similarly, understanding how data is analyzed on the Stall Catchers platform required following the flow of data between human input, the interface and servers, code and its algorithms, as well as the database and developer's interventions.[18]

Taken together, participant observation allowed for the generation of situated knowledge, since "to observe is not to objectify; it is to attend to persons and things, to learn from them, and to follow in precept and practice. Indeed there can be no observation without participation—that is, without an intimate coupling, in perception and action,

---

18 The approach of following actors, commodities, or ideas is especially prominent in multi-sited ethnography (Marcus 2009).

of observer and observed" (Ingold 2000, 108; cited in 2014, 387–388).[19] Through participant observation, I gained insights into the everyday practices of the different actors, of which it was specifically valuable to analyze the interplay and different sociotechnical relations they enter. It also revealed tacit, implicit, and embodied knowledge often difficult to articulate, for example, in interviews (Beck, Niewöhner, and Sørensen 2012, 19).

However, the method of participant observation also has limitations when it comes to phenomena that include globally distributed actors, making specific participant observations impossible, or when it comes to including discursive and historical contexts that also shape such phenomena. My investigation, therefore, draws on qualitative interviews, media analysis, and exploratory methods, such as source code and in-game chat analysis, to cover these shortfalls.

Qualitative Interviews

To gain access to the ideas, perceptions, and values behind practices in line with a "practice of understanding" (Jeggle 1995, 56) typical of a cultural anthropological approach, I conducted qualitative semi-standardized interviews across all three projects. All interviews were guided by a questionnaire, which served as a flexible guide to structure the interview but did not determine the conversational flow (Schmidt-Lauber 2001). I followed the topics my interview partners cared about or brought up on their own accord, including questions in the conversation according to the specific situational flow (Schmidt-Lauber 2001, 176) rather than actively steering the conversation back to a fixed agenda. Interviews, therefore, differed in their form, length, and depth. I talked to participants, researchers, developers, project leads, community liaisons, and other team members involved in HC-based CS projects. The interviews were conducted in English, German, and Dutch.

Qualitative interviews opened the door to certain dimensions of knowledge and perceptions, which, especially in the case of CS participants, would not otherwise have been possible to include in my research. It was through the interviews, for example, that I gained a sense of the importance of a personal connection to Alzheimer's disease for many Stall Catchers participants and how deeply entangled these personal experiences were with their contribution to HC-based CS. The global distribution of the phenomena studied made it effectively impossible to observe participants' playing practices at their physical location, therefore, remote interviews were vital to covering this gap. In-person interviews were invaluable in other settings: I only understood the complexity of the Stall Catchers project at the laboratory by being there and talking with biomedical researchers. Sitting down with them, I learned how much they cared about their research, sometimes because of the same connections to Alzheimer's that the participants described, and the problems they faced. At the same time, I gained access to different layers of designers and developers' imaginations that went beyond the inscribed and programmed ones and were not accessible for me by reading code. Sociotechnical imaginar-

---

19   My approach could also be considered to carry autoethnographic traits insofar as I include my personal experiences, positions in the field and their reflections in my analysis; on autoethnography, see, e.g., Ploder and Stadlbauer (2013); Caivano and Naumes (2021).

ies of HC drive the design and development of such systems but, at the same time, often break down or do not hold in everyday life. Interviews help access these different layers and narrations.

It should be noted that these narrations are self-descriptions that are contextualized in the research partner's sociocultural, embodied, and material or technological environment. At the same time, they are polished external presentations adapted to the interview situation (Froschauer and Lueger 2020, 236). In my conversations with Stall Catchers participants, for example, interview partners sometimes addressed me not only as an ethnographer but also as a team member of the Human Computation Institute. At times, I may have been seen as something of a mediator who would pass on both praise and criticism. Because information is filtered and interview partners apply discursive modes of representation that do not necessarily correspond to the practical mode of being and conscience (Beck 1997, 346), qualitative interviews require a critical approach to the empirical material. The relationship between representation and observable action is complex (Geertz 1973; 1983; Forsythe 2001g, 139). Beck talks about "translation errors" that are to be expected and points to the problem that these can only be partially corrected by comparison with observations (1997, 346). Keeping these limitations in mind, qualitative interviews, particularly in combination with the other methods applied in this research, form a practical approach to HC-based CS projects.

I conducted a total of 64 interviews, of which 49 were oral interviews and 15 written interviews and questionnaires with follow-up questions. Most of these were within the Stall Catchers case study: 28 interviews were conducted with Stall Catchers participants, five with researchers of the laboratory, and five with the Human Computation Institute team members. Thirteen interviews were conducted with Foldit participants[20] and six with Foldit team members and researchers. Finally, two ARTigo participants and five ARTigo team members were interviewed, as well as three representatives of funding institutions in the US that fund or have funded some of the projects. Given the scope of this study, the insights from the latter serve as contextual background information.

Although I had initially planned to conduct a comparable number of interviews with participants for each case study, it proved difficult to find and get in contact with ARTigo participants. Even with the support of the ARTigo team, who allowed me to post calls for participation via ARTigo's Twitter and Facebook accounts, it was difficult to acquire participants. The call was answered by one ARTigo participant. Another participant interview was acquired via the snowball principle. An additional problem was that it was not possible to work directly with the ARTigo platform during these interviews in 2020 because the server was down at the time, which limited the interactivity in some ways. Furthermore, the platform does not include any in-game communication features (such as a forum or chat), making it challenging to analyze participants' experiences and practices by other means. A small written survey in which participants were asked to play ARTigo and share their experiences was conducted in January 2021 with fellow students

---

20   While 28 Stall Catchers participants immediately answered an open call for participation in my research via email, recruitment of interview partners from Foldit participants was slow. In addition to posting a call in the game's forum, I actively sent more than 50 interview requests to individual participants via the game's platform and Discord server.

as part of an explorative study for a computer science master's seminar at LMU Munich. We created a small qualitative questionnaire which we shared in our networks to compare the user experience of different games and because there was no data on the participants' perspective on ARTigo.[21] Due to these circumstances and the resulting limited empirical material, ARTigo was not included as a comparative study to the same level as Stall Catchers or Foldit in my final analysis. However, a new iteration of the ARTigo platform was launched during the later course of my research introducing a new stage of the project and an interesting turn regarding my analysis of how HC-based CS are formed and change over time. I discuss this development in Chapter 6, less empirically than analytically, applying my theoretical concept of intraversions to this development and exploring its explanatory potential using this example.

All but one interview with participants were conducted via video conference services, such as Zoom or Skype, or via a landline due to the geographic distance and the COVID-19 pandemic. Although in-person interviews allow for a richer communication and interaction context, the digital and phone interviews were particularly suitable for this research since they allowed CS participants from all over the world to contribute and even increased the possibilities of participation (Markham 2005, 801; *cf.* Hengartner [2001] 2007, 201).

The majority of participants interviewed were based in the US, followed by five interview participants based in Germany, and three in Belgium. In addition, one interview partner contributed from each of the following countries: Australia, Brazil, the Netherlands, Nigeria, and Singapore. The location of one participant was unknown.[22] The average length of interviews was about one hour, with the shortest being 21 minutes and the longest being two hours and 16 minutes. It is worth noting that interviews with participants of the case studies analyzed are necessarily limited to those who actively volunteered to contribute to my research by responding to my call for interview participants. Hence, the viewpoints represented are those of participants that were indeed willing to share their experiences with Stall Catchers, Foldit, or ARTigo and the motivations that drive them.

The interviews with team members, researchers, and funding institution representatives were mostly conducted via video conferencing services, except for conversations with ARTigo team members and those with the researchers at the Schaffer–Nishimura Lab.[23] Even though I spent three months with the Human Computation Institute in

---

21 A total of ten participants in the age range between 20 and 64 contributed to our written questionnaire. Even though we did not collect personal information in addition to age, it is very likely that most participants had a computer science background and were based in Germany. Five participants had not heard of ARTigo previously, three had heard of it but never played it, and two had played it before our small study.

22 It should be noted that I left it open to my interview partners to indicate their age and gender, as these categories did not form a focus of my research. The age range generally included individuals between around 20 and 75, with a slight majority of participants being over 40. Slightly more participants identified as women across the three different case studies.

23 As in the case of interviews with participants, gender and age distributions are not considered in my analysis. However, I would estimate that the distribution of different genders was roughly representational of the general population among researchers and team members of Stall Catchers

Ithaca, NY, which allowed many in-person conversations, most interviews with its team members were also conducted virtually, as the institute's team is distributed around the world.

All interviews were audio recorded and transcribed. The quotes used in this work were minimally smoothed to improve readability, on the one hand, and protect the privacy of my interview partners, on the other hand. This means that expressions such as *um*, *uh*, *you know*, and *like* as well as word repetitions, if not purposefully repeated, were omitted from the quotes. These expressions can transport important meanings to which cultural anthropology is sensitive, similar to hesitations or pauses while speaking. However, since not all interview partners were native speakers, the usage of such expressions and terms varied across all interviews, potentially resulting in identifiable speech patterns. This was especially noticeable as I translated the non-English interview quotes used in this work.

Another strategy I employed to protect the privacy of research participants is the use of pseudonyms and random gender changes. Pseudonyms were chosen arbitrarily, with the best effort to maintain the overall representation of origin and diversity of real names across all data. Furthermore, due to some roles at the Human Computation Institute and the laboratory being unique, I partly generalized roles—such as "researcher" for different positions including experimentalists, PhD students, postdocs, and developers—and duplicated positions that are unique to an individual to multiple representations of the same research partner in order to make deanonymization more difficult. These obfuscations do not affect the overall results of my research, and I specify the position in cases where the role is relevant. For these reasons, only a few of my research partners' names, whom I refer to by their last names, were not pseudonymized.[24] These exceptions were necessary due to their unique field position and public appearances. Anonymization could, therefore, not be guaranteed. Consent for this was granted by all individuals affected.

Finally, I jointly refer to users, players, and citizen scientists as "participants" since some participants reject the term "game" and do not identify as "players." I also prefer "participants" over "users" and "citizen scientists" to emphasize their active role in shaping the HC-based CS systems.

In addition to the qualitative interviews, informal conversations with Human Computation Institute team members, researchers from the laboratory, and a few participants were of great value to my research. In these cases, I took field notes from which I quote in this work when conversation partners agreed.

While qualitative interviews, informal conversations, and participant observation formed the core methods of my research, it also involved the analysis of existing data, such as external perspectives on the projects studied (e.g., in media articles), self-presentations of the project's teams, and, most importantly, the Stall Catchers source code and

---

contributing to my research. For the research partners of Foldit and ARTigo, the majority seemed predominantly to align with male.

24 The decision to refer to research partners whose names I have not pseudonymized by their last names while using first names as pseudonyms for others is purely pragmatic and not meant to imply any hierarchical distinction between them.

data pipeline as socio-technological foundation and constituent entities of the project. The methods used in connection with these are described in the following. As will become clear, human–technology relations and the sociomaterial assemblage only became understandable in the combination of the different methods which revealed the entanglements of the different human and nonhuman actors.

Experimental Approaches to Infrastructures, Code, and Digital Chat Data

In the course of this research, I analyzed a variety of textual material. This included media articles about the different HC-based CS examples, blog posts, and the case studies' own websites and forums. The websites' content and project descriptions were valuable sources, revealing how the teams wanted to represent themselves to the world. In the course of my collaboration with the Human Computation Institute and due the openness of its members toward my ethnographic approach, I also received access to the institute's workspaces and digital infrastructures, such as its primary communication space on Slack (Slack Technologies, LLC n.d.), the Github repository,[25] databases, Stall Catchers' admin spaces, and, where required for certain collaborative work, computational infrastructure. The analysis of these text-based and infrastructural sources followed an exploratory approach with the aim of including them as supplementary material to support or contrast with other observations and empirical materials.

Particularly crucial was the access to computer code and digital chat data from the institute. I focused on analyzing the code of Stall Catchers, which is a key component in HC-based CS assemblages (Mackenzie 2006, 2) to understand how user–technology relations unfold. Through the analysis of code, the intra-actions of participants and technology can be traced and the underlying and (consciously or unconsciously) inscribed design logics of developers and designers revealed (Koch 2017b, 117).

The analysis of computer code is a relatively new area in social sciences and humanities, with emerging subdisciplines like software studies, digital STS, critical data studies, and CCS (e.g., Fuller 2008; Vertesi and Ribes 2019; Marino 2020; Hepp, Jarke, and Kramp 2022).[26]

Code is multidimensional and, as such, requires a multiperspectival analysis to fully appreciate its sociotechnical embeddedness, its becoming and implications, as well as its

---

25 *Github* (GitHub, Inc. n.d.) is a hosting service for version control using the distributed version control system *Git* and software development, which allows the distributed development of software in teams.

26 Today, ethnographic studies including computer code often refer to these fields, which provide a rich repository of useful and important methodical approaches. However, depending on their scientific situatedness, they follow specific research interests. The CCS, for example, are strongly influenced by and emerged from the field of literary studies. By comparison and despite the expressed need for it, an ethnographic approach that focuses on the practices and meaning-making processes, for example, has not yet been established (e.g., Carlson et al. 2021; Vepřek et al. 2023). In order to change this, the *Code Ethnography Collective*, a group of researchers from mainly cultural anthropology and STS has met regularly since 2021 to discuss ethnographic approaches to computer code (Code Ethnography Collective n.d.).

inscriptions and the various practices associated with it (Vepřek et al. 2023). In my ethnographic research, code analysis played an important role in qualitative interviews and walkthrough sessions (Light, Burgess, and Duguay 2017).[27] Additionally, it was also crucial in participant observations during special events and laboratory sessions. This also extended to my collaborative engagement tasks, such as performance and load testing of the platform. The different approaches, thereby, supported each other because insights from participant observation revealed interesting starting points for focused code analysis. My focused code analysis approach (first discussed in Carlson et al. 2021; Vepřek et al. 2023) builds on the CCS method (Marino 2016; 2020) of reading code in a critical way. That is, as Marino writes,

> to explore the significance of the specific symbolic structures of the code and their effects over time if and when they are executed (after being compiled, if necessary), within the cultural moment of their development and deployment. To read code in this way, one must establish its context and its functioning and then examine its symbols, structures, and processes, particularly the changes in state over the time of its execution. (Marino 2020, 23)

Unlike CCS, however, my research used code analysis as an additional perspective, particularly focusing on the *flow of code in practice*, which is why I call it *focused* code analysis. I selectively analyzed code sections related to algorithms and I/O (input/output) operations. Accordingly, I traced selected I/O operations or sequences of function calls, which I had previously identified as interesting starting points. This way, I followed the flow of the code during its run-time operation, entangled with and in relation to other elements of the sociotechnical assemblage. I included specific code sections in MAXQDA and annotated them similarly to other empirical material, such as interview transcripts, to analyze selected code blocks.

By integrating this analysis with other ethnographic methods, I traced the flow of actions of human–technology relations within and beyond the text. This approach follows an understanding of software "structured as a distribution of agency" (Mackenzie 2006, 19).

Due to the dynamic nature of code, it was only through the combination of the different approaches and methods described in this chapter that I was ultimately able to gain an in-depth understanding of the participant–technology relations as they unfolded in the sociotechnical assemblage. This combination also revealed the different development, maintenance, and use practices and how values, norms, and imaginations of future hybrid human–AI systems were inscribed into the Stall Catchers project and guided its implementation. Since Stall Catchers' source code is proprietary, I do not include actual code samples in this work, except for one instance which the Human Computation Institute kindly granted permission to include. However, whenever necessary to make an argument or provide an example, I describe insights from analyzing the code. An example of this can be found in Chapter 7, which focuses on trust. In this

---

27   I also included the code underlying the data pipeline and ML model at the Schaffer–Nishimura Lab in the form of walkthroughs with researchers and readings on the laboratory's Github repository.

context, integrating code into my analysis enabled me to comprehend how algorithmic mechanisms contribute to preventing cheating and ensuring data quality.

Finally, as a second exploratory method used in this work, the analysis of Stall Catchers' in-game chat data during my fieldwork in 2020, particularly in the midst of the COVID-19 pandemic, provided valuable insights into the project and participants' perspectives. Yet, it required a more exploratory approach because, despite advances in digital text data analysis in the digital humanities and computational social sciences (Lemke and Wiedemann 2016; Franken 2023), qualitative analysis of large datasets like chat records remains challenging for qualitative and inductive approaches. While methodological considerations exist for qualitative approaches to the analysis of written online conversations (e.g., Schirmer, Sander, and Wenninger 2015; Nam 2019), the question of how to make large amounts of data manageable for qualitative analysis remains generally little discussed. Conducting a manual analysis of the entire chat record following the grounded theory approach proved to be arduous, if not infeasible, due to the large amount of data involved, particularly considering its supplementary role in the overall scope of my research. The chat record in 2020 contained around 17,000 messages, spanning a period of around one and a half years. To manage this, I explored a new approach using relational databases[28] for (re)organizing and analyzing the extensive chat data, making it more accessible for qualitative study.

The Stall Catchers' chat, located in the lower right corner of the game's interface (see Figure 1), forms a fundamental aspect of Stall Catchers in that it is the primary method of direct communication between participants. Conversations ranged from play practices and Alzheimer's disease research to questions on game functionalities that experienced participants answered. At the same time, the chat provided a direct communication channel for the participants to the Stall Catchers team and vice versa. Even though not all participants actively engaged in the chat, it was important for the motivation and

---

28   Digital databases are collections of electronically stored information that can be maintained, accessed, and updated via a database management system, which functions as an interface between users or programs and the database. However, databases are more than mere information repositories; they are a consistently organized set of data whose informational patterns allow us to ask various questions—so-called queries—to the data (Quamen and Bath 2016). Among the different types of databases, relational databases are currently the most common form. They built upon a relational model (Schubert 2007, 35). This means that data are stored as so-called relations, which are typically represented as tables with rows (describing an object, such as a chat message) and columns (describing attributes or characteristics of a chat message, such as the sender's name). In these tables, each row describes an object, such as a message in the case of the chat analysis. Each column describes an attribute of the object, such as the sender id or the content of a message. In relational databases, object attributes can reference other objects by referencing their identifiers (typically the so-called primary key) in one of their columns. It is important to mention that databases are not neutral storage media but provide a specific perspective on data, as has been shown by sociologist Christine Hine (2006). Storing data in a structured database keeps it more manageable than simply using spreadsheets or other unstructured documents. The main advantage becomes apparent when it comes to analyzing and transforming data: with a relational database, one can interact with the data in complex and highly specific ways, which allows the answering of certain questions much more easily than a manual or spreadsheet-based approach could. For relational databases, this is commonly done via SQL.

contribution of some, as participants used it for mutual encouragement. Through chat analysis, I uncovered themes that were not mentioned in interviews, such as the playful reinterpretations of research data as artworks (see Chapter 5). Moreover, the analysis of chat messages provided valuable indirect and exploratory access to the field.

Figure 1: Stall Catchers' main UI

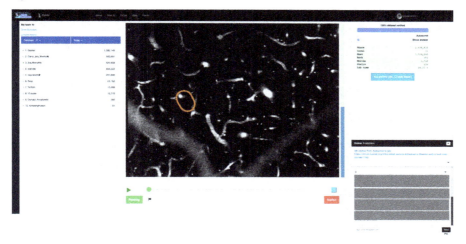

Source: Screenshot taken by LHV on Mar. 24, 2024 (https://stallcatchers.com/virtualMicroscope)

In another work (Veprěk 2023a), I provide a comprehensive discussion of my methodology, including a discussion of the chat format's particularities and addressing the ethical considerations and challenges involved in analyzing digital chat data. A brief summary is provided in the following for the purpose of this section. The process involved the four steps of data acquisition, cleaning, structuring into a relational database, and restructuring for analysis with SQL queries. My approach aimed to conducting a content-based analysis that focuses on the narration of play practices and meanings of Stall Catchers. I extracted sets of conversational contexts involving a specific player in order to gain a deeper understanding of the perspective of some of these individual Stall Catchers players via the chat. The extracts included entire conversations between participants with only minimal interspersed sections where other users engaged in the chat without actively contributing to the conversation of interest. I maintained the conversation dynamics by including a window of messages before and after each message of the user in focus. This way, I changed the superficially linear structure of the data into separate contextually linear excerpts. Occasionally, I revisited the data to isolate topical fragments of discussion or perform other small supporting analyses.[29] In the main analysis, these

---

29   The approach could also be extended to specifically isolate (or exclude) contexts where specific words or phrases were mentioned, making it easier to drill down on specific topics of conversation (in general, or once again involving a particular user). Since the results of a query are presented in virtual tables that can be exported and the queries do not operate on or change the dataset itself, it is possible to return to the dataset at any time and query it with different foci. Such flexibility

extracted conversations were exported into a format suitable for loading into MAXQDA for manual analysis, where they presented an access point to participants' perspectives during gameplay.

Chat analysis, despite its potential for ethnographic research, also has some limitations and challenges that are important to consider. These include the lack (or subtlety) of nonverbal elements in written conversational data, the question of how representative chat data are for the overall participant base, and significant questions of privacy (for a more detailed discussion see Vepřek 2023a).[30] I chose in this work to anonymize the chat data and combine several player identities into one. I also refrained from directly linking any personal interview data to the chat data or quoting directly from the chat in this work. Despite these limitations, if contextualized and included in a multi-perspectival and -methodological ethnographic approach, I expect this approach of restructuring and querying digital textual data with relational databases to have potential for qualitative ethnographic research beyond the specific analysis presented here. It is conceivable to apply this method to other textual data, such as tweets or comment threads, or to design queries that focus not on the textual content but on metadata, such as timestamps, to learn more about the temporal flows of chat communication, or other aspects and phenomena.

After having discussed the methodological foundation and its operationalization of my research, I now turn to the first empirical chapter, in which I analyze the imaginaries behind HC that guide and shape the development of HC-based CS.

---

is particularly helpful for ethnographic and inductive research, where fieldwork phases alternate with analysis and reflection and do not follow a linear scheme.

30   The Stall Catchers chat is semi-public; it is accessible to all registered Stall Catchers participants. Because registration only takes the provision of an email address and accepting the terms and conditions, the chat can be considered almost public. Participants are informed about this in the terms and conditions.

4 Envisioning and Designing the Future

The holy grail for generalized AI is to achieve humanlike intelligence. We have humanlike intelligence, it's called humans! [...] [A]nd what we really need to solve today's problems and tomorrow's problems is superhuman intelligence. So, and I believe the only way to achieve that, at least today, is with humans in the loop. (Michelucci 2019a, 11:22–11:47)

In his talk at the Microsoft Research Faculty Summit with the title "Crowd, Cloud and the Future of Work: Updates from human AI computation," Michelucci argued for HC as an alternative to AGI. Instead of pursuing "strong AI" or AGI efforts to build "human-like intelligence" with machines, HC aims to keep "humans in the loop" to build "superhuman intelligence." Michelucci, who I regard as an HC advocate, together with colleagues at the Human Computation Institute follow such approaches to human-in-the-loop computing by building HC-based CS systems. For this pursuit, Michelucci described going back to the "biggest pictures [he] can think of" (Jan. 21, 2021). The biggest picture for him was the understanding of information processing, which forms the basis of his HC visions and enables the development of HC systems for "human survival" (Michelucci, Jan. 14, 2021). The institute is, therefore, dedicated "to the betterment of society through novel methods leveraging the complementary strengths of networked humans and machines. We strive to engineer sustainable participatory systems that have a profound impact on health, humanitarian, and educational outcomes" (Human Computation Institute, n.d.). With this dedication, the institute places its ethical mission at the center of all its endeavors and daily working practices to develop "superhuman intelligence."

While Michelucci's words cited above do not describe what including humans in the loop to build human–AI systems means in concrete terms, they do provide a glimpse of how HC advocates construct the narrative of HC from AGI as a counterpole. Human computation, as I will discuss in this chapter, is not only understood to be *better* and more *ethical* because humans remain in control, but also more feasible. Michelucci's words represent and carry imaginations of the "human in the loop" in HC which not only shape the discourse on HC but also the development of such systems in general, as well as in the field of HC-based CS. Imaginations are crucial and powerful forces in the formation of assemblages and their desire. Therefore, it is important to disentangle these imaginations to further analyze how HC systems form, taking into account the interplay of future visions, everyday negotiations and associated human–technology relations. The latter are fundamental to HC because human-in-the-loop imaginations do not consider humans and AI in HC as separate but as relations between humans and AI. As I will show in this work, the intraversions of human–technology relations are, hence, always imagined.

This chapter, thus, analyzes the imaginaries from the perspective of HC advocates and developers as fundamental parts of the (everyday) becoming of HC-based CS assemblages and their intraverting human–technology relations. I first unpack the ambiguous term "human-in-the-loop" within HC, drawing particularly from the empirical material collected during my fieldwork at the Human Computation Institute. The analysis of this expression reveals different shifting and partly competing imaginations of concepts, such as "humans," "AI," "computation," or "technology" in HC and the broader field of AI. The analysis is structured around six imaginaries related to "humans in the loop." It should be noted that "human-in-the-loop computing" is a common phrase in computer science and technology development that goes beyond HC. I specifically focus here on its meaning in the field of HC. The first part of this chapter is structured around the individual imaginaries. First, HC forms a counter-imaginary to AGI. This describes the basis of HC endeavors and imaginations. Humans are put at the center of HC achievements in contrast to AGI, where the focus lies on the technological achievements. With humans in the loop, following the imaginary, unprecedented capabilities can be reached that exceed todays AI capabilities. In this overall counter-imaginary of HC, I describe situated human-in-the-loop imaginations which refer to the imagined *human* in the loop—*who* is imagined to be in the loop and *how* are humans imagined in HC?—; GWAPs imagining humans as *players* and loops as *games*; the imagined *loop* between humans and AI as human–AI conversations of the future; and the imaginations of humans in the loop as *crowds* in the loop. Finally, HC is imagined to lead to a future *Thinking Economy*, in which human creativity is unleashed.

By analyzing these visions and imaginations, which go hand in hand with normative principles guiding the everyday practices, it becomes clear that the development of HC systems can be understood as "ethical projects" (Ege and Moser 2021a) striving for a *better* future and human–technology relations as they *ought* to evolve (Jasanoff 2015a, 4). Moreover, it becomes apparent that there are not only overlaps between the imaginaries but also tensions between the HC future imaginaries and existing implementations, which can be explained with cultural anthropologist Genevieve Bell's observation that AI, and in this case more precisely HC, is "both a technology, and an imagination of a technology" (2021, 4).

The analysis of the imaginaries and visions necessarily decontextualizes them to a certain extent and separates them from everyday practice. This is due to my attempt to juxtapose concrete imaginaries with the example of the Human Computation Institute and general perspectives found in literature and the development of the field of HC.

For this reason, the second part of this chapter analyzes examples of infrastructuring that go hand in hand with the imaginations at the Human Computation Institute. Infrastructuring, as I will show, not only materializes these imaginaries but also destabilizes and forms them. Within the realm of HC, these practices can be observed at different levels. On the one hand, it is part of the bigger goal of establishing HC and the vision of hybrid human–AI systems, which I describe as *fundamental* infrastructuring. Here, I discuss the example of Civium, an "eco-system" which is being developed to build a basis for more "sustainable" HC (Michelucci 2019b). On the other hand, infrastructuring is part of everyday practices of maintaining and developing technological systems in general. I provide examples of what everyday materializing and negotiating human-in-the-loop imaginaries looks like from the perspective of the team, using the case of Stall Catchers. However, this discussion of infrastructuring continues throughout the next chapters with a special focus on the development of the Stall Catchers' data pipeline in Chapter 6, since exploration and infrastructuring play a major role in forming, redefining, and materializing human-in-the-loop imaginaries in HC. Because, while the ultimate goal seems to be clear, HC-based CS systems, according to the imaginaries described in the following, must stay at the edge of AI developments and scientific research. Furthermore, the role of humans and technology in these sociotechnical assemblages is not set *a priori* but negotiated along the way, guiding the future development of HC systems and further (re-)forming the visions. Developing HC systems, therefore, depends on future imaginations of human–technology relations (building on the counter-imaginaries), practices of exploration, and path dependencies (Klausner et al. 2015; De Munck 2022) introduced in the past. At the same time, and this will be the focus of the next chapters of this work, imaginaries are negotiated and partly also contested and contingent in the everyday. Inoperable materialities and life cycles of mice unaligned with the research process form or reform these relations just as much as other actors, most prominently the participants, who bring in their own aims and values, and engage in their chosen ways in relations with technology.

The aim of this chapter is to reconstruct the imaginaries driving the development of HC, in a manner akin to Bachman's analysis of the "work on the medium" with the example of Dynamicland (2018). These depictions are presented in a way that is aimed at closely reflecting and analyzing how they were presented by HC advocates, to then be contextualized and critically contrasted with other perspectives in later sections of this work.

My close collaboration with and ethnographic fieldwork at the Human Computation Institute allowed me to thoroughly analyze the imaginations driving the institute and provide a thick interpretation of a concrete example of narratives and normative principles that form and steer the current development of HC systems. While some of the points to be elaborated came to the fore due to questions I asked in interviews, others emerged, for example, from work meetings for brainstorming new features or project ideas or discussing current issues and problems with existing projects. Conversations, particularly during the field research periods in Ithaca, NY, in which I shared my ethnographic observations with Michelucci and other team members, sparked reflections on and comparisons of our perspectives, the questions we were asking in our respective work, and the approaches we took to arrive at answers. It was in these discussions that the imaginations underlying the development of HC became apparent.

The focus on the Human Computation Institute's specific approach to HC, thereby, is especially interesting due to its striving for the development of HC systems generally and not primarily to the solution of one particular scientific problem, as in the example of Foldit, which was specifically developed to solve the problem of protein structure prediction and design. Insights from other empirical material, for example, from interviews with HC advocates such as Bry, will be included in the following to contextualize the institute's imaginaries and further explore and problematize HC visions. Additionally, without claiming completeness, I include literature and HC standpoints presented in media content from the field of HC (and HI) to situate these imaginaries further.

As discussed in Chapter 2, I build upon Jasanoff and Kim's term "sociotechnical imaginary" (2009; 2015) as well as Hilgartner's notion of "vanguard visions" (2015) to think about HC as a counter-imaginary. The aim is to stress how HC advocates are forming their visions in contrast to AGI through boundary work and by explicitly distancing themselves from such AGI endeavors. The concept of sociotechnical imaginaries forms a helpful starting point for analyzing how HC-based CS projects are envisioned due to its focus on the imaginaries' normativity and the materialities that are part of sociotechnical assemblages (Jasanoff 2015a, 19). Some of the imaginaries specific to the Human Computation Institute can even be understood as normative principles that guide the team's work in maintaining existing and developing new projects. However, if we take Jasanoff and Kim's definition seriously, HC imaginations, as opposed to AGI narratives, would be more accurately described by Hilgartner's term "vanguard visions." These visions originate from "sociotechnical vanguards" who are

> relatively small collectives that formulate and act intentionally to realize particular sociotechnical visions of the future that have yet to be accepted by wider collectives, such as the nation. These vanguards and their individual leaders typically assume a visionary role, performing the identity of one who possesses superior knowledge of emerging technologies and aspires to realize their desirable potential. Put otherwise, these vanguards actively position themselves as members of an avant-garde, riding and also driving a wave of change but competing with one another at the same time. (Hilgartner 2015, 34)

While Hilgartner observes that visionaries or groups also compete and support slightly different views for the field of synthetic biology, I obtained similar results in the analysis of HC visions. Vanguard visions have to move and legitimize themselves between powerful sociotechnical imaginaries and, hence, cannot be successfully established solely by argumentative means (Hilgartner 2015, 45). Instead, "hands-on activities are needed to present their futures as feasible and achievable" (Hilgartner 2015, 45).

Among HC's sociotechnical vanguards is Michelucci, whose institute is cited as a source for new HC developments (Lazar, Feng, and Hochheiser 2017, 435–436). The initiative "2014 Human Computation Roadmap Summit," led by Michelucci, earned him and his colleagues the title of "crowdsourcing pioneers, and visionaries" in the *MIT Technology Review* magazine (Emerging Technology from the arXiv 2015). Due to Michelucci's prominent position in the field, focusing on the imaginaries at the Human Computation Institute is a fruitful point of departure for the analysis of HC visions.

The narratives and visions presented are not necessarily shared by all HC and HI practitioners or visionaries. Focusing on a few sociotechnical vanguards, however, allows one to gain a deeper understanding of HC visions, imaginaries, and their directions and scientific programs, because, as Koch writes following Ulf Hannerz' understanding of culture, "[i]ndividuals [...] as producers of culture play a key role in cultural analysis" (Koch 2005, 26). They "are constantly inventing culture or maintaining it, reflecting on it, experimenting with it, remembering it or forgetting it, arguing about it, and passing it on" (Hannerz 1992, 17).

Moreover, HC is becoming an increasingly "organized field of social practices" (Appadurai [1990] 2002, 50; cited in Jasanoff 2015b, 327) that can currently be situated between vanguard visions and widespread sociotechnical imaginaries. Seeing HC as a form of counter-imaginary to AGI helps me to analyze the specifics of its visions and their emergence. I show how these visions not only distinguish themselves from AGI sociotechnical imaginaries but are also related and situated in common perspectives of AI research. After all, as Hilgartner writes, "[s]ociotechnical vanguards seek to make futures, but (to paraphrase Marx) they cannot make them simply as they please; they do not make them under self-selected circumstances but do so using vocabularies and practices already given and transmitted from the past" (2015, 50). To a certain degree, these vanguard visions guide the evolution of HC systems and, therefore, of intraversions of human–technology relations (by which they are also produced), even though they are also contested and negotiated.

Human-In-The-Loop Imaginaries

Human Computation as a Counter-Imaginary to Artificial General Intelligence[1]

At the heart of the human-in-the-loop imaginations of HC are humans, who are put in the spotlight when talking about the achievements of HC systems. While this may seem trivial at first, it is one of the key differences between HC and many strong AI narratives. The latter claim technical achievement and commonly gloss over human involvement altogether or even imagine humans to be out of the picture. For example, it is still usually not mentioned that most AI models have been trained on data annotated by humans, that datasets have been manually cleaned beforehand, or that rules and rewards, in the case of reinforcement learning, had to be set initially; not to mention the adjustments and maintenance work that guide the lifetime of any sociotechnical system. In contrast, HC, one could say, takes the opposite approach in sketching their imaginaries as they tend to hide the system's technical complexity and the AI algorithms operating in the background, and focus on humans when it comes to achievements. This difference is key to the imaginary of HC as a counter-imaginary to AGI. To better illustrate how this unfolds, I first turn to imaginations of AI in general with a recent example.

On May 14, 2022, Nando de Freitas, who is an ML professor and research director at Alphabet Inc.'s AI subsidiary and research laboratory DeepMind, tweeted about Deep-

---

1 First ideas of this section were presented at the 8th International Working Conference of the Digital Anthropology Commission German Society for Cultural Anthropology and Folklore Studies, "Digital Futures in the Making: Imaginaries, Politics, and Materialities" on September 15, 2022, at the University of Hamburg.

Mind's new AI-agent Gato (Reed et al. 2022): "It's all about scale now! The Game is Over! It's about making these models bigger, safer, compute efficient, faster at sampling, smarter memory, more modalities, INNOVATIVE DATA, on/offline, ... 1/" (Nando de Freitas [@NandoDF] 2022). With "the game is over," de Freitas declared that the search for AGI is over with Gato AI. All that is needed now is scale. The "generalist agent" Gato is presented as the solution to a decades-long search for the path to strong AGI. While it surely presents an interesting development in ML and reinforcement learning, the purpose of mentioning de Freitas' tweet does not lie in the question of whether Gato presents the straight pathway to AGI or not. Instead, it serves as an example of a very current manifestation of the beliefs in AGI.

While the latter have been particularly pursued since the early 1950s (Fjelland 2020, 2), imaginations of future machines with "intelligence" have long existed in literature and mythology and even before the term AI itself was coined in the 1950s (Koch 2005, 9; Cave and Dihal 2019). But since then, researchers have focused on developing AI, often striving for AGI with "human-like intelligence" (Fjelland 2020, 2). Despite various attempts to define AI, it, similar to its natural model, cannot be clearly defined (Koch 2005, 48). Koch (2005) has already aptly demonstrated this in her research on the "culturality" of the technological formation of AI in Germany in the 1980s and 90s. Nevertheless, this is still true today, where the term AI is used to refer to various technologies from robotics to NLP and DL and where there are different understandings of what the ultimate goal of AI research should be. One of the most prominent endeavors is the development of AGI or "strong AI." Some advocates of the AGI thesis argue that AGI, utilizing its "human-like intelligence," will subsequently lead to Artificial Super Intelligence (Kurzweil 2005; Bostrom 2016; cited in Peeters et al. 2021, 220).[2]

Even though the implementation of actual AI systems has been lagging, it has been attracting lively interest in public discourse since its early years. Koch attributed the widespread interest in AI to the fact that it challenges humans to reconsider how they perceive themselves in a way that few other technologies do (2005, 48–49). The "scientific attempt to technically reproduce abilities that until then had been regarded as originally human is perceived by many people as a threat" (Koch 2005, 49), leading to an extensive social debate in which participants from various backgrounds contribute (Koch 2005, 49). Similar observations can be made today—still and to no lesser extent—for discourses in, at least, many European countries and the USA.

The journey of AI research has seen ups and downs since its uptake in the 1950s. An example of the latter was the "AI winter" of the mid to late 1980s, which emerged due to overoptimistic expectations of AI's possibilities from AI sponsors, industry, and government in the face of a "failure to deliver systems matching these unrealistic hopes, together with the accumulating critical commentary" (Nilsson 2010, 408). However, with the increase in computing power and advances in, for example, ML and NLP, as well as recent developments, such as the text-to-image model DALL-E and others, AI—and increasingly the discussion of its societal and ethical implications—remains a hot topic of public debate. The "AI winter" has been overcome, and AI is becoming more pervasive in everyday life, demonstrating capabilities in ways that were not expected of these

---

2 Other scholars divide AI research into cognitivist or engineering approaches (*cf.* Koch 2005).

approaches. But as promising as these advances are that let AI advocates dream and announce that "the game is over," as Nando de Freitas did, predictions as to when the idea of AGI will materialize have not yet been fulfilled, and it is highly controversial as to what extent such an AI is even possible.

Artificial intelligence researchers Marieke Peeters et al. (2021) identify three main perspectives in the current broader debate about the future of AI. While the first, the *technology-centric perspective* resembles the endeavors described above, they identify the *human-centric perspective* as the second perspective, "which holds that true intelligence can ultimately be found only in human beings and (potentially) other intelligent living creatures" (Peeters et al. 2021, 219). While AI can provide assistance to humans, it cannot develop "intelligence" itself. The third recognized perspective is the *collective intelligence perspective*, which overlaps with HC and is discussed further below. The field of AI has been accompanied from its inception by narratives and imaginaries that commonly refer to visions of strong AI. Regardless of whether they ever become "true" or not, such narratives—and this is the key point I would like to make here—shape both the expectations and beliefs according to which AI is evaluated, designed, and developed (see also Cave and Dihal 2019, 74). These both utopian and dystopian AI imaginaries are strongly shaped by fiction and politics (Bareis and Katzenbach 2022). Philosopher Stephen Cave and science communication scholar Kanta Dihal (2019) identify four pairs of the most widespread hopes and related fears of AI from a corpus of fictional and nonfictional narratives: these pairs are immortality vs. inhumanity, ease vs. obsolescence, gratification vs. alienation, and dominance vs. uprising (Cave and Dihal 2019). Each hope or fear is connected to a number of either optimistic and utopian or pessimistic and dystopian narratives (Cave and Dihal 2019, 75). These narratives are, of course, not exhaustive but form composing parts of the sociotechnical imaginary of AI that currently prevails in Western countries. Formulating these hopes and fears as pairs illustrates how sociotechnical imaginaries, no matter how widespread, also always face resistance and conflict. This is particularly true in the field of AI, where the promises and their long-term narratives and visions have not been sustained and where there is a mismatch between AI's imaginaries and current implementations (Sartori and Bocca 2022).

Examples for such alternative imaginaries of AI are the vanguard visions of HC. The advocates of HC build their narrative in an active distinction to dystopic AI narratives and AGI visions with "boundary work" (Gieryn 1983). These HC narratives and visions build upon the limitations of today's mere machine-based AI efforts and the disappointments due to unfulfilled promises of such strong AI visions. Sociologist Jens Beckert describes this pattern regarding new promissory stories in general, which always relate to preceding ones and create their own credibility by removing themselves from those "disappointed hopes" (2018, 284).

In a promotional video by the Carlsberg Foundation about HI at the Center for Hybrid Intelligence at Aarhus University (Elmann 2022), physicist and head of the Center for Hybrid Intelligence Jacob Friis Sherson explains:

> When you watch the news, then you get the experience that artificial intelligence is a tidal wave that is coming upon us, so every week there is a new instance of how artificial intelligence has entered and transformed a new domain. And you get the

experience that it will overcome us, that there will be no place in the near future for humanity. And that's what I want to fight, because the reality is the more we try to build into artificial intelligence, the more we see that there is a core of us that cannot be replaced. (Carlsbergfondet 2019, 00:19–00:53)

As Sherson speaks, the camera alternates between him sitting in front of a large screen and different scenes lending a dramatic visual shape to his words, including an angry-looking robot army, a hand taking a queen in a chess game, a fast-running clock, and a skull surrounded by white fog. To construct the narrative, HI takes off from the fears of current AI sociotechnical imaginaries that are circulating in the media and builds on the dystopian images of these imaginaries.

While not all HC advocates aim to combat dystopian strong AI narratives in the same way as Sherson, their arguments share the common notion that the promises of AGI to "supersede humans" (Dellermann, Ebel, et al. 2019, 637) have not been fulfilled to date, and it remains unclear whether these promises will be achieved in the not-too-distant future. Law and Von Ahn write in the preface to their book *Human Computation* (2011) that HC moves beyond the extent of current AI algorithms. Despite advances in automating mathematical problems humans used to solve, the narrative goes, "there are still many problems that are easy for humans to solve, but difficult for even the most sophisticated computer algorithms" (Law and Von Ahn 2011, 2). Human computation, therefore, starts here, at so-called AI problems and by combining humans and machines they can "achieve futuristic AI capabilities today" (Michelucci and Dickinson 2016; cited in Michelucci 2019b).

One of the key differences between HC and strong AI narratives was mentioned at the beginning of this section: the attribution of claimed achievements. While Strong AI narratives claim technical achievement, HC puts humans in the spotlight when talking about the achievements. In the promotional video, Sherson argues that HI is not about "understanding what AI can do and cannot do, it is understanding what we as humans are best at" (Carlsbergfondet 2019, 01:03–01:08), and the challenge is to find the "human place [...] alongside AI" (Carlsbergfondet 2019, 00:58–01:00). By distancing itself from AGI sociotechnical imaginaries and visions, HC builds its own credibility and aims to legitimize the approach as the *right* way to develop "intelligence" that goes beyond humans and computers' capabilities by putting humans into the focus. This allows HC advocates to circumvent the dystopian narratives and fears in public discourse associated with strong AI pursuits and to present HC as "ethical projects" (Ege and Moser 2021a). They are "future-oriented undertakings" (Ege and Moser 2021a, 7) that are permeated by values and norms of the roles that humans should take and what "artificial intelligence" should look like. This is evident in the choice of words, such as "conscientious development," chosen by Bowser and colleagues in the report "Artificial Intelligence: A policy-oriented introduction," published by the US research center and nonpartisan political forum:

Human computation approaches to AI [...] advanc[e] the design of AI systems with human stakeholders in the loop who drive the societally-relevant decisions and behaviors of the system. The *conscientious* development of AI systems that *carefully* considers the coevolution of humans and technology in hybrid thinking systems will help ensure that humans remain ultimately in control, individually or collectively, as systems achieve superhuman capabilities. (Bowser et al. 2017, 11, emphasis LHV)

It is stressed in this statement that humans ought to ultimately remain in control of these hybrid systems which can be accomplished with conscientious development that takes into account the "coevolution" of humans and technology. The understanding of human–technology relations put forward in this quote is reminiscent of a postphenomenological approach that considers humans and technologies existing together and being mutually interdependent (Dorrestijn 2012a, 63).

The Human Computation Institute ethically frames its endeavors in a similar way, placing its ethical mission at the center of all its efforts. I would like to illustrate this mission in more detail here before situating it within other HC and HI endeavors. Michelucci explained during one of our more formal interview sessions, when asked if there was anything else important to him that he wanted to share, how his work was rooted in the "biggest ideas," as cited in the introduction, and how HC could bring humanity closer to the "good" life. "We can use these new capabilities to bootstrap better [...], more humanistic augmentation methods that allow us to solve world problems [...] by doing things that [...] entertain us, they give us a sense of purpose and they just help us thrive as human beings." (Michelucci, Jan. 21, 2021)

As Michelucci's words quoted at the beginning of this chapter make clear, HC is understood to achieve unprecedented capabilities today that exceed mere computational capabilities and current AI technologies and ensure humanistic approaches.

Michelucci and computer scientist and HC researcher Elena Simperl, as editors, open the first issue of the *Human Computation Journal* with the following quote that is attributed to the economist and public servant Leo Cherne: "The computer is incredibly fast, accurate, and stupid. Man is incredibly slow, inaccurate, and brilliant. The marriage of the two is a force beyond calculation" (Michelucci and Simperl 2014, 1). Building on the understanding of computers as "fast but stupid" and a human as "slow but brilliant," the authors set the agenda for the transdisciplinary approach that taps and combines the respective strength of humans and computers to achieve unsurpassed capabilities (Michelucci and Simperl 2014, 1).

However, despite this huge potential, there is a gap between what could be possible with HC and current HC research and implementations. According to Michelucci, the field of HC research today is not innovative enough due to the organization of academia and its inherent dependencies and power relations (Jan. 14, 2021). If more researchers tried new and more innovative approaches to combining humans and machines, it would not only advance HC and bring it closer to solving some of today's biggest problems but also make it more "interesting" for participants, he explained in one of our conversations (Jan. 14, 2021). Michelucci described this as the "duality" in HC. An HC project must be both interesting and "fulfilling" for participants (Michelucci, Jan. 14, 2021).

With his vision of HC, Michelucci might not necessarily represent the perspective of all HC researchers. However, as I learned during my ethnographic fieldwork at the Human Computation Institute, it is also shared with great enthusiasm by his colleagues. The CS coordinator Paul explained in our interview in October 2020:

> I think human computation could do everything [laughs]. […] I think the […] only limit is sort of human imagination of what problem you can take and tackle with human computation and, of course, still the […] existing resistance to nonconventional ways of doing that. Like, you still need to, [it is] still a huge job to convince scientists and founders—like what happened with Stall Catchers, the scientists didn't believe that this could be done but they did later and no funders would believe that this could be done, but then there was one who decided that he will take a risk and fund this nonconventional project so I think that's […] the only limitations for now to tackle any problem with […] human computation. (Oct. 14, 2020)

The spirit of working on something new and "revolutionary" drives the work at the Human Computation Institute, such as that of developer Kate, who explained that working on this project opened their eyes to the relations between humans and machines:

> I'm really excited to work on a project […] like that. I really gained a lot, learned a lot as well, I'm glad I took that chance actually. It made me think of things in a different perspective. I used to think of humans as separate from computers. I used to think people can do, cannot do what a computer can do or the opposite, but the more I've worked on [different projects at the institute], I started to [gain] a different perspective on things like […] the line between humans and machine is getting more blurred and maybe in the future, Stall Catchers or human computation would have been the first step in a neural network or beaming people's brain into computers, I don't know about that [laughs], but maybe that's part of the revolution, that's part of the first step of how we do these things and I'm really happy to have been part of this project. (Nov. 19, 2020)

Sharing this excitement, developer Samuel half-jokingly explained to me in another conversation that combining humans and machines in a way that humans oversee machines could indeed lead to unprecedented capabilities: "[B]y bringing these two together, it's like we're having a god, we're building a god. [Laughs] We are creating a new era god" (Sept. 2, 2021). But in contrast to an AGI that will outperform and rise up against humanity, as one of the fears stated which was identified by Cave and Dihal as prevalent AI narrative prophecies (2019), HC, according to Samuel, presented

> a big opportunity. It […] can prove that AI will […] help humans instead of replacing humans in the future. And HCI [the Human Computation Institute] is proving that AI can help humans instead of replacing them. because AI still has some disadvantages, still has some issues. It's not like we're all in […] a science fiction movie, where AI is destroying everyone. […] HCI has an opportunity to become, […] the best, to become a role model for people I mean in the AI section, AI development. (Samuel, Sept. 2, 2021)

Human-in-the-loop computing in HC is imagined as leading to *good* sociotechnical systems, in which human control is ensured. Due the institute's dedication to the "betterment of society" (Human Computation Institute, n.d.), Samuel explains, it could become a role model for other AI researchers and developers (see above). The mission of the institute illustrates the ethical motivation behind the projects and work at the institute to develop HC systems to develop *better* futures.

While the Human Computation Institute's vision might be very specific and personal, the counter-imaginary of HC and HI as ethical projects can also be found in one of the first HI papers of the current research trend by human-centered AI researcher Ece Kamar (2016a; 2016b). Referring and connecting to HC advances, Kamar argues for "hybrid intelligence"[3] systems as AI systems that are guided and supervised by humans to "prevent the mistakes and failures that would be caused by an AI system working alone" (Kamar 2016a, 4070). Rather than "designing AI systems that function alone, we should focus on hybrid systems that can benefit from partnership with humans" (Kamar 2016a, 4070), which can lead to a "virtuous improvement cycle for the system to continuously learn from" (Kamar 2016a, 4070). These, according to Kamar, can go beyond the current limitations of mere machine AI (2016b, 25). Kamar's HI definition clearly has great similarities with common definitions of HC. However, Kamar specifically refers to HC as crowdsourcing and emphasizes the fruitful environment of such crowdsourcing or HC platforms since they "function as testbeds for data collection and experimentation" (2016a, 4070) for accessing "human intelligence."[4]

Dellermann et al. (Dellermann, Calma, et al. 2019; Dellermann, Ebel, et al. 2019) further define HI as "socio-technological ensembles" in which both the human and AI "continuously improve by learning from each other." (Dellermann, Ebel, et al. 2019, 640) The paper "Mapping Citizen Science through the Lens of Human-Centered AI" (Rafner et al. 2022), which was initially published as a preprint with the title "Revisiting Citizen Science Through the Lens of Hybrid Intelligence" (Rafner et al. 2021), has been written by multiple authors and visionaries in the field of HC. It can, thus, be understood as an attempt to come to a shared understanding of the fields despite these "incompletely aligned views" (Hilgartner 2015, 35), for example, regarding the definition of HI, and what differentiates HC from HI. Divergent understandings are not uncommon for new or young research fields that have not stabilized into one prevalent narrative that most advocates can agree

---

3   While Kamar's article definitely describes an early take on HI, it is not clear when the term was first coined.

4   For Kamar, crowdsourcing platforms are "testbeds" for HI (2016a, 4070). Including the crowd in online CS projects also creates the starting point for exploring HI at the Human Computation Institute and the Center for Hybrid Intelligence at Aarhus University (Elmann 2022). In an article in the *Human Computation Journal*, the CS researcher and ecological informatics specialist Greg Newman elaborates on the opportunities of "Citizen Cyberscience" for HC: "Some citizen science projects introduce new human computation techniques or engagement modalities, thus directly contributing to a growing body of human computation methods. In this way we can see citizen science as applied human computation, a platform for human computation research, and a body of work that may innovate in the human computation space. As an example, citizen science often generates platforms for citizen engagement that can be used in human computation research" (2014, 108). Platforms such as Stall Catchers, Foldit, and ARTigo can, hence, be understood as laboratories for HC research. The game environment and framing of the project also plays an important role in the intraversions of human–technology relations due to affordances of the platforms as games, for example, which invite participants to play around and find ways to engage with the platform and game that go beyond the developer's intended designs.

on. Nevertheless, HC and HI visions seem to be able to agree on several shared understandings, the most prominent of which are the imaginations of alternative futures of AI to widespread AGI and strong AI sociotechnical imaginaries that place humans at the center of their approaches. While acknowledging that discussing these narratives collectively can result in oversimplification and generalization, examining them as an interconnected yet diverse bundle can provide a deeper understanding of the nuances and specificities of both HC and HI. As Forsythe has aptly put it regarding her research on informatics and AI:

> While the beliefs held by actors within a given scientific setting or community are neither identical nor altogether stable over time, neither are they completely reinvented from one event to the next. Scientists construct local meanings and struggle to position themselves within a cultural context against a backdrop of more enduring beliefs, at least some of which competing actors hold in common. After all, in order to negotiate at all or successfully to frame actions or objects in a given light, there must be overlap in the interpretive possibilities available to the parties concerned. ([1992] 2001b, 2–3)

The overlap in HC lies specifically in the attempts to imagine AI futures with humans at the center and in control. While this utopian perspective is reminiscent of traditional techno-optimist imaginaries, according to which automation and AI will free humans from undesirable work, the HC imaginary as a counter-imaginary to AGI endeavors focuses on the "co-evolution of humans and technology," so that the former retain control over the latter. "In-the-loop" refers not only to *being* in the loop but also to being *in control* of the system, as opposed to other uses of this term, such as in crowdworking applications, where humans are in the loop in the sense of directly training an ML model by annotating data.

Shared Paradigms

Even as advocates position HC as a counter-imaginary to other AI pursuits, they remain open to the possibility that AGI may become possible at some point in the future. This is evident in Michelucci's words quoted above, in which he states that the only way to achieve "superhuman intelligence [...], at least today, is with humans in the loop" (Michelucci 2019a, 11:40–11:47). The insertion "at least today" indicates that Michelucci does not completely distance himself from the idea that AGI might someday be possible, similar to Dellermann, Ebel, and colleagues, who speak of the "near future" (2019, 637).

Despite the boundary work of HC to distance itself from AGI visions, the visions share much in common. Following the cultural studies scholar Toby Miller, "[e]very cultural and communications technology has specificities of production, text, distribution, and reception. But the utopias and dystopias of successive innovations share much in common. As private excitements and public moral panics swirl, they also repeat" (2006, 6). Although HC refers not only to previous innovations but also concurrent developments in AGI endeavors, Miller's observation about the commonalities of utopias and dystopias of successive innovations can also be observed for these fields. This is not surprising given that many HC researchers and advocates commonly have a background in computer and/or cognitive science. To understand the intentions behind the design of HC systems, it is, nevertheless, important to discuss these parallels.

I here want to briefly mention two of the perspectives and paradigms that form parallels between HC and strong AI approaches. First, imaginaries of AGI (both utopian and dystopian) and HC have in common that they view these developments (AGI and HC) as having an impact on humanity as a whole. They describe phenomena that do not only solve specific problems but have a large-scale influence on humanity and its problems in general. In the case of HC, this can be seen in the example of Michelucci's vision and in the quotes that introduce the new *Human Computation Journal*. In their introduction, Michelucci and Simperl not only quote Cherne but also founder of evolutionist theory Charles Darwin (1859): "In the long history of humankind (and animal kind, too) those who learned to collaborate and improvise most effectively have prevailed" (cited in Michelucci and Simperl 2014, 1). By invoking Darwin and Cherne in this way, they frame HC as a revolutionary phenomenon with possibly existential implications. This way of framing parallels popular imaginations of the rise of AGI, for example, as physicist, cosmologist, and president of the Future of Life institute[5] Max Tegmark writes in *Life 3.0. Being Human in the Age of Artificial Intelligence*: "In comparison with wars, terrorism, unemployment, poverty, migration and social justice issues, the rise of AI will have greater overall impact" (2017, 37–38).

Second, both HC and AI approaches generally build upon information processing as their fundamental paradigm. Information processing, which interprets cognitive processes as processes of information processing, is considered the key paradigm underlying AI research (Ahrweiler 1995, 15; cited in Koch 2005, 45; see also Simon 1996; Turkle 2005a; Becker 2023). It is also fundamental to HC, as Michelucci writes:

> [T]he construal of computation as being equivalent to information processing seems to best fit the practical context of human computation. In HC, "computation" refers not just to numerical calculations or the implementation of an algorithm. It refers more generally to *information processing*. This definition intentionally embraces the broader spectrum of "computational" contributions that can be made by humans, including creativity, intuition, symbolic and logical reasoning, abstraction, pattern recognition, and other forms of cognitive processing. As computers themselves have become more capable over the years due to advances in AI and machine learning techniques, we have broadened the definition of computation to accommodate those capabilities. Now, as we extend the notion of computing systems to include human agents, we similarly extend the notion of computation to include a broader and more complex set of capabilities. (2013d, 84, emphasis i.o.)

Though Michelucci argues for a broad understanding of cognitive processing (see below), the examples provided still refer primarily to abstract, mental processes.

---

5   The Future of Life institute is a nonprofit organization with the mission of "steering transformative technology towards benefitting life and away from extreme large-scale risks" (Future of Life Institute, n.d.).

This understanding of information processing can be contrasted with Hutchins' understanding of computation and distributed cognition.[6] Even though Hutchins' *Cognition in the Wild* was published in 1995 and his critique of the cognitive science approach might not be applicable to all current understandings, his elaboration on the problems of approaches that focus on information processing seems relevant here:

> The model of human intelligence as abstract symbol manipulation and the substitution of a mechanized formal symbol-manipulation system for the brain result in the widespread notion in contemporary cognitive science that symbols are inside the head. [...] And while I believe that people do process symbols (even ones that have internal representations), I believe that it was a mistake to put symbols inside in this particular way. The mistake was to take a virtual machine enacted in the interaction of real persons with a material world and make that the architecture of cognition. (1995a, 365)

The problem, following Hutchins, is that an approach to cognition or intelligence as information processing makes it difficult to combine it with action, which is fundamental to an understanding of distributed cognition as process. "History and context and culture will always be seen as add-ons to the system [if cognition is reduced to information processing], rather than as integral parts of the cognitive process, because they are by definition outside the boundaries of the cognitive system" (Hutchins 1995a, 368). Such an understanding also has consequences for the imagination of the human in the loop, as I will discuss below. Although only briefly discussed so as not to go beyond the scope of this research, the two examples show how HC is, to a certain degree, situated within the broader field of AI research and, thus, rooted in the same foundational paradigms and perspectives with endeavors from which it distances itself.

The Imagined *Human* in the Loop

Who is the human in the loop? While the use of the term *human* in the sociotechnical imaginary of HC implies a very broad understanding, a closer look at the human-in-the-loop imagination reveals that the *human* in the loop in HC narratives and literature does not refer to developers or designers of HC-based CS systems but specifically to users and participants who are invited to contribute by the system designers. In the examples of Stall Catchers, ARTigo, and Foldit studied, however, the participants who are *in* the loop are not necessarily those who are *in control*, even if they play an active role in forming the systems, as will be discussed in the next chapters. Instead, the projects and sociotechnical systems have been initiated and created by researchers and developers, who, in the end, have decision-making power over the overall project.

Human-in-the-loop generally refers to a paradigm or approach within computer science research fields such as ML (Monarch 2021; Mosqueira-Rey et al. 2022). The US De-

---

6   For further discussion of the information processing paradigm, see, for example, Ahrweiler (1995), Koch (2005), Turkle (2005a), Becker (2023). Unlike Hutchins or literary theorist N. Katherine Hayles in her book *Unthought: The Power of the Cognitive Nonconscious* (2017), my research does not seek to define "cognition" but to examine how it is conceptualized in the field I investigate.

partment of Defense defined it in 1998 as a "model that requires human interaction" in their *Modeling and Simulation Glossary* (Under Secretary of Defense for Acquisition Technology 1998). More recently, it has been defined as "an interaction between humans and an artificial intelligence (AI or machine), with the goal of improving the machine's AI" (Rueckert and Riedl 2022). Even though usually not explicitly stated, human-in-the-loop is understood as a way to model these interactions, which are defined by the designers and developers of such systems or interactions. With *human*-in-the-loop, authors using the term are most often not referring to themselves. They are referring to the imagined users. It should be noted that humans in the loop can also be, for instance, medical doctors using a system in the example of medical AI systems (Kieseberg et al. 2015). During my second research stay in Ithaca and when discussing this observation of the human-in-the-loop approach, Michelucci agreed: "I'm a human in the loop but not in a way we often talk about it" (fieldnote, Nov. 9, 2022). Rather, the participants in today's HC systems are those in the loop who might later be replaced by automated processes as computational capabilities advance. Designers, developers, and researchers set the terms and decide what process should be automated and where participants should be invited to solve a specific task. In this sense, human control over AI and hybrid systems currently does not typically take place "in the loop," but as control "of the loop." Therefore, to better understand sociotechnical assemblages, it is crucial to consider the role of the designer and developer in the development and evolution of HC-based CS projects. Only then is it possible to contextualize the imagined *human* in human–technology relations of the future and analyze how the imagined humans actually relate to these. That is, how they actively engage and bring in their own ideas of how they want to be human in human–technology relations in HC-based CS.[7] This observation of the human in the loop also implies not only a relational but also a processual perspective, focusing on both the becoming of the assemblage and the continuous processes of territorialization, reterritorialization, and deterritorialization, which are "mutually enmeshed," following Deleuze and Guattari: "It may be all but impossible to distinguish deterritorialization from reterritorialization, since they are mutually enmeshed, or like opposite faces of one and the same process" (Deleuze and Guattari 2000, 258). This chapter, therefore, discusses both the imaginaries of HC advocates and the work of designers and developers on materializing them, while, at the same time, changing the imaginaries.

The imagination of *who* the human in the loop is already alludes to *how* humans are imagined by HC researchers and the projects that they create. The humans in the loop are not imagined as the researchers or experts in HC. But, in collating their nonexpertise responses, the information they provide can be computed into "expert-like" information or output. At the same time, "human intelligence" (Law and Von Ahn 2011), "cognition" (Michelucci et al. 2015, 2), "human intellect" (Hartman and Horvitz 2013, xi), and the "human processing power" (Von Ahn 2005) have remained central to the image of the human since

---

7   In fact, following a relational understanding of technology or Louise Amoore's approach to "cloud ethics" (2020), "the human in the loop is an impossible subject who cannot come before an indeterminate and multiple *we*" (2020, 66, emphasis i.o.). This implies that we must focus on the relations and entanglements of the various human and nonhuman actors, which I attempt to do in this work.

the very beginning of HC, as I have shown in the previous sections. The idea of human processing power is linked to the broader idea of information processing, which does not only refer to computers, but applies to computation in general, which is performed by a wide range of organisms, including humans, animals, and even bacteria (Michelucci 2017a). According to Michelucci's introduction to the *Handbook of Human Computation*, computation, as understood in HC, includes computer algorithms and numerical calculations but, at the same time, "embraces the broader spectrum of 'computational' contribution that can be made by humans, including creativity, intuition, symbolic and logical reasoning (though we humans suffer so poorly in that regard), abstraction, pattern recognition, and other forms of cognitive processing" (2013b, xxxviii). Human computation then refers to such information processing processes "that derive from the computational involvement of humans in simple or complex systems" (Michelucci 2013b, xxxviii). Michelucci elaborated on his understanding in one of our interviews in early 2021 as follows:

> [W]hen I think about human computation [...], I'm really thinking about information processing in the large. We have a lot of different ways to do information processing. We have human humans. We have other animals. [...] [W]e have [...] other kingdoms that can do information processing. And then we have inventions. We have machines we've created that help us do information processing. So, I'm always interested to look at this entire collection of things and [...] this is our tool kit, our tinker toys that we can build with. What can we build with these things to create new capabilities that didn't exist before and that can help us survive. (Jan. 14, 2021)

Michelucci has referred to this holistic approach to thinking as "organismic computing" (2013c) in earlier work.[8] By considering not only humans in contrast to technology but also nonhuman entities, this approach is reminiscent of naturecultures and more-than-human perspectives in the humanities and social sciences research.[9]

The idea of HC here does not detract from the question of how to automate human tasks with machines or AI, for example, but describes them—including other nonhuman life—as different "information processing" tools that can be combined into something that exceeds the individual capabilities. The combination of these different elements can, in the institute's understanding, make possible the survival of humans, which depends on the survival of the entire ecosystem and world (fieldnote, Nov. 9, 2022). Humans in the loop are, thus, computational elements in HC systems that can solve a specific problem presented, for instance, an image recognition problem in the example of CAPTCHA, Stall Catchers, and ARTigo, that cannot currently be solved by *other* information processors, such as computers. Moreover, humans are imagined as creative problem solvers whose creativity can be freed and enhanced by HC. Finally, humans in the loop in crowd-based HC systems are imagined as aggregate individuals who can be networked together in the

---

8 In Michelucci's understanding, HC and "organismic computing" are not two separate concepts but "organismic computing is an instantiation of human computation" and "multiple human computation methods can be implemented in" different parts of organismic computing systems (fieldnote Aug. 19, 2022).

9 On naturecultures and more-than-human approaches, see Chapter 1, footnote 3.

loop. Before delving into these aspects, it is important to consider a human or user HC imagination specific to HC systems designed as GWAPs, as in the case of the examples studied. This imagination connects the image of the user with the image of the loop, since in GWAPs, according to Dabbish and Von Ahn's definition, the user is imagined as a contributor who must be entertained and participates due to their interest in playing games, not solving AI problems (2008). These user images play a fundamental role in the design and actual implementation of HC systems in the field of CS, which is why I here want to briefly discuss how and why the examples studied were designed as GWAPs.

Imagining Humans as Players and the Loop as a Game

The examples of Stall Catchers, Foldit, and ARTigo, while addressing very different scientific problems, have in common that they are designed as computer games, and, as such, share several game features. Most prominently, all of the games include a scoring feature so that participants can earn points and compete against each other on the leaderboards. The projects are not accidentally designed as games. In fact, many online CS projects are not game-based but rather designed as more classic scientific studies, where participants fill in surveys or collect data that they upload to a dedicated platform (e.g., the Great Influenza Survey [Land-Zandstra et al. 2016] or iNaturalist [n.d.]). In the field of HC, however, the idea of developing human-in-the-loop systems as games has been part of the HC idea from the very beginning.

Von Ahn introduced "human algorithm games" or the more often referenced term "games with a purpose" in his aforementioned doctoral thesis. Instead of following "traditional approaches" concentrating on the improvement of software to solve problems that computers are not yet able to solve, Von Ahn aimed at "constructively channel[ling] human brainpower using computer games" (2005, 3). Games present a fruitful environment for this goal because, according to Von Ahn, they offer the incentives required to make humans contribute to computational systems (2005, 11). "In every case, people play the games not because they are interested in solving an instance of a computational problem, but because they wish to be entertained" (Von Ahn 2005, 12). Von Ahn thereby builds on previous research, arguing for the success of gamification, or making "work fun," and the need to design enjoyable UIs but transfers this to using game-design to solve AI problems. These "human algorithm games" could be thought of as algorithms in which humans perform the computational steps. "Instead of using a silicon processor, these 'algorithms' run on a processor consisting of ordinary humans interacting with computers over the Internet" (Von Ahn 2005, 11). In his thesis, Von Ahn points out the importance of asking the following question when assessing such games, which is also one of the normative principles guiding the development of HC systems at the Human Computation Institute:

> "[C]an a computer play this game successfully?" If the answer is yes, then the game is of little or no utility because the computer could simply play the game to solve the computational problem. If the answer is no, then the game has truly captured a human ability and the time invested by the players may provide a useful benefit. (Von Ahn 2005, 81)

In contrast to the normative principle expressed by Michelucci that it is unethical to have humans perform a task that computers can solve, the question here is more about the "utility" of the human algorithm game for a given AI problem. The understanding that people contribute to such games because they "enjoy the game [...], in turn producing more useful output" (2005, 76) is at the core of Von Ahn's idea. The ARTigo project, developed just a few years after Von Ahn's doctoral thesis, built directly on his understanding of GWAPs. This approach seemed particularly attractive for collecting large amounts of image tags, since, as the team had calculated, paying student assistants to do the work would not have been affordable (Bry, Mar. 2, 2020). As a result, ARTigo was designed as a casual game that can be played occasionally, either for a very short time between subway stops or for a longer period of time. In a similar way, Stall Catchers was also designed as a casual game that can be played without much time commitment (Vepřek 2023b). The idea behind Stall Catchers' game design builds on the user image the Human Computation Institute had in mind when developing the system:

> I thought it was going to be 30 something people. I thought it was going to be the casual game that people in the workforce, [...] 20 something, 30 somethings, they're going to be standing in line at the bank and they're going to be catching stalls because it gives them a sense of purpose it's like a kind of fun, casual game. (Michelucci, Jan. 14, 2021)

The team also saw the advantage of designing it as a casual game, rather than a crowdworking task, in that annotating the data in a game format allows one to build up a "self-sustaining community" (Michelucci, Jan. 21, 2021). They viewed the game approach as advantageous due the possibility of educating people: "[I]t feels like we're making a bigger difference not just in analyzing the data but in educating people, in building a sense of community and purpose around finding a treatment for Alzheimer's disease" (Michelucci, Jan. 21, 2021).

In the example of Foldit, crowdsourcing the human participants' contributions in a crowdworking context, also did not seem feasible due to the complexity of Foldit, which would require training crowdworkers to perform the task. Developer Daniel explained in our interview:

> [T]here is a huge learning curve. And if you try crowdsourcing this, there is a lot of overhead costs and up-front costs to teaching the crowdsourced workers how to play and [...] how to solve the task. And, in doing so, they are not very motivated to learn this huge thing. Imagine if you were being paid some small amount to learn a different language so that you could translate some paragraphs. That is not very interesting, not very motivating, it seems like a [...] huge cost up-front just to do some task. And so, by framing it as a game, there is a reason, there is an intrinsic motivation to learn how to play, there is a reason to stick around and continue doing it, [...] there is a sense of long-term investment. (Jan. 24, 2020)

A game-design would make the complex matter of protein design— which exceeds the complexity of crowdworking microtasks—more accessible as players would be more motivated to invest time and effort in learning to play Foldit. In addition, as Gidon, a team member and one of the co-creators of Foldit, explained, the reason for designing it as a game emerged from the possibility of tapping an existing large user base:

> [O]riginally the motivation for the game was that people spend a lot of time playing games already and games kind of engage people in solving problems and so maybe we could make a game that takes all of the time and effort that people spend solving problems in games and put that towards a real-world problem. (Jan. 31, 2020; *cf.* Miller et al. 2019)

This understanding aligns with the GWAP idea formulated by Von Ahn and Dabbish (2008) and reflects the user image the Foldit team had in mind when developing the project. The aim was to address "gamers," people already spending a lot of time playing games. As Foldit researcher Brian Koepnick explained in an interview with the *German Video Game Awards* (*Deutsche Computerspielpreis*), interest in and enjoyment of games can also lead to an interest in science: "Gaming is a great way to get people excited about science. The 'average citizen' has a lot to contribute to scientific research, but most people don't want to read a textbook or even get a PhD. If you can solve a scientific problem through play, it becomes accessible to nonscientists." (Koepnick 2020)

In a paper in which the authors analyze how learning and motivation frameworks can be utilized to improve player experience in Foldit, game scholar Josh Aaron Miller and colleagues note, based on their work, that Foldit, in fact, attracts participants from both the gaming and CS community (2019, 8).[10]

However, as I will show in Chapter 5, for many of the Stall Catchers and Foldit participants I interviewed, it was not the game design that brought them to the projects but the role Alzheimer's disease played in their everyday lives. Some of the participants with a personal connection to Alzheimer's disease even rejected the project being called a game. Specifically in the example of Stall Catchers, the team was well aware of this fact and acknowledged it. Nevertheless, and coming back to the imagination of the *human* in the loop and the imagination of the *loop*, it can, thus, be summarized that the human in the loop in HC design in the field of CS is often first envisioned as a gamer or someone interested in playing games who can be incentivized to contribute to the computational systems through entertainment. The *loop* in GWAPs, therefore, is imagined and designed

---

10 Additionally, GWAPs, as Von Ahn had already noted in his dissertation, are not suitable for every AI or scientific problem. In the example of Foldit, game design seemed to fit naturally with the way protein folding had been addressed in science for a long time. Foldit team member José explained: "[T]he way that we think about proteins and protein folding, it actually lends itself to competition pretty easily. So, the way we think about proteins already is a very, it's a kind of competitive paradigm. [...] The way that we think about the problem is very much game-like already" (Jan. 22, 2020). One example of the game-like and competitive paradigm is the biennial *Critical Assessment of protein Structure Prediction* (CASP) experiment which has been taking place since 1994 (University of California, Davis, n.d.). Even though it is framed as an experiment, it represents science as a playful approach to the world (Dippel 2020). It is organized as a protein structure prediction challenge and resembles a sports tournament in which participating research groups computationally or experimentally try to come up with structures for certain amino acid sequences (University of California, Davis, n.d.).

as a game. In the field of HC in general, the imagination of the loop further refers to how humans and technology or AI should collaborate in future HI systems.

Human–Technology Conversations of the Future

The role of humans in HC systems is described in the current HC literature as computational contributors (Michelucci 2013d, 84) or assistants *for* AI systems (Kamar 2016a, 4071).[11] Authors such as Quinn and Bederson draw on terms from crowdworking applications to define computer, worker, and requester as the roles that exist in any HC systems (2011, 1410). These roles, although the order in which the roles are performed may differ from application to application, are considered consistent (Quinn and Bederson 2011, 1410). Taking the case of reCAPTCHA, for instance, and focusing on the workers and the computers in Quinn and Bederson's example, the authors describe the relation between computers and workers as follows: "[A] computer first makes an attempt to recognize the text in a scanned book or newspaper using OCR. Then, words which could not be confidently recognized are presented to web users (workers) for help" (Quinn and Bederson 2011, 1410). Here, the worker assists the computer by solving the text that the computer cannot recognize. They are solving the same task. In a different scenario, workers provide image labels that are subsequently aggregated computationally to delete irrelevant labels (Quinn and Bederson 2011, 1410). In this case, the tasks performed by the workers or the computer are different, while they are working together toward the same goal (even though the workers might not be aware of this goal). The point is that these different role assignments and relations are assumed to remain the same throughout the life of a project or system.[12] However, as I will show in later chapters, the roles of humans and the computer, or more precisely software or AI, in HC-based CS are not fixed and consistent but are, instead, constantly in flux and subject to reshaping and rearrangement over time.

Similar to Quinn and Bederson's second scenario described above, Michelucci envisioned humans and computers in the loop as working together to solve different tasks. Human–technology relations—or *the loop*—were imagined as a "conversation" in which the roles and tasks were distributed depending on the respective abilities of the computer and human. Imagining current and future relations between humans and technology, Michelucci explained in one of our meetings, using the example of a new, more complicated version of chess than the common one, that "it's our scrappiness[,] our ability to [...] very quickly figure out, [...] which part of the search space is not worth exploring" (Jan.

---

11   It should be noted that developers sometimes also view AI as serving as an assistant to humans, depending on the approach taken, for instance, when a computational problem requires human assistance, as in the examples described here, or when humans receive help from AI technology in solving a problem, such as in the example of AlphaFold in Foldit (discussed in Chapter 6). In either view, developers and designers consider these role distributions to be rather fixed at the system level; I argue, however, that they are typically dynamic and intraverting.

12   Notably, some of the literature on HI disagrees with such a static understanding and advocates for a dynamic understanding of roles, tasks, and relations in hybrid human–AI systems (e.g., Akata et al. 2020; see also Chapter 6).

21, 2021). The loop between humans and computers, he continued to explain, could then be imagined as follows:

> [L]et's say I'm a human player of chess and I have one hundred and fifty possible moves ahead of me, I can immediately exclude 90 percent of those because I know they would be pointless or most likely. Intuitively I just know and from all my experience with chess, there's no point. So, I've narrowed it down to maybe five moves. So now if a human does that, the human can go to the machine and say, I think these are the best possible five moves to be considered here. And then the machine can now search out that space, which has been substantially narrowed. It can use its brute force computation to search out possibilities. And the machine can put a bunch of possibilities in front of a human and say, OK, what do you think about these? And the machine says: Here are one hundred possible moves of the human says, no, no, no, no, these seven. Now try these seven and so on, and they can have this sort of conversation where they help each other. (Michelucci, Jan. 21, 2021)

Here, human–technology relations are envisioned as a "conversation" or "partnership" (Michelucci 2017a) in which the human first delimitates the search space for a problem for which the AI then makes suggestions that can again be narrowed down or corrected by the human in a dialogue. In the description above, the human retains decision-making power within their sociotechnical relation with the AI. For "making this all work," Michelucci argued, it is important that "our automated systems [...] adapt to humans as opposed to getting humans to adapt to our automated systems" (Jan. 21, 2021). This idea of getting the AI to adapt to humans and not vice versa goes beyond the dialogue described above. The humans in the loop should decide how they want to engage in such systems, for example, by dancing or writing poetry, and to get there, "we just need to figure out the mappings, I think, somehow," Michelucci explained (Jan. 21, 2021). Such an approach to hybrid human–AI systems, according to Michelucci, differs from an approach that focuses on "information processing efficacy," in that it follows a human-centered and "ethical" approach:

> [W]e often [...] look from the standpoint of information processing efficacy, so we say, OK, well, what are the computers best at? What are the humans best at? And then how do we combine those complementary strengths in the most effective ways? But what's most effective isn't always what's most enjoyable for a person. There are organic aspects to this, and you don't have to motivate a machine-based computer, but you do have to keep a human interested somehow if you're going to be ethical about it and [...] I think those things are very intertwined. [...] I mean, if the fundamental question is about how do we get humans and machines to work best together, I think that's evolving because the machines are getting more human-like in the things they can do. So that's a moving target. I mean, we're always going to be reengineering that *as* we discover new techniques for AI, for example, and as we discover new techniques for getting humans to work together and humans and AI to work together as well. (Michelucci, Jan. 14, 2021)

The "ideal" human–AI or human–technology relations in HC systems are not always the most efficient or effective solutions. Allowing humans to decide and set their own terms for how they want to be involved in such sociotechnical systems is, in this sense, described as an ethical choice. I, therefore, consider the development of HC systems to be an ethical project in which "beneficial" takes precedence over engineering/technical optimization. Moreover, as Michelucci explained, what is considered the best combination of humans and machines is constantly evolving. Acknowledging that there is a gap between the future imaginations of a self-chosen human–computer conversation and current instantiations of HC and its human–technology relations, Michelucci concluded with: "So, it sounds kind of crazy and how could that ever work? How could writing a poem solve a problem? But I actually think it's not so far-fetched" (Jan. 14, 2021).

Considering HC-based CS systems and how humans are currently involved in projects like Stall Catchers, there seems to remain a discrepancy between the idea of placing humans—and their primarily human attributed ability, creativity—at the center of these hybrid systems and the need to abstract problems and the tasks of humans in HC systems in a machine-interpretable way. This abstraction, however, almost inevitably leads to the alienation of people and to the replaceability of the individual. This reduction of the human seems to form a dilemma of crowd-based HC-based CS projects that the Human Computation Institute team is aware of. In an attempt to solve this dilemma, if only partially, they created the category "The humans of Stall Catchers" on the institute's blog where they present interviews and stories of participants and team members (Human Computation Institute, n.d.b). While there are only ten portraits of participants at the point of writing, these introduce, for instance, the participant's background and motivation for contributing to Stall Catchers.[13]

Linked to this imaginary of a conversation between humans and AI at the Human Computation Institute is the self-imposed "oath" or normative principle that guides development and working practices (Michelucci and Egle [Seplute] 2020). The oath is adapted from the Hippocratic Oath, which was named after the ancient Greek physician Hippocrates, although its root and authors remain uncertain (Shmerling 2015). "It represents a time-honored guideline for physicians and other healthcare professionals as they begin or end their training. By swearing to follow the principles spelled out in the oath, healthcare professionals promise to behave honestly and ethically" (Shmerling 2015). At the institute, this oath is customized to "First use no humans" (Michelucci and Egle [Seplute] 2020). In a blogpost, the institute discussed an ML challenge it organized together with the data science crowdsourcing platform Driven Data (DrivenData Labs, n.d.) and the company and MATLAB[14] creator MathWorks (The MathWorks, Inc., n.d.) to see whether ML engineers could develop ML models to solve the analysis task currently performed by humans in Stall Catchers. In this context, the post described the adapted Hippocratic Oath as follows:

---

13   By contrast, in the example of the ARTigo project, participants do not even have to register to contribute and are, thus, sometimes not identifiable beyond their IP address.
14   MATLAB is a programming language and platform for programming and numeric computation.

> [I]f there is a way to solve a problem using machines, we should not ask a human to do it. [...] This might seem counter intuitive given that our mission is *"dedicated to the betterment of society through novel methods leveraging the complementary strengths of networked humans and machines"*. But we believe that sometimes action with the best of intentions can have costs that outweigh the benefits. If there is a job that machines can do, we think it would be unethical to waste volunteer human cognitive labor on that job, when there are other, more pressing societal needs that require the unique mental faculties of the magnificent human mind. (Michelucci and Egle [Seplute] 2020, emphasis i.o.)

The aim is not to develop *any* kind of conversation between humans and AI. Instead, humans should only be part of HC systems when the problem cannot be solved by computational systems. This oath reveals the contradiction inherent in HC, that is, the commitment to an ethic that values human labor on the one hand, and the simultaneous treatment of humans as a functional element on the other. Nevertheless, the normative principle implies a constant reevaluation of the possibility of automating the task performed by humans in Stall Catchers, as in the ML challenge mentioned above. As I show in this work, it also implies that Stall Catchers and HC systems in general must be built as systems that remain open to new human–technology relations with the "moving target" (see Michelucci's quote above) and to the development of computational solutions to tasks previously performed by humans. HC systems, therefore, can never present complete or finished products but must remain at the edge of AI developments. This is where the imaginaries play a driving role for the intraversions of human–technology relations.

From the imagined human–computer conversation described above that is ethical and enjoyable for humans, and the imaginaries' consequences for HC development at the Human Computation Institute, I return to the question of who is in the loop in the next subchapter, but this time, focusing on how crowds—rather than individuals—are imagined to be in the loop.

Crowds in the Loop

Although HC systems do not always or necessarily involve a collective or crowd of humans, the imagination of the human in the loop as a "crowd in the loop" is a fundamental idea behind the HC-based CS examples studied. Here, the understanding is that the power to build HI lies in the combination of machines with (digitally) interconnected humans (Human Computation Institute, n.d.). In this way, HC is linked to both the idea of the "wisdom of crowds" (Surowiecki 2005) and the concept and research field "collective intelligence."[15] The term "collective intelligence" gained popularity in the 2000s, in part

---

15   In the *Handbook of Collective Intelligence*, in which the editors and organizational theorist and management scholar Thomas W. Malone and computer scientist Michael S. Bernstein aim to form the new interdisciplinary field "collective intelligence," they define "collective intelligence" following Malone, Laubacher, and Dellarocas broadly as "groups of individuals doing things collectively that seem intelligent" (Malone, Laubacher, and Dellarocas 2009; cited in Malone and Bernstein 2015, 1). They do not specify further what they mean by "intelligence" to keep it adaptable to different understandings of "intelligence" and point to the perspectivity with "seem" (Malone, Laubacher, and

with the journalist James Surowiecki's publication *The Wisdom of Crowds* (2005; *cf.* Malone and Bernstein 2015, 6). The wisdom-of-the-crowds approach is a fundamental aspect of all crowd-based HC systems. Building on examples ranging from computer algorithms to stock prices and votes, Surowiecki defines four characteristics for a group or crowd to be "wise:"

> diversity of opinion (each person should have some private information, even if it's just an eccentric interpretation of the known facts), independence (people's opinions are not determined by the opinions of those around them), decentralization (people are able to specialize and draw on local knowledge), and aggregation (some mechanism exists for turning private judgments into a collective decision). (2005, 10)

With these characteristics, Surowiecki argues, the likelihood of an accurate crowd answer is high as each individual person's errors will cancel each other out (2005, 10). The characteristic of diversity of opinion implies that individuals do not have to be experts on the problem in question. In an interview with Neil Savage for the *Communications of the ACM* (Association for Computing Machinery) magazine Adrien Treuille, computer scientists and one of the creators of Foldit, reflected on the astounding insight of Foldit's functioning: "[l]arge groups of non-experts could be better than computers at these really complex problems," he argued (Treuille in Savage 2012). "I think no one quite believed that humans would be so much better" (Treuille in Savage 2012).

The calculation of crowd answers in Stall Catchers is also built on the assumption of nonexperts contributing to the game: "we don't assume that every catcher is as accurate as a trained laboratory technician. Instead we use 'wisdom of crowd' methods that can derive one expert-like answer from many people" (Michelucci 2017b). Humans in the loop in HC are, thus, not imagined as individual experts, capable of solving the problem at hand but as crowds of nonexperts capable of producing "expert-like" answers that are better than computational results.

However, the "wisdom-of-the-crowd" approach, while emphasizing the collective effort of combining human efforts, still relies on the isolated individual human in the crowd, as Surowiecki's second characteristic ("independence") states. Furthermore, if we follow Surowiecki's definition, these individuals are then aggregated in a way that their "private judgements [are turned] into a collective decisions" (Surowiecki 2005, 10, see above). The "crowd" describes an imagination of aggregated individuals not interacting, communicating, or even aware of each other. This imagination of the crowd, however,

---

Dellarocas 2009; cited in Malone and Bernstein 2015, 1). Moreover, with "acting," they constrain that "intelligence to be manifested in some kind of behavior" (Malone and Bernstein 2015, 3) that emerges from individuals or groups that act together. Michelucci defines "collective intelligence" in the *Handbook of Human Computation* in a similar way but with specifying "intelligence" as "problem-solving." "A group's ability to solve problems and the process by which it occurs" (2013d, 84). "Collective intelligence" can, hence, be understood as a subdiscipline of HC or HI, since the latter focus on the combination of crowds, groups of people, with computers and AI, while "collective intelligence" could also refer to groups of people or biological systems without any computational elements.

does not completely align with those of Foldit's and Stall Catchers' teams, who, instead, invest in building "communities."[16]

Even though participants in crowd-based HC-based CS at the Human Computation Institute are considered the same as classification providers, the scientific purpose of the projects requires careful evaluation of the individual contributions. The "wisdom-of-the-crowd" approach generally does not account for individuals who do not contribute with the best of intentions. In order to ensure that no harmful input occurs, algorithmic control functions are implemented in Stall Catchers. At the same time, these control functions take into account that user input may vary; it is assumed that humans do not always perform equally well. The control mechanisms will be described in detail in Chapter 6. In a nutshell, while humans are currently doing the annotating of research videos because there are currently no algorithms that can solve this task, it is still an algorithm that evaluates how good the human input was in each case by measuring the participants' "sensitivity." At the same time, individual participants' responses are weighted according to their sensitivity level to the research data. Thus, although the actual task cannot be solved computationally, the ability of humans is, nevertheless, continuously reviewed by algorithms and the skill level dynamically adjusted if necessary. Humans are integrated into these systems in a way that is primarily technical, reflecting the previously discussed understanding of humans in the loop as information processors. Stall Catchers relies on users processing information differently (regarding other humans) and accounts for fluctuations in individual performance in near real time (fieldnote, Nov. 16, 2022).

In this sense and summing up the three imaginations of the human in the loop, they build upon crowds consisting of individual input providers or information processors whose annotations and skill levels are evaluated, rated, and weighted by computer algorithms. This contrasts with social scientific understandings[17] of individuals as "embodied, socially situated, and 'cultured' human beings" (Beck 2012, 136). While computational models cannot yet take over the analysis, the analysis results are only created in human– or crowd–technology relations by the combination of different participants through computer algorithms based on statistical methods.

From these human-in-the-loop visions, I now turn to the HC imaginary of the *Thinking Economy*, to which HC is understood to lead and in which the idea of the human in the loop is understood to fully unfold.

---

16   For example, for both Foldit and Stall Catchers, this includes creating ways for participants to connect, such as through forums or in-game chats.
17   Similar observations have been made by Forsythe with the example of the understanding of knowledge when studying knowledge engineers who "treat knowledge as a purely cognitive phenomenon. Knowledge, to them, is located solely in the individual mind; expertise, then, is a way of thinking. This contrasts with the social scientific view of knowledge as also being encoded in the cultural, social, and organizational order. Given the latter view, contextual factors are seen to play a role in expertise, and knowledge appears to be a social and cultural phenomenon as well as a cognitive one" (Forsythe [1993] 2001h, 52).

Humans in the Loop in a Future *Thinking Economy*

The potential of HC is considered to be particularly significant in the field of CS, which has played an important role in HC research since its beginning.

In our interview, Bry, who was one of the principal investigators (PI) of the ARTigo project, editorial board member of the *Human Computation Journal*, and contributor to the *Handbook of Human Computation*, saw the potential of HC and CS as "tremendous" (Mar. 2, 2020):

> So, I'm completely convinced that in many, many [scientific] areas, such as ethnology for example, human computation, crowdsourcing, citizen science have a *huge* future, but even more than that. I believe that this is the approach of the future in science, [...] in the cooperation of people, also at the workplace in general. The future of human work is either caring for people or creativity *and* traditional professions, which will remain, but will be so optimized by software that compared to today only a fraction of people will do it. [...] There is nothing better than human computation to tap into creativity. Science lives from people without expertise coming up with ideas that experts don't have. [...] So, I believe the future of work will be in the direction of human computation, citizen science. (Bry, Mar. 2, 2020)

According to this vision, HC enables creativity and particularly supports "out of the box thinking." In the example of science, this is achieved by including perspectives of people without specific training in a particular field. According to Bry, the focus on care and creativity will change work in general, also in the scientific field at universities (Mar. 2, 2020). He further explained that "[y]ou don't need this tight frame, these working hours, a permanent room and so on. People work much more creatively when they are freer. And one will get this freedom. [...] And the complement to that will be more creativity. This will make this freedom affordable *and* justify it. (Bry, Mar. 2, 2020) With HC, Bry argued, most routine tasks will become more efficient and largely automated, freeing people in all sorts of jobs to think creatively, such as the working conditions now enjoyed by a few professors. The potential of HC is imagined to be endless; HC will unleash humans from undesired work so they can focus on their creativity. When I asked him if there were also limits to what HC could do, Bry declared that while there would be some limits, he did not currently know what they would be, "we will see" (Mar. 2, 2020).

The understanding of the future of work enabled and shaped by HC is shared by Michelucci, who uses the term *Thinking Economy* to describe the future he envisions. While Bry focuses on the creativity of humans which cannot be replaced by computers, Michelucci centers "problem-solving" instead:

> I think the bottom-line is, I think we're scratching the surface of how people can be involved in science. And I think science, what is science? To me it's problem-solving. So, what we're basically saying is: Science is a methodology that helps us do problem-solving effectively and we can get lots of people involved in that process in many different ways and citizen science is helping us begin that process of involving people. And frankly, I think, that's what people are going to be doing. I mean, 50 years from now, factories will be automated. So, we don't need people working in

factories, I mean, basically we're just going to have, like okay, we've created all these technologies, we've created a bunch of problems [...], we have problems we didn't create but we didn't fix, and we need to solve them. So, we just need all the people to be spending their time on the problem solving. So, this is a way to do that. (Jan. 21, 2020)

Michelucci envisions a future where, science, i.e., problem-solving, will be a collaborative effort involving not only professional scientists but potentially anyone whose work may be automated in that future. In addition to (scientific and societal) problems in general, problem-solving will need to address the problems created by humans while developing technologies.

In contrast to marketing and information management researchers Roland Rust and Ming-Hui Huang, who argue that we currently live in the *Thinking Economy* (Rust and Huang 2021) and are in the midst of the development of a "Feeling Economy" (Huang and Rust 2019), Bry and Michelucci locate the *Thinking Economy* in the future. In this development toward a *Thinking Economy*, the COVID-19 pandemic was "sort of a catalyst for the Thinking Economy" (Michelucci, Jan. 14, 2021). It has "accelerated the shift [...] that I envision toward a Thinking Economy. And so now that it's sort of greased those wheels, there's even more opportunity" (Michelucci, Jan. 21, 2021). In 2020, with the pandemic causing lockdowns across many countries, more and more people started to connect online and engage in CS. Online CS projects in general but especially those focused on coronavirus research, such as Foldit, Folding@home (The Folding@Home Consortium (FAHC), n.d.), and Rosetta@home (University of Washington, n.d.), which are all HC-based projects, experienced a huge upswing (Vepřek 2020). "[B]ecause of this, it'll be easier to get people who are already engaged online to start [...] to participate in human computation systems. And I think people do it already without knowing it." (Michelucci, Jan. 14, 2020) The potential of HC is understood to be infinite by both Michelucci and Bry. In this imaginary, HC will enable people in the future *Thinking Economy* to spend their time with problem-solving in creative and enjoyable ways. The chosen modes of engagement should then be integrated into the sociotechnical system, in which the computational parts are supposed to be built around these individual engagements.

The HC imaginary of the *Thinking Economy* provides insights into the thinking in which HC-based CS are embedded and how the human-in-the-loop visions are understood to ultimately materialize and unfold. As I have discussed earlier, there is a discrepancy between the human-in-the-loop ideas and current implementations. In the *Thinking Economy*, this gap is understood as being closed, and it is, thus, both a utopian aim for HC development and a contrast foil to current implementations. These efforts flow into and partly define HC-based CS's intraverting human–technology relations.

Weaving Together the Imaginaries

Taken together, the imaginaries of HC presented, including the imaginations of the human in the loop (and the loop itself), though not exhaustive, describe the narratives, understandings, and visions driving HC-based CS development as open systems at the edge of AI and scientific research and, therefore, at least to some extent, the intraversions of human–technology relations. According to these imaginaries, HC systems are to be developed in such a way that they allow people to participate in a self-determined way and leave room for human creativity. Humans and AI in HC imaginaries are partners in a dialogue in which the AI is an assistant to the humans. But, as I will show in the following chapters, the roles of the AI or computer and humans in HC-based CS games are not fixed. Instead, they change continuously and intravert over time due to properties that may be inherent in HC assemblages and external forces acting upon them.

Moreover, there is a mismatch between the HC imaginaries and current implementations. I argue that it is precisely this mismatch that plays an important role in the intraversions of human–technology relations because the very gap to full HI that this mismatch represents is what drives developers and scientists to push the limits of the system further toward their envisioned goal. Human computation advocates and developers play important roles in forming these systems and their human–technology relations. However, their imaginaries also face resistance and conflict. Participants and researchers with their own aspirations and motivations, and computational models not meeting scientific requirements, for example, interfere with these visions in everyday life.

Human Computation visions of creating "hybrid thinking systems" (Bowser et al. 2017) with humans in the loop remain largely abstract. The question of how humans and/or human intelligence could and should be involved remains at the core of current research efforts in the HC field in general and is often negotiated on a case-by-case basis. The practices of designing and developing HC systems that are shaped and created by the prevalent visions and counter-imaginaries (while, at the same time, forming and creating them) do not generally follow established procedures or frameworks for combining humans and AI.[18] Instead, these "future practices," to borrow sociologist and cultural theorist Andreas Reckwitz's (2016) term, consist of continuous experimenting, sketching, prototyping, and infrastructuring. By introducing path dependencies, these practices not only materialize imaginaries but also constrain them in a particular direction. I now turn to these practices, which, along the way, negotiate questions of practical human involvement and the distribution of roles between AI and humans to realize enhanced performance and desirable futures.

---

18   The *Handbook of Human Computation* can be considered a first attempt to address this gap (Michelucci 2013a). However, on the level of software architecture or development, for example, developer Kate of the Human Computation Institute explained that she could not rely on existing code to build HC-based CS as she would when working on other software projects, such as e-commerce platforms. For the example of Stall Catchers, Kate argued that "Stall Catchers is something unique in a sense that you don't really have a lot of people that have worked on similar types of projects so there are not a lot of open-source tools that you can use if you want to do something. So, most of the stuff I had to code from scratch" (Nov. 19, 2020).

Imagining as Practice: Infrastructuring and Experimentation

Imagination, not only but especially when visions have not risen to a hegemonial or sociotechnical imaginary in the understanding of Jasanoff and Kim (2009; 2015), does not happen in intangible space. Instead, it relies on "hands-on" (Hilgartner 2015, 45) practices and exploration. In the field of HC, the question of *how* superhuman capabilities can be achieved by combining humans and machines or AI in new ways, or how imaginaries of AI overcoming humans can be fought against to find the "human place" in the future (Carlsbergfondet 2019, 00:58-1:00), is negotiated alongside the development through infrastructuring and experimenting with sociotechnical systems.[19]

I apply the term "infrastructuring" to focus on the material-semiotic practice and stress how infrastructures are constantly in the making (Niewöhner 2015, 5; *cf.* Bossen and Markussen 2010). Every sociotechnical system, no matter how established and common, requires continued infrastructuring. While infrastructuring practices are, thus, part of every computational project or software platform development, I here want to discuss two different forms of infrastructuring I observed at the Human Computation Institute. Just as "infrastructures operate on differing levels simultaneously" (Larkin 2013, 330), infrastructuring takes place on different levels. Because HC systems are at the edge of AI and scientific research, they often cannot build upon existing frameworks, and developing HC requires very "fundamental" infrastructuring. I first turn to an example of *fundamental* infrastructuring pursued at the Human Computation Institute before discussing instances of *everyday* infrastructuring in the Stall Catchers project. While I focus here on infrastructuring related to the Stall Catchers platform performed by team members of the Human Computation Institute, infrastructuring related to the development of the data pipeline in the laboratory that prepares the data to be sent to the platform is analyzed in Chapter 6.

Infrastructuring Toward "Sustainable Human Computation"

Infrastructuring can be understood as part of the bigger goal of materializing and establishing HC and the vision of hybrid human–AI systems. One such infrastructuring endeavor pursued by Michelucci and colleagues at the Human Computation Institute is to make the development of HC systems more "sustainable" (in the sense of self-sustainable as opposed to environmentally sustainable). This is one of the specific aims of their Civium initiative, "an integration platform and commerce engine for sustainable human computation" (Michelucci 2019b). Michelucci described the idea of this platform to be built in an article on *Medium* from 2019:

> Civium is an operating system for a new class of supercomputers powered by computer hardware and cognitive "wetware" that will enable us to build, improve, and deploy transformative human/AI systems in support of open science and innovation.

---

19 In this sense, HC practitioners share with AI practitioners studied by Forsythe that their goals are very broad and the "meaning and appropriate scope of 'artificial intelligence' [, or in this case HC,] are subject to ongoing negotiation" (Forsythe [1988] 2001a, 76).

> It is also a bazaar, for sharing, trading, and finding the widgets and services we need to create and sustain the capabilities we seek, breathing new life into unsupported projects and platforms. Ultimately, we believe Civium has the potential to seed a new thinking economy that rewards the uniquely human cognitive abilities needed to tackle our most pressing societal issues. (Michelucci 2019b)

Civium's goal of building a foundation for future HC systems and currently unsupported projects originated in part from the "sustainability problem in citizen science" (Michelucci, Jan. 21, 2021). Most HC-based CS projects are developed by nonprofit research or academic institutions and are, therefore, highly dependent on funding. If funding ceases, most projects cannot be sustained (*cf.* Miller et al. 2023). Addressing this problem is important and valuable to Michelucci, as he explains, since these projects

> are serving [...] a valuable purpose in society. So, then it is a question of well, who finds it valuable and why aren't the people who find it valuable paying for it?! If there's a need for this, if there's no need for it, there's no need for it! But it exists because there was a need in the first place, and somebody is benefitting from that need. So, who is benefitting and why aren't they getting sustained?! (Jan. 21, 2021)

From this starting point, Michelucci, together with collaborators, began to identify where the problem was coming from:

> I realized that part of the problem is that citizen science, like many research activities, has become isolated in our broader economy in that there are these sort of cultural divides between academic research and industrial research and that there's an opportunity there. [...] So even though there may be distrust … there may be exploitation, there might be, in other words, reasons for distrust. (Michelucci, Jan. 21, 2021)

Civium was conceived with the aim not only of enabling projects to overcome the sustainability problem but also of enabling individual users—from CS participants to developers, researchers, and other actors in the future *Thinking Economy*—to decide for themselves the conditions under which they wish to participate in human/AI systems (Michelucci, Jan. 21, 2021). "And we use our actual human computation methods to enable these things" (Michelucci, Jan. 21, 2021). The idea also aimed to reflect on and address the power hierarchies that exist in "top-down" online CS projects. Here, designers, developers, and researchers currently set the terms under which participants can choose to contribute to scientific research. This does not mean that participants do not play an active role in shaping the projects and cannot resist inscribed meanings or intended usage (see Chapters 5 and 6). However, even when participants are invited to co-shape the projects and their involvement by providing feedback to the developers, in the case of disagreement or differing priorities about which features to implement or which bugs to fix (*cf.* Miller et al. 2023), it is the project teams that decide, not the participants. All they can do then is decide whether to leave or to stay. By contrast, within Civium, Michelucci envisions that participants can decide whether they're comfortable with that role and if they're not, they don't have to adopt that role, or, maybe they feel like they should be getting something back that they're not and they can advocate for themselves more easily to do that [...] in so far as ... human cognition becomes a more critical resource for executing scientific research. I think that gives the individual contributors leverage which can provide pressure against hierarchy to say, hey, you need us! (Jan. 21, 2021)

Additionally, Michelucci explained, Civium is intended to solve the problem of "duplication" in CS, referring to the fact that CS projects are often built from scratch, rather than building on existing mechanisms and approaches, because the infrastructures required are missing (Jan. 21, 2021). Even if "one-size doesn't fit all, [...] there are ways to do community outreach, there are ways to build communities. And there are people who are very good at it. So, if these communities exist already, do you need to spend two years building one from scratch?" (Michelucci, Jan. 21, 2021).[20] Michelucci's vision was for Civium to become an "eco-system" that combines all these functionalities and becomes "a marketplace and an integration platform and [...] a policy engine so that people can set terms" (Jan. 14, 2021) to enable sustainable HC. As such, Civium is a "matter that enable[s] the movement of other matter" (Larkin 2013, 329) and is itself a sociotechnical system building on HC. As Larkin formulates for infrastructures, "[t]heir peculiar ontology lies in the facts that they are things and also the relation between things. As things they are present to the senses, yet they are also displaced in the focus on the matter they move around" (2013, 329).

At the time of writing, Civium is still under construction. The Human Computation Institute and external collaborators have started working on implementing the structures of and for the platform, such as an "experimentation toolkit" that allows the creation of sandbox versions of online CS projects to run experiments without affecting the actual 'live' project and the scientific research behind it (Vepřek, Seymour, and Michelucci 2020). To this date, however, Civium, which is supposed to provide the infrastructure for HC, is "a neat idea and it looks like we're doing a few interesting things related to that but it's not a thing yet" (Michelucci, Jan. 21, 2021). Here, the "duality" of infrastructures described by Larkin (2013, 329) becomes apparent, since infrastructuring itself is dependent on resources to become technology supporting systems.

This brief excursus shows how infrastructuring not only materializes but also shapes the not yet stabilized visions of HC through the introduction of "path dependencies" (Klausner et al. 2015). By promising to facilitate HC development, these path dependencies also constrain the possibilities of what is possible to imagine. Just as the development of Civium's modules or experiment toolkit is shaped by Stall Catchers as the institute's

---

20   In addition, building on experiences of applying for ethical review for HC-based CS projects in the US, which to date is not specifically tailored to emerging fields, such as online CS or AI research, Civium should include a new approach to ethical review to ensure that research and CS are conducted ethically and that no participants are harmed. During my fieldwork and collaboration with the institute, I contributed to the discussions and analysis of the current Institutional Review Board or ethical review approach in the US and how it could be improved to better fit new emergent research and application fields (Vepřek, Seymour, and Michelucci 2020; Vepřek 2022b).

main project, Stall Catchers itself was inspired by and built upon the existing online CS project Stardust@Home (Westphal et al. 2005; Stardust@home, n.d.).

Stardust@Home is an online CS project launched in 2006 after NASA's Stardust mission returned with a collection of interstellar dust embedded in an aerogel collector. It invites participants to search for micron-sized interstellar dust particles in images, respectively, short videos consisting of image stacks, which are based on the data collected. The data is presented in a customized virtual microscope, similar to the Stall Catchers platform. In fact, Stall Catchers' UI was built from the Stardust@Home's UI, including the virtual microscope, and even its code served as a starting point for Stall Catchers. "Who knew stardust and blood vessels could be so similar?," asks Stall Catchers' website (Human Computation Institute, n.d.c). By drawing on the existing CS platform in the field of astronomy, the way in which stalled blood vessels in Alzheimer's disease research are to be identified was pathed. And since Stall Catchers then served as the base model for future HC-based CS projects, such as the Human Computation Institute's Dream Catchers project on sudden infant death syndrome (Ramanauskaite 2020) and other future projects to be built on Civium, specific ways of analyzing research data and engaging participants guide the development of future HC-based CS further. As in the general example of evolutionary histories of machines described by computer scientist Iyad Rahwan and colleagues (2019, 481), parts are reused in different contexts which constrains future performance but also enables new innovations. Or, as assemblages, which are always to be thought of as multiples, extend, they also change (Deleuze and Guattari 2013, 7). To understand how path dependencies emerge and guide developments, it is important to trace both the emerging (counter-)imaginaries and their situatedness as well as the infrastructuring that materializes and shapes these narratives and visions. Civium presents such an infrastructuring project that, if completely realized, will probably path the way and form of future HC-based CS development.

Puppies in Stall Catchers: Everyday Infrastructuring

Shortly before the official start of the Catchathon on April 29, 2021, at 7:00 pm CET, several participants, including the new AI bot GAIA and myself, had problems accessing Stall Catchers. Instead of its UI, a screen with a picture of a puppy looking at the user expectantly and an error message saying "Error. Sorry...Please try again." appeared (see Figure 2). At 6:25 pm CET Michelucci sent a "mayday" to Stall Catchers' lead developer. Something had gone wrong with the activation of the dataset. While two team members were busy setting up the live streaming for the kickoff event and entertaining those participants who had already joined the Zoom meeting for the kickoff, Stall Catchers' lead developer, the bot creator, Michelucci, and myself were trying to figure out where the puppies came from (fieldnote Apr. 28, 2021).

Figure 2: Puppies in Stall Catchers. Error page

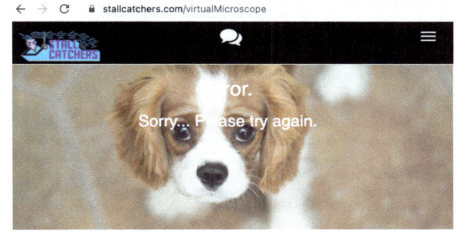

Source: Screenshot taken by LHV on May 17, 2021 (https://stallcatchers.com/virtualMicroscope)

While a moment like this was exceptional in the sense that the Stall Catchers team was constantly working to prevent such situations, such breakdowns are not uncommon in software and technology design in general. Things and processes fail; breakdown is "a condition of technological existence" (Larkin 2008, 234). Even if a process works well most of the time, it can fail for various reasons. In the example of the Stall Catchers puppies, these reasons included data format issues in the dataset to be activated. After creating workarounds and debugging the problem, the puppies went away. But shortly thereafter, a new problem arose that required the team's full attention and troubleshooting once more. Finally, after two turbulent hours, the Catchathon was underway, AI bot GAIA was catching stalls, and participants were able to access the platform and event pages. On the same day and one of the institute's main slack channel, Michelucci summarized this incident as follows: "Well that was exciting [face with tears of joy emoji] We've learned to expect the unexpected. Within minutes of the event starting the new dataset appeared to crash the server, but then we got that working in time, and then GAIA Bot errored out, and thanks to quick team support we got that working again." (Apr. 29, 2021)

This example, while not an everyday experience, illustrates how "put[ting] out fires" (fieldnote Apr. 29, 2021) and dealing with the unexpected is part of HC-based CS design, experimentation, and maintenance. Not only did the introduction of a new element (GAIA) cause problems but also routinized processes, such as activating a dataset, failed. Infrastructuring is a crucial part of the everyday maintenance of HC-based CS projects.

With limited time and financial resources, the Human Computation Institute's team not only had to put out fires at special moments, such as a Catchathon, but also often had to interrupt the day-to-day development of new features, work on Civium, or other new projects. The institute's developers (at most times there was one, sometimes two developer(s) working for the institute and the developers changed over the course of my research), for example, were not only full stack developers but also responsible for the entire troubleshooting, maintenance, and operational processes, as developer Kate described in our interview:

> I basically oversee everything about the platform, making sure that everything is running smoothly. So, development of course, getting new features and debugging any issues with the platform but also making sure that the infrastructure is running smoothly. So, anything related to the servers, the database, things like that. (Nov. 19, 2020)

It was not unusual for the institute's developers (as well as other members of the institute) to abandon their carefully prepared work plan to intervene: "[S]ometimes I have some intervention I need to deal with [if] Stall Catchers is having some issues. I have to … wake up in the night and try to debug it—sometimes, not all the time," explained developer Samuel (Sept. 2, 2021). In these moments, acute measures and actions were required that relegated the imagination of how HC-based CS should be developed to a secondary concern until the problem was resolved. Since these interventions were not infrequent but rather common and recurring, they not only pushed back set timelines but also influenced the imaginations of future systems and human–AI collaborations (an example of the latter will be provided in Chapter 6).

Keeping the system up and running was critical but not the only practice of everyday infrastructuring. I would like to focus here on two brief examples of such practices which also demonstrate how the imaginations of HC-based CS as ethical projects are continuously materialized and performed in the everyday. I see these practices as infrastructural work because of their essential role in creating a participant base and, thus, a foundation for the projects. The examples are 1) creating a seamless and effortless experience for participants, and 2) being transparent and responsive to participants. They show that for the HC system to *work* in the CS domain, from the perspective of the Human Computation Institute team, it is not only necessary to continuously work on the infrastructures as described but also to present the project to participants in a meaningful and enjoyable way.

The first example of the creation of a seamless and effortless experience relates to game design:[21] HC-based CS projects rely on game design to motivate participants to

---

21   The example also relates to practices of developing HC, which are also shaped by contingencies and uncertainties that are specific to it due to its human-in-the-loop approach. While one could say that, in a simplified way, software developers generally write code for it to be executed by computers, in HC, humans perform part of the computation. As Gray and Suri have described using the example of ghost work: "Normally, when a programmer wants to compute something, they interact with a CPU [Central Processing Unit] through an API defined by an operating system. But when a programmer uses ghost work to complete a task, they interact with a person working with them through the on-demand labor platform's API. The programmer issues a task to a human and relies on the person's creative capacity—and availability—to answer the call. Unlike CPUs, humans have agency: they make their own decisions. While CPUs just execute whatever instruction they are given, humans make spontaneous, creative decisions and bring their own interpretations to the mix. And they have needs, motivations, and biases beyond the moment of engagement with the API. Given the same input, a CPU will always output the same thing. On the other hand, if you send a hungry human into a grocery store, he or she will walk out with

complete what are often monotonous tasks and to present complex processes in a simple way. Therefore, the role of the game interface is central. Human Computation Institute team member Paul explained in our conversation that the experience of the participants had to be "as seamless, effortless as [...] possible" so that participants were not "burdened with [...] the whole interaction" (Oct. 14, 2020) taking place behind the game interface. By "whole interaction," he was referring to the complex participant–software interplay required to perform data analysis valuable to the scientific research in the laboratory. Creating such a "seamless" experience involved practices of not only ensuring the platform's availability but also introducing new game-specific features or organizing special events and playful challenges that motivated participants to continue contributing. With these efforts, the team aimed to move the human–software interplay that generates crowd answers into the background of the participant experience. In my conversations with Stall Catchers participants, it was, thus, not surprising that HC itself was only mentioned in very few conversations.

Being transparent and responsive to participants was also important for the Stall Catchers team. It aimed to be as transparent as possible about the decisions made at the institute, the science in the laboratory, and the scientific processes and progress behind the projects (fieldnote Oct. 14, 2022). The institute saw itself as a mediator between scientists and participants, for example, by explaining and translating the scientific research and sharing it through blog posts (fieldnote Oct. 19, 2022). They also sought to be transparent about design decisions, problems, and mistakes they made in relation to the platform or project: "[W]e showed our human side right away and [...] we made ourselves very humble in front of the community and we let the community know that that's how we were coming to the relationship with them, with this deep sense of humility," explained Michelucci (Jan. 14, 2021). Besides communicating with participants through blog posts and public events such as "hangouts," where participants could engage in conversations with the developers and scientists, this principle also includes being "responsive" (Paul, Oct. 14, 2020) by monitoring the different communication channels and answering questions from participants. The same principle was also shared by the Foldit team whose developers, for example, shared a "good faith agreement" about transparency:

> I think that as long as we as the developers really have a good faith agreement that we are going to try and always make it fun, and we are always going to try to be as transparent as we can for players about the science that we're doing. And then it's their choice entirely as to whether they want to participate. We just hope that they want to continue participating. (Hugo, Jan. 28, 2020)

---

a dramatically different bag of groceries than if they were not hungry. In exchange for this impetuousness and spontaneity, humans bring something to work that CPUs lack: creativity and innovation." (2019, xiv) Thus, this defining characteristic of HC itself introduces new challenges in developing the human–software system due to the unpredictability of human engagements (see Chapter 5). These uncertainties and the resulting contingencies not only interfere with grand HI imaginaries and project development plans but require continuous attention, immediate response, and practices beyond software development.

Nevertheless, for CS games, there remains an unresolvable tension between the principle of being transparent to participants and not revealing too much about the game mechanics to ensure the scientific data quality. I will address this "tricky balance" (fieldnote Oct. 14, 2022) that must be maintained in Chapter 5.

Between Counter-Imaginary and Infrastructuring

In this chapter, I explored the visions and imaginaries of future human–technology relations or combinations building "hybrid thinking systems" that underlie HC systems and drive their design and development. I showed how advocates present HC as an alternative to strong and general AI endeavors. They distinguish themselves from strong AI narratives by positioning HC as a counter-imaginary to such pursuits and emphasizing the importance of the human-in-the-loop approach, which they see as crucial to achieving capabilities beyond purely computational ones, while mitigating the dangers voiced in dystopian AGI imaginaries. Despite these definitory efforts, however, HC shares common paradigms with such AI efforts. At the level of visions and imaginaries, then, HC is also rooted and situated in the very reference points from which it seeks to distinguish itself.

Furthermore, I discussed different human-in-the-loop imaginations that describe *who* is envisioned to be in the loop and how the *human* is imagined, how the *loop* is envisioned as a game and as human–AI conversations of the future, and the imaginations of *crowds*-in-the-loop. Eventually, I explored the imagination of HC as leading to a future *Thinking Economy* unleashing human creativity. While these imaginations form and inform the everyday design and development of HC systems, as "social practices" (Jasanoff 2015b, 323) they simultaneously rely on infrastructuring and experimenting. How these imaginaries or vanguard visions are materialized and explored through examples of everyday practices of designers and developers was the focus of the second part of this chapter. Mackenzie described for code that

> [c]ode understood as a collective imagining seems a long way from code as a program of instructions for a machine to execute. However, practices of imagining are not purely mental operations; in no way does imagining reduce to a detached, abstract fantasy. It constitutes collective relational realities. Software attaches different localities to each other because it diffuses relations between them. The composite texture of software is reliant on unfinished exchanges between code and coders. (2006, 138)

Similarly, HC systems emerge in practices of designing, developing, and maintaining concrete sociotechnical systems, and such practices not only fill the idea of combining humans and AI to achieve superhuman capabilities with hands-on examples and meaning but also renegotiate and transform these imaginaries.

Yet, imagining and infrastructuring HC does not only take place within AI discourses, but design and development of HC systems are just as much shaped by their own everyday becoming in the interplay of all human and nonhuman actors involved. Often, they change and adapt through serendipitous discoveries or other coincidences, through object potentials and timely moments seized by actors or through the human–technology relations unfolding in (and forming) the sociotechnical assemblages. While this chapter focused on how HC is imagined and HC-based CS systems are designed to "stabilize practice," the following chapters will also focus on how these systems or assemblages are "destabilized" through (creative) practice (Beck 1997, 296).

5 Multiple Meanings and Everyday Negotiations: Play/Science Entanglements

From the imagination, design, and infrastructuring of HC systems, I now turn to how HC-based CS systems unfold in everyday life. In this chapter, I will explore how HC assemblages emerge in the everyday practices, the participants' interests and backgrounds, the entanglement of science and play, and how the visions and imaginations of HC themselves are influenced and reconfigured alongside the becoming of HC-based CS projects. Following Beck, I turn to the contingencies of HC-based CS sociotechnical systems as "use complexes" (1997, 350). I discuss how different motivations, interests, aims that drive participants, as well as the software's affordances (Gibson 1979; Bareither 2020a) and action potentials emerging from the human–technology relation relate to and sometimes challenge the systems imagined and designed by developers. I analyze how participants realize the object potentials of Stall Catchers in relation to algorithms (Beck 1997), including the shared meanings and values that constitute appropriate practices and modes of engagement. Here, the interferences (Dippel and Fizek 2017a; 2019) of play and science play an important role and create a productive space. It is within this space, including the object potentials and different motivations, that human–technology relations in HC-based CS unfold and constantly change, since human–technology relations in HC-based CS are always situated and embedded in the sociotechnical assemblages they simultaneously create.

The chapter is structured as follows: to briefly familiarize the reader with the two examples, Stall Catchers and Foldit, two short notes provide introductory snapshots of the examples from the participants' perspective.[1] For a condensed description, I combine the perspectives of different participants in a given project into one. In doing so, I create a fictive, ideal-typical description that is inspired by "ethnographic portraits" (Gutekunst and Rau 2017) and "cultural figures" (Wietschorke and Ege 2023). I then approach the interferences, starting with the observation that, in the case of Stall Catchers, some participants do not even accept the categorization of the CS project they are contributing to as a game

---

1   ARTigo will be discussed and described in more detail in Chapter 6.

in the first place.[2] I next analyze how participants ascribe meaning to the systems in their own ways and how they are included in the participants' everyday lives, which may not always align with the imagination of the human in the loop described in the previous chapter.[3] Although some participants reject the description of the projects as games, it is precisely the entanglement of science and play that opens up the space for Stall Catchers and Foldit to emerge, in which these adoption and meaning-making processes take place. This space is not without friction but contested due to different understandings and logics of science and play which create tensions that can only be partly resolved and impact the formation of HC-based CS projects. Understanding how play and science interfere is also crucial because it is in this space that human–technology relations unfold and continuously intravert (see Chapter 6). Adoption, meaning-making, and the changing relations depend not only on the intentions, motivations, and values of participants or other human actors but just as much on the materialities and nonhuman entities with which they engage and the coincidental and "timely moments" (Mousavi Baygi, Introna, and Hultin 2021, 431) that can be seized. Furthermore, new potentials for practices are opened up by the assemblage and its relations and their embeddedness in the context of science and play. In the last part of this chapter, I provide and discuss examples of such play practices that go beyond those intended by the system's design.

A Snapshot of Foldit

A rendering of a protein structure and a brief explanation of the project welcomes users to Foldit's website: "Foldit is a revolutionary crowdsourcing computer game enabling you to contribute to scientific research." (Center for Game Science [University of Washington] et al., n.d.a) The "About Foldit" section gives more details on its aim and how it works:

> Foldit is a one-of-a-kind **protein folding computer game** developed by university scientists. By playing Foldit, you can contribute to advanced research on human health, cutting-edge bioengineering, and the inner workings of biology. Foldit is **free to play** and not-for-profit. Discoveries made in the game are published in peer-reviewed research journals, and Foldit players are always credited for their contributions. Every week, Foldit scientists post new puzzles focused on the latest

---

2   This seems to differentiate the example from other games, where in most situations, players or gamers at least initially decided to play a game, and they mostly do so in their leisure time in contrast to their working time—if we exclude professional gaming activities where players make their living from playing and promoting games, for example. The point is that in the majority, it is accepted that something *is* a game.

3   It should be noted that the participants' perspective in this chapter is not shared by all participants but serves as an example of how such systems can be adopted in different ways. In the case of Stall Catchers, for example, even though I focus here on participants who do not call it a game, other participants actively identify as "players" or "gamers," as the following quote from Elisabeth when describing the main idea of Stall Catchers shows: "I'm a gamer … but contributing to a valuable purpose" (May 9, 2020). The arguments are, thus, situated and partial (Haraway 1988).

problems in protein folding. Read on to learn about ongoing research in **protein design** to treat diseases like influenza and COVID-19, **small molecule design** to invent new drug compounds, and **protein structure solving** to map the molecules that drive biology. (Center for Game Science [University of Washington] et al., n.d.b, emphasis i.o.)

While the first version of Foldit, launched in 2008, focused only on protein structure prediction and design, in 2023, participants can also work on small molecule design and protein structure solving problems. Especially when their gameplay leads to significant discoveries, Foldit participants are actively recognized as scientific contributors by the researchers who use their output. By February 2023, nine scientific papers had been published with Foldit players listed as authors (Cooper, Khatib, et al. 2010; Cooper, Treuille, et al. 2010; Cooper et al. 2011; Foldit Contenders Group et al. 2011; Khatib et al. 2011; Eiben et al. 2012; Khoury et al. 2014; Horowitz et al. 2016; Koepnick et al. 2019).

Figure 3: Foldit overview UI after login

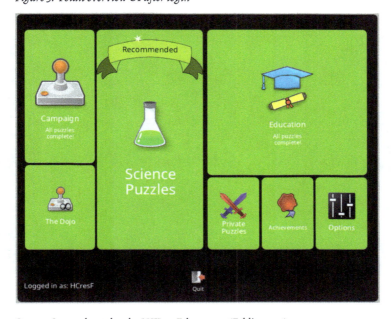

Source: Screenshot taken by LHV on Feb. 2, 2023 (Foldit game)

In order to access and play the game, users must first download and install the software on their computer and create a user account. After logging in (or playing "offline," without an official user account but with limited play experience) and starting the game, the main interface (see Figure 3) appears with different options. New participants can learn how to contribute to Foldit by completing the 34, as of March 2023, tutorial puzzles in the "Campaign" mode, which introduces different tools and essential aspects of the game.[4]

Once the tutorials have been completed, or once participants feel comfortable enough to start working on the actual "Science Puzzles," they can choose the puzzle they want to contribute to from a list of currently active puzzles of varying difficulty. This list always includes puzzles specifically designed for Foldit beginners and a "Revisiting Puzzle," i.e., a puzzle that has already been solved in Foldit but can be completed again (Foldit Wiki 2019). Additionally, there are different puzzles, such as "Design Puzzles" or "De-Novo Puzzles,"[5] which run for a limited time, usually a week. Participants can earn points and compete by working on these puzzles.

The fictional participants Muhammed and Taylor have been contributing to Foldit daily for the past five years. Before starting a new puzzle, Muhammed carefully reads its description, type, and objective, as his play approach varies depending on the problem type presented. Today, he wants to try out the new design puzzle. After clicking on a new puzzle, the main game interface for the puzzle appears (see Figure 4).

Figure 4: Foldit main game UI

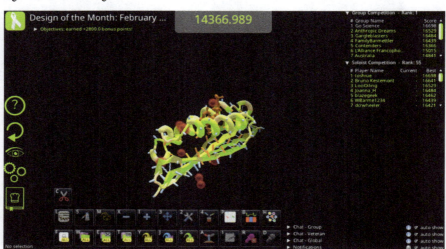

Source: Screenshot taken by LHV on Feb. 21, 2023 (Foldit game)

The initial 3D protein structure is displayed in the center of the screen and can be rotated by clicking and dragging the mouse. At the bottom are several small buttons for different tools. Some are manual tools, such as "cut" or "delete," while others are programmed and perform some automated operations on the protein structure. Examples of the latter are "wiggle" and "shake," which automatically search for better positions of

---

4   "Education" mode was created for use in the classroom; it includes different introductory puzzles with more biochemical information than the introductory puzzles in "Campaign" mode.

5   De-novo puzzles are characterized by the fact that only the primary structure (the amino acids sequence) is given at the beginning (Foldit Wiki 2017e).

the backbone, the protein's main structural framework,[6] and sidechains[7] (wiggle) or the sidechains only (shake) (Foldit Wiki 2017c; 2017d).

Two leaderboards, one for team-based competition and one for individual participants, are in the top right corner. Muhammed's score for his current puzzle is displayed in the center above the protein structure. In addition, a menu on the left side of the screen contains a help menu and, importantly, the "cookbook." The cookbook includes so-called recipes, which are scripts that automate certain tasks. Muhammed first inspects the protein structure from different angles to find out which structure could meet the puzzle objectives provided[8] before he starts "hand folding" the secondary structure, i.e., making manual changes to the protein structure by, for instance, placing cuts or dragging parts. He decides, for example, how many helices and sheets he can add, and how many segments they should have. He then proceeds by changing the structure provided according to his design choices. Once he likes the shape, he uses the "mutate" tool, which changes the amino acids of mutable segments that are mutable (Foldit Wiki 2017a); this can be followed by another round of rebuilding and correcting sections that did not score[9] well. After that, and if the protein structure seems stable enough to him, he starts using recipes to increase his score and further optimize the protein. In this late game stage, he mostly observes what the automated script does, but sometimes, manual intervention is required to locally optimize a particular section.[10]

While Muhammed prefers to solve puzzles as an individual player, Taylor enjoys playing together with others, which is why he joined a group. They share their designs and approaches, help each other out when they get stuck, and sometimes just chat in the in-game chat while they play. Writing new recipes or improving others' is Taylor's favorite activity in Foldit. Contributions to Foldit can, thus, vary and include folding proteins manually (also called "hand folding"), applying automated scripts, or writing them. Compared to this variability, the "official" task in Stall Catchers is more straightforward.

A Snapshot of Stall Catchers

Visitors of *stallcatchers.com* are first presented with a short video clip introducing Stall Catchers as a CS game (Human Computation Institute 2017). The video shows a child and (presumably) their father sitting on a couch together playing Stall Catchers on a tablet

---

6   The backbone is the chain of amino acids which is linked together via peptide bonds (Foldit Wiki 2018a).
7   Sidechains are the chemical groups or shapes that are attached to the backbone of the protein (Foldit Wiki 2020a).
8   Objectives are guidelines for folding protein structures in Foldit. If a design generally meets the puzzle objectives provided, participants receive extra points increasing their score (Foldit Wiki 2022).
9   The scoring function in Foldit is based on the Rosetta software for protein modeling and generally indicates how well the protein is folded (Foldit Wiki 2018b).
10   This simplified and general description illustrates how participants engage in Foldit for this research and does not necessarily reflect it in its full complexity. More details on Foldit gameplay and how participants, together with automated tools, solve protein puzzles are provided in Chapter 6.

computer, followed by shots of seniors walking and dancing. A voice-over narrates the scene and captions summarize the key points: "We are fighting Alzheimer's, a disease affecting ~44 M people worldwide" (Human Computation Institute 2017, 00:03–00:08) A world map appears with pictures of people all over the world. "Catchers worldwide are analyzing real data, telling apart flowing & stalled vessels in the brains of mice to speed up Alzheimer's research at Cornell University," the caption says (Human Computation Institute 2017, 00:09–00:19). The video shows images of the vascular network of a mouse's brain, mice in a cage, and researchers working in the laboratory. Before viewers are invited to join the Stall Catchers community, the director of the Human Computation Institute, Michelucci explains, "As a global community, we will work together to find a cure for Alzheimer's disease" (Human Computation Institute 2017, 00:20–00:25). An orange "Join now!" button on the right side of the video frame invites visitors to join the ~50,000 "catchers" (as of February 2023) who are already registered with the game. As visitors scroll down the website, they learn more about Alzheimer's disease and how Stall Catchers is helping to speed up research. After registering, participants are directed to the main UI of Stall Catchers. A tutorial guides them through the interface, also called a "virtual microscope" (see Figure 5). Once the tutorial is complete, catching can begin.

The fictional participants Luis and Fiona both contribute regularly to the project. While Fiona often plays on her tablet, Luis prefers his desktop computer. After accessing stallcatchers.com and logging in, Luis immediately jumps into the flow of "catching stalls." The "virtual microscope" includes a video frame in the center of the interface where participants analyze short blood flow videos, resembling a look through a microscope (the research data presented has, in fact, been cleaned and transformed in several steps to make it easier to analyze). Inside the frame, the videos are actually augmented in the form of a small orange circle that indicates the specific area to be annotated. Below the frame is a slider allowing Luis to manually scroll through the video at his own pace. There is also a play/pause button next to the slider and an option to enable autoplay, which lets the video loop indefinitely. However, when using the slider manually, Luis sometimes stops at a specific time point and moves the slider back and forth to examine a dark spot more closely. The vessels, shown as white lines on a black background, move[11] or rather fade in and out of the video frames. To determine whether the vessel is stalled, he must closely follow the white pixels flowing through the encircled area, searching for black pixels that do not appear to be moving, which can indicate "stuck" blood cells and, therefore, a potentially stalled vessel. If he identifies the vessel as stalled, Luis clicks on the red "stalled" button below the slider on the right side. He is then prompted to indicate the exact location of the stall in the circled area by clicking on the position in the video frame. If the vessel is flowing, he clicks the green button below the slider on the left. After submitting his answer, he receives automated feedback. In the case of so-called "calibration movies" that test the participants' skill level, the feedback indicates whether he was right or wrong. In the case of an actual research video of the current dataset, the feedback message asks Luis to redeem points later after a crowd answer has been calculated. Luis pays close attention to the blue bar on the right side of the video frame which indicates

---

11   The vessels "move" depth-wise, i.e., either toward or away from the screen, rather than off toward any one side of the frame.

his skill level. A horizontal bar in the top right corner indicates the progress of the crowd analysis of the active dataset. Luis is fully focused on the repetitive task of analyzing the short videos lasting several seconds. In doing so, he inadvertently ignores some aspects of the game, such as the leaderboard on the left, with the usernames and scores of the top ten participants or the in-game chat in the bottom right corner.

Figure 5: Stall Catchers' main UI with the "virtual microscope"

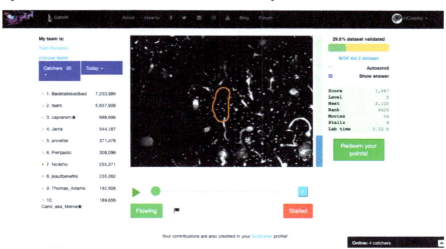

Source: Screenshot taken by LHV on Dec. 20, 2019 (https://stallcatchers.com/virtualMicroscope)

Fiona, by contrast, particularly enjoys the company of her fellow participants and loves to chat with them while playing Stall Catchers. She is also eager to climb the leaderboard and compete with others. Therefore, Fiona often participates in special events, such as Catchathons, because they add a little variety to catching stalls every day. For example, if there is a "double points" hour during the night in her time zone, she sets her alarm to get up and collect extra points. Fiona participates in Stall Catchers primarily for fun and to unwind after a long day at work. She also likes the project because she can do something good and valuable in her spare time. Compared to conventional mobile or mini games, such as Candycrush,[12] playing Stall Catchers is still meaningful in that she is helping to advance research while having a good time.

Contributing to Cope With Everyday Life

The two snapshots of Stall Catchers and Foldit, even if they serve primarily as introductory illustrations, show how both systems, intentionally designed as games, invite different modes of contribution, such as focusing on the "task at hand" of folding proteins

---

12   Candycrush was commonly referred to as a counterfoil to Stall Catchers by participants and team members.

or classifying vessels, or concentrating on the social and competitive game aspects.[13] In the previous chapter, I discussed why these projects were designed as GWAPs in the first place. On the one hand, in the case of Foldit, protein folding in science was thought of as a competitive and game-like endeavor.[14] On the other hand, it was assumed that people would be more motivated to contribute to something that was fun and that they enjoyed. Designed as a "casual game," participants could play Stall Catchers for a few minutes at a time and, thus, using what would otherwise be idle time, such as waiting for the bus or standing in line somewhere, to do something valuable and fun. The representations of users as "gamers" and "people in the workforce" were, therefore, initially "inscribed" in the project's design (Akrich 1995; Fischer, Östlund, and Peine 2020).

When I started interviewing Stall Catchers participants in the Spring of 2020 and asked them about Stall Catchers as a "game," I was initially surprised by reactions such as the following, which I encountered from several participants: "I still don't get the game aspect, I don't even want … I don't, I don't need that, no. That's not got any relationship to why I play it," explained Akin (May 11, 2020). Even though Stall Catchers was designed as a game, some participants rejected this notion. Although to a lesser extent, I noticed similar attitudes in the case of Foldit. For both projects, in written responses to my questions, some participants (Gordon, Jul. 14, 2020; Aram, Feb. 28, 2021; Ada, Mar. 1, 2021) used quotation marks when talking about the "game" or "playing" or emphasized the term when talking about it, adding "the game, if you will" (Alyssa, May 14, 2020; see also David, Mar. 4, 2021 discussed below). Participant Asher explained in our conversation that its design as a game was not a motivating factor for them to contribute to Stall Catchers: "It is technically a game. I don't view it as a game. […] I guess. I'm not doing it for the gaming aspect. [LHV: How would you describe what you're doing?] [I'm] just contributing. Just contributing towards knowledge." (May 20, 2020)

Asher did not "necessarily categorize it in [their] head as a game" (May 20, 2020), and similar to others (i.e., William, May 7, 2020; Elle, May 13, 2020; Noemi, May 14, 2020), the points and scores did not "matter to" (Asher, May 20, 2020) them. Julia mentioned she did not even "understand the point system," and that she had not "really looked into that" (May 11, 2020). So, I was initially surprised that the system's design as a game was not a given for some participants but a contested term. Such statements sparked my interest in understanding how these participants understood their participation, why they contributed, and what meanings they ascribed to Stall Catchers, especially those that did not align with the inscribed meanings and, hence, the visions, values, and norms that the designers and developers implemented in the system's design. In the following, I, first, stay with the example of Stall Catchers to better understand why some participants rejected the classification of the project as a game. To do so, I focus on their personal motivations,

---

13 In a post on the Stall Catchers forum from September 2020, replying to a new participant's question about the idea of Stall Catchers, Michelucci explained the reason for calling Stall Catchers a game and acknowledged that it was enjoyed by some participants while being ignored by others: "We call it a game because we have added game elements – like a score and leaderboards, and we sometimes run competitive events. Some 'catchers' (Stall Catchers players) really enjoy competing for spots on the leaderboard, and some catchers completely ignore it" (Michelucci 2020).

14 See Chapter 4, footnote 10.

and, in particular, the meaning of Stall Catchers for those who participated because of a personal connection to Alzheimer's disease.

Personal Connections

As discussed in Chapter 2, there is a rich body of literature on participant motivation in CS projects and CS games. Although the literature points to the heterogeneity and dynamic nature of motivations, which also depend on the participants' professional as well as sociocultural and economic background, the most commonly identified reasons include an interest in contributing to "real scientific research or to an important cause such as the environment or health" (Land-Zandstra, van Beusekom, and Koppeschaar 2016, 3), an interest in the research topic of a project, enjoyment, the chance to learn about a particular research area and community, and social reasons, such as connecting with others who share similar interests (Land-Zandstra, van Beusekom, and Koppeschaar 2016, 3). More specifically, research has shown that the primary motivations in CS *games*, although play and fun may be motivating factors, are not in the game itself but rather in the scientific contribution involved in participating (e.g., Curtis 2015; Miller et al. 2019). The results of my research generally confirmed these findings. However, the motivations listed only capture some of the motivations I encountered during my fieldwork on Foldit and Stall Catchers. Specifically for Stall Catchers, more than ten of my interview partners named another key motivating factor: a personal connection to Alzheimer's disease, either because of close friends or family members who suffered or had suffered from the disease or because they were caregivers or feared developing the disease themselves.[15] Even though there may be some selection bias at play, in that participants who chose to share their perspectives in an interview or otherwise contribute to my research, are likely to be among the more active or deeply motivated participants. That is, the participants interviewed are not necessarily a perfectly representative cross-section of the large Stall Catchers participant base. However, written entries in the "dedication" section of the project website indicate that this motivation seems to apply to a significant number of participants in addition to those who actively participated in my research. On the "dedication" page, participants can publicly dedicate their contribution to someone or something (Human Computation Institute, n.d.a). In fact, most of the around 300 entries are dedicated to a close person or family member who was currently suffering from Alzheimer's disease or whom they had lost to the illness or another form of dementia. For those participants and interview partners who have a personal connection to the disease, the motivations and meanings of Stall Catchers go beyond those commonly cited in the literature.

One of my first conversations with Stall Catchers participants was with Elle, who joined Stall Catchers shortly after its launch and was brought to the project because of her father's Alzheimer's disease. She described that,

---

15   I did not directly ask interview partners about a personal connection to Alzheimer's disease. In the cases mentioned, it was brought up by the participants themselves. It is, therefore, possible that even more of the Stall Catchers participants who contributed to my research have a personal connection to Alzheimer's disease but chose not to share it with me.

> I guess as soon as I heard about it, it gave me a sense of […] it's too late for dad, but […] it might, […] if I can do this and maybe feel a little bit less helpless […], I could perhaps, by participating, I might help […] to find a cure or prevention […] that might prevent other people from suffering […] the way dad has […]. [M]y initial motivation was, here is something practical I can do to help with research. […] Basically, the whole motivation really stems from my personal connection that […] I'm seeing and still living, 'cause my father is still alive and watching him deteriorate and yeah just […] going through what we're going through with his illness and just wanting to help. (Elle, May 13, 2020)

Although Elle was not able to contribute "as much as I wish I could" due to her daily schedule and life being "a bit overwhelming," contributing to Stall Catchers helped her feel "a little bit less helpless" in her everyday life, of which Alzheimer's disease formed a permanent part (May 13, 2020). Similarly, in an interview with the Human Computation Institute published on the institute's blog, Michael Landau, one of Stall Catchers' most active participants, who is deeply invested in promoting Stall Catchers, explained, "I like to play the game because it makes me feel less powerless as I sit and watch my mother's mind slowly being wiped out" (Landau 2018). While contributing to Stall Catchers helped Elle and Michael Landau feel less helpless or powerless, other participants engaged in the project out of fear of one day suffering from the disease themselves after family members had been affected by it. This fear or risk was often shared by the entire family, as in Olav's case:

> On my mom's side of the family, my grandfather and grandmother […] both have—or my grandmother passed away a few years ago, but they both have Alzheimer's. […] [A]nd so […] certainly my mom—because both her parents had it—[…] my mom has never gotten any test to figure out for sure if she has the genetic […] marker that means she's likely to get it but, […] certainly, […] there's that concern. And so, she does everything she can, but obviously, Alzheimer's research […] for those reasons […] is very important to all of us […], and even, I guess, me, frankly 'cause I'm obviously related. (May 21, 2020)

Comparable motivations were shared by some of the participants of Foldit, which addresses Alzheimer's disease amongst other diseases such as cancer or HIV/AIDS (Center for Game Science [University of Washington] et al., n.d.c):

> Like many others, I started playing Foldit while having primary caretaking responsibilities for my parents. One died from Alzheimer's, the second later from Lewy body dementia and Parkinsonism. Even if Foldit is working on different problems at any given time, it helps that there is hope it will address some of the many medical problems at some point. I discovered many "team mates" and Folders from other teams were also providing caretaking for family members. (Ada, Mar. 1, 2021)

During her time caring for her parents, "it was nice" for Ada "to be able to unwind with Foldit around spending time with them" (Mar. 1, 2021). The hope of finding a cure for dementia to save his mother was the main motivation for David to contribute to Foldit, which he discovered while researching medications.

> [B]ecause my mother had dementia. And then I started looking for medicines, I couldn't find them on the Internet, and then I arrived at Foldit, and I thought it would be a very good idea to participate. To try and find a drug for dementia myself. So, the connection is important […] also intrinsic motivation. (David, Mar. 4, 2021)

Despite the vast differences in design, development, and functionality between Foldit and Stall Catchers, the personal connection to Alzheimer's disease was a primary motivation for participants of both projects.

In his 2008 essay "Play Theory," play theorist Brian Sutton-Smith describes play as a form of protection: "play as we know it is primarily a fortification against the disabilities of life" (2008, 118). He refers to how play goes beyond "life's distresses and boredoms and, in general, allows the individual or the group to substitute their own enjoyable, fun-filled, theatrics for other representations of reality in a tacit attempt to feel that life is worth living" (Sutton-Smith 2008, 118). While Sutton-Smith may not have had GWAPs and CS in mind when writing these lines, and though participants contested the classification of Stall Catchers or Foldit as games, I argue that it is, in fact, *because* the projects move between science and play that they can become so meaningful to participants. Stall Catchers and Foldit can have a "healing function" for participants who contribute because of their personal connection to Alzheimer's disease, corresponding, perhaps explicitly but at least incidentally, to what Sutton-Smith describes for games: "[p]lay was always intended to serve a healing function" (Sutton-Smith 2008, 124). However, the healing function in the examples studied goes beyond the "healing function" that Sutton-Smith ascribes to all games because the fact that Stall Catchers and Foldit address the very disease that affects the participants is crucial to them. This understanding of the healing function goes hand in hand with anthropologist Veena Das' conception of "the everyday as a way of inhabiting the very space of devastation yet again" (2020, 58). Das applies this understanding of the everyday to "think of a politics of the ordinary as a stitching together of action and expression in the work of bringing about a different everyday—I call this the birthing of the eventual everyday from the actual everyday" (2020, 58). Although Das' analysis emerges from her fieldwork in India, where she studied low-income urban families, this understanding can be very informative in analyzing the role of these HC-based CS projects for caregivers or close family members of people with Alzheimer's disease. For Das, the everyday is restored and wounds are healed through and in the everyday itself. What cannot be said, for example, because it is too painful, becomes expressible again in the everyday. Individuals caring for a family member with Alzheimer's disease and who feel powerless over the disease can contribute to research and empower themselves with Stall Catchers.

They also find themselves in an understanding and empathetic "community," as Stall Catchers participant Alyssa explained to me (May 14, 2020). Challenging experiences, such as the loss of a loved one to a disease like Alzheimer's, or the powerlessness others experience in the face of the disease, are part of the everyday. It is precisely the ordinariness of Stall Catchers and Foldit, their situatedness in the everyday, and the possibility to contribute whenever and for as long as they want that helps participants deal with these challenges and feel less powerless. The games can be easily integrated into daily routines. To quote Stall Catchers participant Akin: "[I]t's always there. It's there at 3 AM, it's there whatever time I have available, it's there" (May 11, 2020).

The projects give participants the feeling that they can do something about Alzheimer's disease while dealing with it. For these participants, unlike others, it is not about contributing to just any scientific study but specifically to Alzheimer's disease research. While it remains unclear if and how their participation in Stall Catchers will lead to a treatment or cure for Alzheimer's disease, the project opens a "horizon"[16] for the participants where their current situation and the state of medicine does not. Stall Catchers, in this sense, can be seen as providing a horizon in the face of a currently incurable and deadly disease in two ways: On the one hand, Stall Catchers introduced a new way of analyzing research data that would not have been possible or feasible otherwise. On the other hand, it can be understood as a form of "horizoning" (Petryna 2022) for participants to cope with everyday life which is marked by a deadly disease. Contributing to HC-based CS is a moral practice with which they participate in the ethical projects. For them, the ethics of the project is not in optimizing HC. It is in fulfilling their perceived responsibility to do something about Alzheimer's disease.

Because of their connection to the disease and the meaning of Stall Catchers as a way of coping with the everyday, the project, like Foldit, is not just fun and not just a game to pass time to them but a serious endeavor.[17] Foldit participant David, therefore, suggested simply calling the project "citizen science" or "gamification of science" because they both contain the word "science:" if the word "is included, then it's already an improvement compared to 'game.' Because by 'game' a lot of people also think of shooting games [...] for example, just games you play to kill time" (Mar. 4, 2021). Similarly, long-term Stall Catchers participant Alyssa would prefer for Stall Catchers to be thought of as research, which is how she describes what she does to family and friends:

> I started in [...] 2017, and I think by ... maybe Spring of 2018 I started calling it "I do research for Cornell University," and that's how I have described it: "I do research for Cornell University," and [people would answer] "Really?! What do you do?" And then I describe the game. The game. And I tell them that it's a game but that in essence, it's doing research for Cornell University for Alzheimer's. (Alyssa, May 14, 2020)

Not only participants with a personal connection to Alzheimer's disease used quotation marks to refer to the projects as "games". When I encountered the use of quotation marks

---

16 Anthropologist Adriana Petryna (2022) thinks about horizons regarding climate change, where "'horizon work,' allows experts and the public to find other meaningful points of reference from which to imagine how to organize a response to the current crises before we lose the capacity to respond" (2022, 3). Petryna introduces the term "horizoning" as a "conceptual device for thinking about and responding to complex futures" (2022, 5).

17 In her book on VDC, Holohan describes similar reasons for contributing for Folding@Home participants, who are "motivated by the personal helplessness of watching loved ones die of diseases for which the cure is still being sought" (Holohan 2013, 20).

in a written response about Foldit, I asked Aram why he had used quotation marks to refer to the game.

> For me, Foldit is not really a game but a simulation tool. And the players are rather creative researchers than gamers. Many Foldit users don't play for points, but they use Foldit for protein modeling and to implement their own ideas. I think this application goes far beyond a game. In essence, this is why I had put the "". Foldit is more than a game. (Feb. 28, 2021)

For John and Gordon, Stall Catchers was not a game but rather work (John, May 7, 2020; Gordon Jul. 14, 2020). Gordon argued that the "game" is more of a cover for the work previously done by paid scientists, which is why he wanted to be paid for his contribution to Stall Catchers:

> Also, I wish there was some way I could make some money playing the game. I know that it's just supposed to be a volunteer game, but I do think that people put a lot of work into it, and it is really work and not a game, so people should be compensated for their time. Why should people work for free? The researchers don't work for free, so why should the players be expected to work for no pay? It's not a game. It's work made to look like a game, and everyone knows that that's the case, so why pretend otherwise? I'm not saying that it doesn't have value. I fully understand the value of the "game." But I don't think it's fair to call it a game when at some point, I'm sure people were paid money to complete the same task. Just because they have created a facade to make it look like a game, that doesn't mean that it's really a game. (Jul. 14, 2020)

Gordon began contributing to Stall Catchers on a daily basis about six months after it was launched but recently stopped participating because he "got bored with it" (Jul. 14, 2020). Digital media cultures theorist Tiziana Terranova has argued that such forms of "free labor" are inherent to the Internet and the digital economy (2012). While play/work interferences can be considered inherent to games, the relation of CS to or as work, and whether CS participants should be paid for their contributions, have been the subject of scholarly and public debate (e.g., Liboiron 2019; Robinson 2019). However, Gordon's wish to be paid was unique among the examples in my empirical material. This does not mean, of course, that there are no other participants who would appreciate financial compensation, but it was not actively brought up in conversations by participants. For them, in addition to the ethical motivations and coping strategies discussed, it was more important that their contributions were meaningful.

Meaningful Contributions

It was of utmost importance to the participants to feel that their contribution was worthwhile and meaningful. This became particularly apparent in situations where something did not work as needed, such as server outages that made the platform inaccessible, when research data would not load properly, or when the data they were asked to annotate was of poor quality and, therefore, difficult or impossible to interpret correctly. Stall Catchers participant Ellen explained that "if you get the bad quality pictures all the time, you lose interest because it's, you don't feel that you are doing something valuable" (May 19, 2020). She added that "because it's, you just fear that your work is meaningless" (Ellen, May 19, 2020). Kamon shared this perception, since one was forced to simply guess the answer if images were too grainy, "[b]ut that is not so fun [...] because, then you end up asking yourself: What am I really going to contribute in this dataset [...] if every player makes a guess?" (May 15, 2020). These moments were described as the most frustrating, as the purpose of their participation was not clear to them. It is not unique to HC-based CS games that gamers become frustrated when their game does not work as intended. However, an important difference seems to be that while glitches in other video games are merely disruptive because they interrupt the flow and prevent players from being immersed in the game, in CS, this "immersion" is tied to the perceived real-world purpose.

The understanding of CS games as a way of coping with everyday life can also be applied to other overwhelming situations and experiences in which people feel helpless, such as in the context of the COVID-19 pandemic. Although the meaning of contributing to CS as a form of coping is not necessarily generalizable to all different kinds of CS or individual experiences, comparable forms of engagement in CS projects could be observed during the onset of the COVID-19 pandemic in Spring 2020, when CS projects dedicated to advancing research on the coronavirus or a drug and vaccine against it experienced a considerable increase in participants. In order to better understand this increase, I conducted an exploratory study in which I contacted people via the snowball principle who had started contributing to the Folding@home project (The Folding@Home Consortium (FAHC), n.d.), a distributed computing project for simulating protein dynamics (Vepřek 2020). Participants could contribute to coronavirus research by donating their computing power to the project through downloading and running the software. Six contributors participated in my study by completing a written questionnaire. Here, contributing to such a project was, for instance, described as a way of "doing something" about the crisis. "I felt that I was currently not doing anything to help with the current crisis, and this felt like an easy way to help" (Francis, Mar. 27, 2020), one participant explained in writing. In times of lockdown and social distancing structuring the daily lives of many people not working in jobs of "systemic importance," participating in CS projects such as Folding@Home or Foldit allowed them to feel like they were really "doing something" to contribute to the fight against the pandemic in a meaningful way. Framing contributions to CS as *forms of coping* enriches our understanding of participants' motivations and goals. This perspective goes deeper than the science-focused perspective of doing good or simply enjoying a game. Focusing on the participants' meaning-making processes reveals how they renegotiate and extend the inscriptions and design of the systems.

With regard to GWAPs, participants contribute to science or some form of data analysis and collection "as a side effect of playing the game" (Von Ahn and Dabbish 2008, 60). However, this analysis of the participants' perspectives shows that for many participants it was the other way around. From their point of view, they contributed to science and knowledge production, and the gameplay was the side-effect: "[Y]ou can make games out of it too. But the main thing is to hopefully advance the research," explained longtime Stall Catchers participant Caitlin (May 5, 2020). To Noemi, "it's not important [...] to compete except that I do want to contribute, and that's [...] my personal goal" (May 14,

2020; see also Jeshua, May 8, 2020; William, May 27, 2020). Based on the analysis of my empirical data, I argue that, in this case, a more suitable description for these *games with a purpose* would be *purpose with a game*. Contributing to "real" science gives participants a sense of purpose, and their understanding of "games" or "playing" refers to a "field" (Bourdieu 1985) of the everyday that does not include seriousness and purpose in this sense. At the same time, the fact that the projects are designed as games is crucial here, as it is the interplay of science and play that forms the basis and opens up the space that allows different forms of meaning-making and adoption.

The "game" is the basis here. It enables participation in scientific research with a low entry threshold and allows participants to actively do something. As Stall Catchers team member Paul aptly describes, "there may or may not be a [...] result [...] that will help you, but you're still doing something to move things forward in fighting this disease, and [that] just gives people the [...] drive to keep going and [...] not give up, 'cause otherwise there's nothing else you can do." (Oct. 14, 2020)

The points, the game score, and the leaderboards—even if participants do not participate to compete—help to experience the feeling of progress: "anything that can help measure and feel like there's progress" (Maya, May 13, 2020). In the following, I turn to the specific entanglements of game design and science in HC-based CS, which, as I will show, create object potentials and affordances that open up the possibilities for new relations between participants and software to emerge, and, hence, for intraversions to occur.

A Phenomenon Between Play and Science

Designed as GWAPs, HC-based CS games should, from the designers' perspectives, not only serve the means of the game itself but also contribute to the solution of a particular computational or scientific problem, or to the training of an AI model (Von Ahn and Dabbish 2008, 60). However, as I have shown above with a focus on participants who care about Alzheimer's disease, participants assign varying meanings to their involvement in HC-based CS in their daily lives. Based on these meanings, which can range from a leisure activity to a meaningful contribution to science, work, or even a moral obligation, they position themselves differently. As a result, they stabilize and destabilize the assemblages in different ways.

Game and play theories have long shown that games and play are not merely fields of fun and enjoyment but can involve demanding and complex activities that are more akin to "work," or that games can include repetitive tasks that can make a game tedious and unpleasant at times (*cf.*, e.g., Stevens 1980; Sutton-Smith 2001). There are always interferences between play and work. Terms like "playbour" (Kücklich 2005; Lund 2015) and "laborious play" (2019; *cf.* Abend et al. 2020), along with "playful work" (Abend et al. 2020), have been introduced to describe this phenomenon, emphasizing play "as an act of drawing or blurring boundaries. It is not a given, but an active achievement of all actors involved, including non-human actors like interiors, or hard- and software" (Abend, Fizek, and Wenz 2020, 8). Dippel and Fizek, in their pioneering work on the field of CS, where "science" is an additional pillar in play/work interferences, state that: "[c]itizen science games may be perceived as laborious playgrounds, placed between the two poles of *ludus* and *labora*, oscillating between qualities associated previously with leisure or pastime and with productive or useful time" (2019, 256, emphasis i.o.). Playful and productive aspects are interrelated in GWAPs (Turner 1995, 56; *cf.* Abend, Fizek, and Wenz 2020). According to Fizek, the success of CS games lies precisely in "[t]he immediate leverage of a playful and pleasant activity with a socially productive outcome, the element of competition in a large collaborative environment, and the feeling of belonging to a community with a common goal." (Fizek 2016)

My aim is not to further define play, work, and science but to use the terms "play," "science," and "work" as discussed in the field by my research partners. In this sense, the purpose of this work is not to directly contribute to or give a new definition of play, work, or science. Rather, the purpose is to gain a better insight into how different understandings of play, work, and science and their inscriptions in HC-based CS are interwoven and how this opens up the space for intraversions of human–technology relations. In addition, I focus on play and science as the main fields. In the field of HC-based CS, "work" and "science" are blurred. While data analysis had been part of the working routines in the laboratory for which researchers were paid or which were considered part of the study program, it was framed as "citizen science" in the context of the game platform Stall Catchers. Here, from the beginning, participants contributed for reasons other than financial reward (as discussed above). In this sense, contributing to Stall Catchers was not considered "wage labor" or *paid* work even though the task itself was sometimes framed as "work," as I will discuss below. Nevertheless, the overarching reference points in the field were "play" and "science" and, thus, form the focus of my analysis. I consider these terms as tools for the boundary work performed by different actors.

As Sutton-Smith puts it, "[s]omething about the nature of play itself frustrates fixed meaning" (2008, 80). Just as researchers do boundary work by defining what "scientific" knowledge production should look like, understandings of what play is and what it should look like on a digital platform inform the individual perspectives, producing boundaries and resistance. In both examples, Stall Catchers and Foldit, the playful approach was not part of the professional scientist's daily working practices but was delegated to volunteer participants, creating a boundary between the tasks performed by professional scientists in the laboratory and the tasks performed by volunteer participants. Play is, thus, differentiated from professional, scientific practice;[18] the playful approach is understood, or intended, to be incidentally helpful. In Chapter 6, I illustrate this further with the example of biomedical engineering researchers who considered inviting Stall Catchers participants to help with data preparation. The researchers felt that such a contribution to data preparation should not be gamified, unlike the task on the Stall Catchers platform. This was because it would involve the participants in the scientific process of data curation (fieldnote, Oct. 21, 2023).

These entanglements of play and science, and the different expectations, visions, and meanings of various actors, form a productive space that is not frictionless but contested

---

18 In fact, as shown in Miller et al. (2023), CS games often face the problem that game design is not directly funded by scientific grants, leading to missing resources.

due to different understandings and logics of the fields of science and play, creating tensions that can only be partly resolved, and that affect the everyday formation of their sociotechnical assemblages. The HC-based CS projects, like the intraverting human–technology relations, must, therefore, be considered multiples (Mol 2002b). Referring to Annemarie Mol's research on how atherosclerosis is enacted (2002b), Mol and Law write that "[i]n practice, if a body hangs together, this is not because its coherence precedes the knowledge generated about it but because the various coordination strategies involved succeed in reassembling multiple versions of reality" (2002, 10). Just as a coherent body is the result of different coordination strategies, HC-based CS form a 'coherent' project through the diverse and continuous attempts of all actors involved to assemble the project. These attempts do not always create the desired "seamless space"[19] (Vertesi 2014). Instead, different and divergent modes, logics, and interests converge and coexist in both productive and tense ways. As Mol and Law observe:

> *Often it is not so much a matter of living in a single mode of ordering or of "choosing" between them. Rather it is that we find ourselves at places where these modes join together. Somewhere in the interferences, something crucial happens, for although a single simplification reduces complexity, at the places where different simplifications meet, complexity is created, emerging where various modes of ordering (styles, logics) come together and add up comfortably or in tension, or both.* (2002, 11, emphasis i.o.)

The interferences of play and science are essential to the creation of the HC-based CS assemblage and the participants' experience in the first place, even though some participants may object to calling Stall Catchers or Foldit games. Play and science combine comfortably and in tension at the same time. In what follows, I aim to provide concrete examples of how these play/science interdependencies create the spaces in which participant–technology intraversions unfold. With respect to the mutually supportive or productive science/play entanglements that I turn to first, I focus on the following examples, which mainly, but not exclusively, concern the Stall Catchers project: 1) new approaches to scientific findings with "out-of-the-box thinking" as part of play, 2) making a "boring" analytical task bearable, and using competition and points as motivators and short-term rewards along a lengthy scientific process, and 3) legitimizing play with scientific purpose and making games meaningful.

---

19 Sociologist of science and technology Janet Vertesi (2014) borrows the term "seam" from critical studies in Ubiquitous Computing. To approach heterogeneity and complexity, she suggests adopting this vocabulary of seams to "consider the constraining nature of infrastructures at the same time as it observes how actors skillfully produce moments of alignment between and across systems: not fitting distinct pieces together into a stable whole but *producing fleeting moments of alignment suited to particular tasks with materials ready-to-hand*. Rather than moving to the macro view of a meta-infrastructural analysis, [the vocabulary] must hold our focus steady on the micro: actors' observable, reportable activities as they wrestle with many infrastructures' limitations and possibilities to bring them into moments of alignment." (2014, 268, emphasis i.o.) The concept allows one to focus on the (micro)practices and efforts of different actors to align various elements, since "seamlessness cannot be assumed" (Vertesi 2014, 274). Instead, it is continuously created and maintained.

Mutual Supportive Science/Play Entanglements

Out-Of-The-Box Thinking

As a novel mode of biomedical or biochemical practice, for example, CS games open new horizons in knowledge production through the "level of creativity in games" (Dippel 2019b, 248), especially when current technology and established practices fail to meet prior expectations. Their playful encapsulation of scientific problems allows participants without any prior specific domain knowledge "to contribute to [...] science through a side door," as Foldit participant David put it in our interview (Mar. 4, 2021). Using the metaphor of a side door, David referred to the fact that CS participants might not rely on established scientific approaches to solve a scientific problem which is presented to them. In the Foldit example, the team saw "out-of-the-box thinking" as particularly promising and could be stimulated in a playful environment:

> [W]hen people are maybe in a playful [...] mindset they are more willing to try things that they wouldn't try and look at things creatively and not be as afraid of failure and that kind of thing and so, those would all be good things, I think, for people who are playing Foldit or other citizen science games to try new things, to have fun. Cause in a way, that's kind of [...] one of the core motivations for doing things in any case's way is they have to come up with new things that maybe someone who is a biochemist might not have, try out something new. Maybe a little bit unusual, look at a problem in a different way. (Gidon, Jan. 31, 2020)

Games invite creative approaches that can allow CS participants to come up with new ideas that have not yet been tried in the conventional settings of academic or professional biochemistry.[20] Following Dippel, "[t]he old modes of production may still be reflected in the way many games 'work,' and old 'traditions' may still be at work in many video games. However, new worlds, envisioning other ideas of society, are in the making" (2019b, 248). For Foldit participants like Aram, its design as a game necessarily abstracts away the scientific content to a certain extent and allows participants to keep a distance from the serious scientific background:

> I consider the puzzle rather as a separate optimization task. [...] Nevertheless, I follow the context (blog, newsletter, etc.) regularly to stay up to date. But I understand many details only to a certain extent since I focus more on the optimization task than on the context. This helps me to keep a certain distance which I consider positive. I also think it's good that Foldit offers this option, because it's not mandatory to have mastered biochemistry to be able to keep up well enough in Foldit. Foldit is abstracted enough for that (by being presented as a game with bonuses, scores, [...] which I consider as positive. (Aram, Feb. 28, 2021)

---

20   It should be stressed that regarding the case studies investigated, this specifically refers to Foldit and not so much to Stall Catchers or ARTigo where the task to be performed by participants is rather straightforward. However, Stall Catchers participants still find creative ways to engage with the platform, as discussed in the subchapter "Adaptations and practices beyond design."

Viewing protein folding as an optimization problem rather than a biochemical one also opens the door to unconventional approaches in current scientific work. This perspective on protein folding was also shared by Foldit researcher José: "[E]ven just thinking about it, when you learn about it in biochemistry class, it kind of feels like a game where you just have to [...] find the lowest energy fold" (Jan. 22, 2020). This makes the task in Foldit interesting for participants with various interests, ranging from playing games and solving puzzles in general to computer science, mathematics, or biochemistry. Compared to Foldit, the task in Stall Catchers is more straightforward. This made the science and play entanglements important for other reasons.

Making a "Boring" Task Enjoyable and Keeping the Motivation Up

The interferences of play and science were not only considered to support new creative approaches to scientific problem-solving and out-of-the-box thinking but, especially in the case of Stall Catchers, were essential to making a repetitive task bearable and enjoyable. Participant Ellen explained that the task of

> Stall Catchers itself is tedious. So, it's not really fun but what makes it fun is the way they did it, the way how they [are] getting this rating. Not just the rating is important because it [...] keeps you on your toes. [Laughs] So you are not getting lax because [...] somebody is catching you if you're not [keeping up]. (May 19, 2020)

It was the game design and the competition[21] that made the task enjoyable in the first place: "I thought they did a really good job of keeping it like a fun, lighthearted competitive energy in the face of a *really* boring job" (Maya, May 13, 2020). To make the task more interesting, the merging of the playful setting with the scientific background of Stall Catchers was crucial. The playful setting creates the conditions for participants to keep going and differentiates Stall Catchers from a mere image analysis task.[22] Participant Maya explained further: "[T]he human nature is you want to, you wanna get the right answer. Like you want to be right and then when you're not right you're like ooch ... and they did a good job of not being like 'No! That's wrong!' [laughs] So I thought they did a very good job of having it be playful." (May 13, 2020)

This sentiment was also expressed by participants who did not see their contributions as play, explaining that they did not care about the points but only focused on the analysis. Olav, who contributed to Stall Catchers mainly to do something about his own risk of developing Alzheimer's disease in the future, explained that "if there was less of a game aspect to it, I think it would be for more people and probably me too, [...] a little bit harder to [...] stay motivated [...]. [B]ut it is kind of fun, and that's what [...] continues to bring me back, I think" (May 21, 2020). As a game, Stall Catchers was so engaging that it even risked becoming addictive to Olav: "It's slightly like good games, [...] It's a bit [...] of a

---

21 For participant Caitlin, "competition" in Stall Catchers did not refer to plain competing against each other but to a "friendly competition" (May 5, 2020), where it is okay to pass each other on the leaderboard because they are all working toward the same goal of advancing Alzheimer's disease research, as participants reminded each other in the in-game chat.

22 This is not to say that some participants did not take an active interest in the task itself, which was fun and sometimes challenging for them (e.g., Elle, May 13, 2020; Noemi, May 14, 2020).

challenge which you can then become slightly, [...] not in a bad way addicted to like, oh, just get one more level. Something like that. Those are what make video games, board games, et cetera, fun" (Olav, May 21, 2020). Being addicted to Stall Catchers, however, is not a bad thing for Olav, as it is still for a good cause. Similarly, Foldit participant David sees addiction as part of any "good game:"

> Part of a good game is also the ... in a sense addicting the user to the game. And I think they succeeded with Foldit in this regard because there are often players who say, oh, well I don't need it anymore, I don't do anything anymore. And after a while they come back because they miss it. And I've had that myself sometimes that I thought, okay, I'm having an off-day today and I don't really feel like doing it anymore, so I pull the plug and I don't do it anymore, and then someone comes along with a story about a family member with dementia and I think, yeah, that's why I did it! And the next day I'm back modeling again. (Mar. 4, 2021)

Stall Catchers participant Kamon, who joined the project in 2019, also contributes daily, filling up most of his evenings. With a smile, he admitted in our conversation that Stall Catchers was indeed addictive to him. Sometimes, he said, his partner would remind him of their presence when they felt Kamon was spending more time with Stall Catchers than with them (Kamon, May 15, 2020). "[B]ut I still try to succeed, I'm ranked [...], and I still try to keep that [...] place. So, if I've gone on vacation for a while and I see I've dropped, I still do a few extra hours to keep that [laughs]" (Kamon, May 15, 2020). The competitive aspect keeps participants coming back to improve or maintain their ranking list position.[23] In the case of Foldit, the daily competition and score also kept long-term participant Lucas motivated to keep contributing, given that the scientific rewards were not common: "The scientific rewards take a very long time to come and they're few and far between. Hearing that we've actually made a contribution to science is a very rare reward and so having a reward that you can go after every time you sit down, I think is important" (Mar. 17, 2022). In addition to their purpose as a short-term reward, points allow participants to contribute without understanding the scientific problem's full complexity or the sociomaterial entanglements behind the projects. In Foldit, some participants focused only on optimizing their game score without knowing, for example, how it was calculated or what biochemical processes were involved in protein design. Similarly, in Stall Catchers, participants do not need to know how crowd answers are calculated, and contributions are evaluated and weighted to contribute. The points they receive for analyzing a video indicate their performance. Hence, the game mechanics play a crucial role in breaking down and hiding the scientific and algorithmic complexity. In fact, any reference to the notion of HC itself was remarkably absent in most of the interviews I conducted with participants.

In the example of Stall Catchers, however, the game features described were mainly seen as the only aspects of Stall Catchers that would make the task playful: "yeah [, it's]

---

23   An educator who used Stall Catchers in class and contributed to my research declared the scoring aspect and leaderboard to be particularly important for children participating in Stall Catchers as part of their classes (Ren, May 18, 2020).

not very playful unless you look at the score aspect of it and I think that can get people on board and be like 'uh, neat!'," summarized participant John (May 7, 2020).

Caitlin enjoyed "playing my way back up the skill bar. [...] [T]he first few times I got up there, of course, I didn't stay there very long and, so when I would make a mistake or two and drop back down on the skill, I much enjoyed that process of climbing my way back up to the top" (Caitlin, May 5, 2020). However, after contributing to Stall Catchers for several years, Caitlin's skill bar was almost always at the top, which is why she did not "really have the fun of climbing my way back up the way I used to" (May 5, 2020).

> I'm always there, right? Yeah, and if I do miss, you don't drop very far cause the longer you're at the top when you eventually do miss, it doesn't drop very much for the first miss. If you miss twice, then it's gonna plummet, but the first time you miss, you merely drop, [...] and after a couple of calibrations you're back up at the top again. (Caitlin, May 5, 2020)

While Caitlin and other participants, came up with new play practices that went beyond the coded game features (see below), other participants expressed that they found the game less attractive over time (Gordon, Jul. 14, 2020). Most participants agreed that Stall Catchers was only "medium fun" (Elisabeth, May 9, 2020) compared to other "dumb games" (Elisabeth, May 9, 2020); playing Stall Catchers for a longer period of time exhausts the game's features. Special events, such as the Catchathons mentioned above, were, therefore, essential for keeping participants engaged over a long time and bringing new participants on board. They interrupted the regular game flows and brought the online game into the physical presence (in pre-COVID-19 times) in the form of live meetups in libraries, schools, or even pubs, as explained in a blog post on the institute's website (Egle [Seplute] 2021a). Participants were invited to join the challenge of analyzing as much research data as possible in a given time frame.[24] In particular, the final hours of such events tended to be well attended due to the "double points" that could be earned. In April 2021, I attended the final hour and the related online "hangout" of one such event, where participants could meet the team and complete the challenge together. These meetings would allow participants to pose questions about Stall Catchers and the Alzheimer's disease research behind it. Researchers from the Schaffer–Nishimura Lab often joined the hangouts, as did representatives from the BrightFocus Foundation (BrightFocus Foundation n.d.), one of the longtime funders of Stall Catchers and the CS platform SciStarter (Scistarter.org, n.d.). The hangouts typically ended with the team reading out the challenge's statistics and leaderboards and announcing the winners. The statistics included the number of videos annotated, and how that related to the number of laboratory workdays saved. For example, Michelucci explained in the live hangout on May 1, 2020, that in the *cabinfever challenge*, over the 30-day challenge, 214,000 videos were annotated, of which 156,000 were actual videos from the current dataset, resulting in approximately 205 days of laboratory work (fieldnote, May 1, 2020). Mapping the

---

24 These time frames spanned from 24 hours, as in the example described in the introduction, or a few days, to an entire month, such as in the case of the "cabinfever challenge" that took place in 2020 during the COVID-19 pandemic (Egle [Seplute] 2020a).

analyzed data back to comparable laboratory time was proof of the scientific purpose and legitimization of the playful contribution. In the interviews with participants, these special events were described as having a "motivating" (Kamon, May 15, 2020) effect on participants. To Ebby, it was motivating that "everybody is catching stalls" (May 8, 2020), and to Elle, special events, "where they're really trying to encourage participation that usually gives [her] a little bit of a kick" to continue participating on a more regular level (May 13, 2020). Some competitions even offered prizes for participation. For a long time, this was Gordon's motivation:

> I started playing Stall Catchers about six months after the program started. I used to play about an hour a day, and then sometimes up to three hours during competitions, but recently, I haven't played it all. I'm kind of embarrassed to admit it, but I got bored with it. It was really difficult for me to stay motivated. They used to have competitions where you could win prizes, and that really kept me motivated. In competitions, I won a trophy, a T-shirt, and a mug. [...] Once they stopped giving out prizes, it was very hard for me to stay motivated. [...] I need to have some goal to work toward. At one point, I was one of the best players, but just racking up points was not very motivating for me. Also, they allowed the two top players to accumulate so many points that it became absolutely impossible to catch them even if I worked all day and all night, so I realized it would be almost impossible to advance to number one. (Gordon, Jul. 14, 2020)

While points and competition can incentivize players to engage during a specific timeframe, special events also help reinvigorate interest and engagement among participants who might have forgotten about Stall Catchers, inviting them to come back, as in the case of Sophia (Apr. 28, 2020).

As mentioned above, the "double points" hours that occurred during the last hours of the challenges were important to participants (e.g., Elisabeth, May 9, 2020). When bonus points and "hangouts" occurred at the same time, Elisabeth preferred to focus on annotating videos: "Although I was interested in participating in the recent 'live hangout' events, I did not participate because they were held while bonus points were offered. I was motivated more to earn points than to interact" (Elisabeth, May 9, 2020). For some, accumulating points even faster with double points increased their motivation and the perceived value of their research contribution. Stall Catchers participant Alexandra explained:

> During the recent cabinfever challenge I was especially motivated during the double points day on Fridays as I wanted to see how much I could compete against myself in getting the points and number of videos watched up. It made me feel like I was doing something and making a difference when suddenly the numbers went up. I could see the results of watching thousands of videos. I was thinking *wow*. (Alexandra, May 9, 2020)

The importance of double points, challenges, daily scoring, and leaderboard features for the playfulness of Stall Catchers and as motivators for contribution shows how different meanings and understandings of Stall Catchers are interwoven. For some participants, who rejected Stall Catchers' game design but relied on it for their daily contributions, this led to sometimes conflicting attitudes, as in Daan's example:

> I like the point system in that it gives me an idea of how I'm doing. It does cause me, [...] when I get one incorrect where the expert has also reviewed it, and I think it's stalled and the expert says, no, it's flowing. And I look at that and it's like, oh, I just missed all those points. But deep down, it's also [...] I didn't get that one right. So, I didn't really contribute that time as well as I should have. [...] Whenever I'm doing a job, I try to do the best I can and darn it, I got that wrong. But that sensation doesn't last for too long [...]. I don't care about the points, but it is a nice way to keep track of how you're doing. So yeah, I'll look at the points now and then, and I'll laugh, and I'll go, meaning let's put [it as] fun, and I'll just continue trying to amass them, but really, it's did I get this one right? Did I get that one right? And when I'm just participating in adding to the crowd's evaluation of it, I feel like, okay, good. Some of the people agreed with me. Some of them didn't; I'm adding my voice to it. Overall, when you put all of us together, we'll nail it. One way or another, we'll get it right. (Daan, May 26, 2020)

Even though Daan did not actively care about the points, they allowed him to understand "how [he was] doing" in terms of the goal of contributing to the scientific purpose of the project, and he experienced disappointment when he missed them. However, what really mattered to Daan was not the points he missed but that he did not "contribute [...] as well as I should have" (Daan, May 26, 2020). Together, the aspects described create a playful experience of a serious situation, as Stall Catchers advocate and founder of the Memory Café Directory[25] initiative Dave Wiederrich put it in a blog post on the project: "Make no mistake. Calling this a game IS NOT trivializing the important work taking place inside this 'game' wrapper" (2019). Participants used the gamification elements to motivate themselves to engage in ethical practices and to position themselves as ethical subjects.

Legitimizing Play

At the same time, the scientific background and "real-world" impact of Stall Catchers are just as important as the game features in motivating participants to continue playing. In addition to the personal motivations described above that ascribe meaning to the project, participants need to know that they are contributing to "real" science, that the data is "real," and that the results could have a "real-world" impact. Learning about the developments in the science behind Stall Catchers is a key motivator for Elisabeth, who has been participating in Stall Catchers since 2016: "It is important to me because feedback continues to fuel my purpose and motivation to participate. The game administrators do a good job providing this information via ongoing website updates" (May 9, 2020). This perspective is shared by many other Stall Catchers participants who contributed to my research. In the example of Foldit, for some participants like James, the

---

25   Memory Café Directory is a platform that provides resources for individuals with Alzheimer's disease or dementias and their caregivers but primarily informs about "memory cafés," spaces to meet, share information, and learn about resources for support (Memory Cafe Directory n.d.).

scientific purpose of the game also helps to legitimize their play and the time they spend playing:

> It is usual to have a rationale because in my family [...] they have the impression that I am crazy about this game, I'm busy with it every day. And I then need to have a reason, and I say, "Yes, but it is still useful!" Right? So, for example, now I say to everyone "Yes we are working on the coronavirus." [...] But it is usually to justify that [...] we do it because we enjoy it and then you have to justify it to others because it's a game [...]. Usually the kids, they are grown up now but they say "Come on, you're occupied with the video and we were not allowed more than an hour [...]." So, they laugh, and I say, "Yes, but it's useful." (Feb. 11, 2021)

Contributing to Foldit is useful and valuable because of its scientific purpose, which distinguishes the project from other games and social networks. Participants also described gaining recognition from their families and friends for their contributions to and successes with Foldit and Stall Catchers. Alyssa described her "friends and family cheering [her] on" when contributing to Stall Catchers (May 14, 2020). For David, co-authoring Foldit-related publications was rewarding and an important recognition, which he also used as proof to show that he was doing something "meaningful" when contributing to Foldit:

> [T]he family is really proud that I contributed to that and that we published those papers, that is an important reward for me, that it is recognized somehow, that there is a result and I also like it when my name is on the paper, like, okay, here is proof, I can also use it for my work, that I am doing something meaningful. Other than saying, you're playing games. (David, Mar. 4, 2021)

The scientific purpose is essential to enjoyable gameplay, making it meaningful. At the same time, it serves to legitimize participant's hours of play in front of their friends and family.

Play/Science Frictions

Despite the productive power of the play/science entanglements to create a space in which Stall Catchers and Foldit thrive, it is not an uncontested and frictionless space. Rather, different tensions emerge between play and science, which the actors involved have to deal with, try to work around, or accept. In the following, I will discuss five examples of such tensions between play and science, which were recurrent in the empirical material. While some apply to CS in general, others are specific to HC-based CS games, which help to better understand how these systems form in everyday life. These tensions can be observed at different levels, from the source code level to the discursive level. The examples are: 1) the "balancing act" between software design for scientific accuracy and efficiency versus games, 2) the goals of science versus the goals of games, 3) the uncertainty of science versus the rigidity of games, 4) the hierarchies between play and science, and 5) the different meanings of "success" for science and games.

A "Balancing Act"

At the level of the software, i.e., the source code, of Stall Catchers, the tension between play and science unfolds indirectly as a question or "balancing act," as Michelucci described it in one of our meetings in October 2022 (fieldnote Oct. 12, 2022), of optimizing for efficiency of the analysis vs. optimizing for playfulness. This balancing act needs to be reevaluated and sometimes enacted, for example, when algorithmic changes need to be made to improve system performance for an upcoming Catchathon, as in April 2021. Although, organizationally, it was routine for the Human Computation Institute to organize special events in the form of Catchathons, from the technical side, extra testing—particularly performance testing of the platform—had to be done to ensure a smooth event (see also Thanner and Vepřek 2023). The institute expected a large number of participants in the Catchathon and wanted to ensure that the platform could handle this large crowd annotating data simultaneously. Contributing to these testing efforts as part of my collaboration with the Human Computation Institute during this period allowed me to go beyond my focused code analysis and gain insight into the "balancing act" between play, scientific quality, and efficiency at the software level.

Designed as a game, Stall Catchers was implemented in a way that made the experience enjoyable for participants. This implied that the implementation would not prioritize optimizing the efficiency of the data analysis process over optimizations that made the overall system more satisfying for participants. To illustrate this, it is instructive to consider a simplified process of how data analysis in Stall Catchers could be implemented most efficiently in terms of the individual steps that need to be performed. For example, if a dataset consisted of 500 videos, they could be organized into a simple task queue from which one video at a time could be selected and presented to a participant. Once the participant has annotated the video, it could be removed from the queue so that it shrinks until no video remains to be analyzed. At that point the system's intended annotation task would be complete. This would also mean that the game would end for the participant(s) at that point, at least until a new set of data is uploaded. A cascade of many such queues could be used to gather multiple annotations for each video. However, even then, not every participant would be required to annotate every video, since only a certain number of answers are needed to calculate the final crowd answer per video.

Even though this process would meet the main requirements of Stall Catchers' core data analysis, avoid redundancies, and be relatively easy to implement, the actual data analysis process followed in Stall Catchers is implemented quite differently. This is largely because Stall Catchers participants are not supposed to experience periods where there are no videos to analyze, for example, because all the videos in the current dataset have been annotated. Instead, once a participant has answered all research videos, the video selection algorithm begins to randomly reselect videos from the current dataset. Accordingly, some videos are analyzed more often than necessary to calculate the crowd answer. The purpose of this is not analytical but to keep participants engaged and the game going. Interestingly, and although not addressed by the Human Computation Institute's team, this example of keeping participants engaged seems to be in tension with the institute's normative principle or Hippocratic oath described in the last chapter. According to this principle, humans should not be involved in a task if not necessary. But they also need to be kept busy.

However, a progress bar on the right side and at the top of the video frame indicated the actual progress of the analysis of the current dataset, so when the analysis was officially complete, participants often began asking for new data via the Stall Catchers chat. Annotating videos after the science was complete was considered less meaningful by participants. Participant Akin described these times of waiting for new datasets to arrive as annoying:

> [Whenever the participants are] all way through with this [data]set, you gonna have to wait a while to get another set loaded up, I felt a curious sense of loss [laughs] [...] and yeah a down feeling that then began feeling a little aggravated as I checked back a couple of more times and still no Stall Catchers, just a minor prickle of annoyance, but then it came back so. (May 11, 2020)

Another example of the trade-offs at the source code level that I observed was the routine for selecting the next video to be presented to a participant, along with the possibility for the participant to redeem points. The corresponding algorithm for selecting the next videos was quite complex, mainly due to its game-related features and the aim of allowing researchers to get an early look at the data trend. It had to consider, for example, how many other participants had already annotated a research video in order to generate crowd answers for individual videos in a more data-efficient manner, i.e., to avoid collecting redundant answers. The earlier crowd answers existed for videos; the earlier researchers could observe whether there was a trend toward more or fewer stalls in the data before the dataset was fully analyzed (fieldnote Oct. 12, 2022). However, they had also implemented the "redeem points" feature. Participants received a reduced number of points at the time of submitting an answer for annotating a research video, i.e., a video for which an expert answer did not yet exist. They were prompted with an automated message to "redeem later!" Once enough participants had annotated the same video, a crowd answer could be calculated. At this point, the "Redeem your points" button turned green for all participants who had annotated that particular video, allowing them to receive the actual amount of points the system allocated to them for the annotated video (the specific amount depended on whether or not their answer matched the other participants' answers). Depending on how many participants were actively analyzing data during a certain period of time, the crowd answers could sometimes be calculated within a few seconds. However, as Michelucci explained, the points redemption mechanisms were slowed down by design to increase the amount of time people spent playing Stall Catchers. If it took about half an hour to redeem, he argued, participants might be more motivated to keep playing until they could redeem their points (Michelucci, fieldnote Oct. 12, 2022). Michelucci described how they tried to be transparent about this game mechanic to the participants. At the same time, however, it was important to prioritize videos with a few annotations to give researchers an early glimpse of the data trend. These two requirements conflicted, necessitating a balancing act. As a result, the complexity of the algorithm for selecting the next video increased to accommodate the ensuing requirements and to implement a "happy medium" between "depth-first and breadth-first" searches over the dataset (Michelucci, fieldnote Oct. 12, 2022). Together, these constraints resulted in a longer runtime of the required database queries due to lower efficiency at the algorithmic level. This reduced performance was not necessarily noticeable to individual participants but could become a problem at scale, i.e., if too many participants requested new videos at the same time.

Finally, to ensure that the data quality met the scientific requirements, the crowd answer calculation took into account the individual skill level of each participant in order to weight their answers in the calculation of the crowd answers to which they contributed. As I describe in Chapter 6, the participant's skill level was continuously evaluated with so-called "calibration movies" for which the correct answer was already known. These videos were regularly included and presented to participants between research videos, with the frequency also depending on the individual skill level of the participant. The lower a participant's computed skill level, the more often they had to answer calibration videos, with the side effect of slowing down the main analysis of the current research dataset for scientific data quality. As noted above, these source code and algorithmic/system design trade-offs did not necessarily result in a worse game experience. However, they show how different requirements of play and science influence the implementation of HC-based CS projects and demand continuous balancing acts.[26] Finally, although transparency about these implementation considerations was considered important by the team, trade-offs regarding transparency about the game mechanics and crowd-answer calculations were necessary *because* Stall Catchers was designed as a game.

Goals of Science versus Goals of Game

> The way the score is set up, it encourages you to try to get every little fraction of a point that you can to get higher on the leaderboards. Whereas in practice it's better for us [the researchers behind Foldit] to just have the general shape that you can come up with and then we can optimize it on our own later. We can run those computations so the players are kind of, I wouldn't say wasting time, but they are putting a lot of time into what they call the late game, right? They are putting a lot of time into that refinement process, whereas we are more interested in the early and mid-game of them just coming up with the general shape and trying out a lot of different solutions. (Daniel, Jan. 24, 2020)

As this quote by Foldit team member Daniel illustrates, similar to Stall Catchers, the goals of the scientific research behind Foldit are not perfectly aligned with the goals of the game's mechanics. While Foldit's scoring function motivates participants to collect as many points as possible and optimize their specific protein designs to maximize the score, the scientists working with the resulting protein designs are more interested in the "general shape" and discovering a wider variety of approaches to protein design. Many of the participants I spoke with, especially frequent players like James, were well aware of this fact:

> I [know about] it, but [...] we're still competitors so .... we like the competition too [...]. [W]e remain in competition and also a motivation for many of us. It's being at the top, staying at the top also. That's why you're still playing and always playing because

---

26 Similar balancing acts could be observed regarding the data distribution of stalled and flowing vessels on the Stall Catchers platform.

when you stop playing for a month, you're out again. It's a kind of [a] reputation to keep or something like that. But yes, and we also know that the last three days [of a puzzle] maybe are no longer useful for science, but we do that just to be the first, to be well placed [...] They [the Foldit team] have tried systems [...] to force us, to make some designs. But that attempt didn't work out well. (Feb. 11, 2021)

The tension between the game's scoring and its underlying scientific value was well-known. Nevertheless, the game mechanics' affordances and their own scores remained important motivators for participants. Similarly, the design and implementation of Stall Catchers as a game affords different practices—such as the accumulation of points—than, for example, "dry" analysis tasks or experiments, where playful practices such as tinkering or even modding[27] would not be afforded in the same way.

During the period in 2020 when I interviewed the Foldit developers, they were trying to find a new approach to the game's design that would be less in conflict with the scientific value of the contributions. Team member Daniel explained:

[T]he game doesn't do a lot to motivate playfulness. In fact, sometimes the game mechanics we have in place work against playfulness. For example, we have a score system to motivate players to try to get the best score and that is still the best way that we have of telling players "this is good, this is bad," because that score is derived from actual like chemical energy formulas. So, this score is a measure of how likely is it that it would actually fold this way in nature. With the caveat that sometimes that's not true. There are certain edge cases where you can be getting a higher score, and your shape is just unrealistic for nature. And so, when the scientists are looking at player's solutions, they will glance at it and even if it's a good score they might throw it out if it is unrealistic. And so, one thing that we're doing is trying to find ways to adjust the score function to make sure that we are meeting those edge cases. (Jan. 24, 2020)

Their concerns about the scoring system related not only to the fact that the late-game score did not always accurately reflect the scientific value of a solution but also to the difference in gameplay when participants focused primarily on scoring. Such a focus negatively impacted players' "playfulness," i.e., the degree to which they would "play [...] around" (Daniel, Jan. 24, 2020) with the protein structures, as participants would think less creatively. But this out-of-the-box thinking, which bypasses the "deterministic" (Charlotte, Feb. 5, 2020) approach of computational solutions and established scientific approaches, was seen by the team as the greatest value of human contributions to Foldit: "Foldit players are very good at exploring outside of the box and exploring ideas that we wouldn't probably think about" (José, Jan. 22, 2020). According to long-time participant

---

27   Practices of modifying computer games are also termed "modding" and are generally considered "an important part of gaming culture as well as an increasingly important source of value for the games industry" (Kücklich 2005). On different practices common in gameplay that go beyond the game practices intended by design, see Carlson and Corliss (2007). I return to such practices in the examples studied below.

David, when players focus on scoring, their creativity is directed toward accumulating points rather than finding novel solutions:

> And I think the disadvantage of playing for points is that people become very creative to increase their score by consulting sources where there is an example. There are examples of this, even in puzzle comments where people say: Oh, but in PDB [Protein Data Bank] you can see that model and you can download it and [...] then you have all those distances and then you can put all those distances, you can put in the length of bands, and you can put that in a script and then you have a very good copy of the original and then you have scored high on the puzzle but that, it completely ignores the point of Foldit. (Mar. 3, 2021)

David lamented that some participants, instead of using their own creativity to come up with a protein structure, would visit the Protein Data Bank (Worldwide Protein Data Bank (wwPDB), n.d.), a database that contains 3D structural data of proteins, and copy characteristics of protein structures into Foldit to get a high score.

The goals of Foldit and Stall Catchers go beyond the objectives of their 'games' and sometimes even conflict with them. However, the games also afford practices that undermine the idea of the overall projects, practices that could be understood in terms of conventional gaming as either skillful play or perhaps even cheating. In the example of Stall Catchers, some participants adapted their behavior according to the temporal flows of the game's algorithms to maximize their rewards at maximum speed (see below). These practices challenge the possibilities of the system that are typical of gameplay and become possible in the intra-actions between participants and software, but they do not necessarily align with the designed or intended play-flows of the game. When the Stall Catchers team became aware of such new practices, their first reaction was to evaluate how these play behaviors affected the scientific accuracy and data quality to determine if such behaviors could harm the system's purpose. If the practice was found not to impact the science negatively, it was accepted and sometimes even supported by the team, as it would speed up data analysis.

Similarly, Foldit participants tried to maximize their score in accordance with the scoring algorithms in ways that were not always helpful to the scientific problem. Here, as team member Daniel described in our interview, the developers and designers tried to respond to this practice by limiting the player's options and improving the scientific precision of the software (Jan. 24, 2020). However, there remains a gap between the software's understanding of proteins, and what proteins look like in the "real lab," as put in one of the monthly Foldit newsletters (Dev Josh 2021b). In the weekly Foldit newsletter of August 27, 2021, for example, Foldit's game designer explained that while "Foldit likes it when the strands on the edges of your sheets are blue hydrophilics [...] [,] in the real lab, those edges are too floppy without some sticky oranges to pull the edges into the core of the protein" (Dev Josh 2021b).

These examples show how the goals of the game can diverge from the scientific goals, requiring additional effort from the HC-based CS teams to ensure that the scientific goals of the games are met.

Uncertainty and Unpredictability of Science versus the Rigidity of Game

Scientific processes in (biomedical) laboratories are characterized by uncertainty, unpredictability, and contingencies at various levels, from the results of experiments to failing materialities, and the life cycles of mice. Sometimes, experiments do not go as planned and must be repeated, and things often fail. This understanding of science is widely shared, not only by STS researchers (Law and Lin 2020, 1) but also by the biomedical researchers in the laboratory: "[b]ecause it is science, you [...] don't just expect it works [the] first time" (Jada, Oct. 27, 2021). In the laboratory, research was practiced around and with uncertainties and failures. The Human Computation Institute also communicated this uncertainty to participants, such as in the example of a "dreamathon" event (for another CS game run by the Institute), where the director reminded participants in a blog post about the dreamathon's results:

> Before we dig into our initial findings, let me start with the usual reminder that all research is uncertain! At this point we are just taking an initial look at how much your labels agree with the experts on the training images (the ones where you got a "correct" or "incorrect" answer). So while you read the below, please keep in mind that these results are not final, they are based only on the training images and can actually change substantially after we look at the entire dataset. (Michelucci 2019c)

Despite these efforts to communicate the uncertainty of science and to manage expectations accordingly, laboratory members described at least some degree of conflict or "disconnect" between Stall Catchers and the laboratory's research. This was primarily due to the need for Stall Catchers to be "functioning" at all times, which was perceived to be in contrast to scientific work:

> [I]t doesn't go very smoothly at times. [...] Like research in general. Yeah, I think that's also part of it. [...] cause in research in general, [...] things fail [...] more often than they work. [...] So, I think when you try to put something more rigid on top of research, like Stall Catchers—it's not rigid, but it's something that's established, functioning, working really well. And then you have it trying to support it with research, which is like, oh [laughs], so it kind of there's ... I don't know if it's a disconnect or different expectations. I think it's similar to like in the corporate world, things kind of tend to run more smoothly than in research where you're trying things that are probably not going to work. And you will constantly run into problems. So then when you're trying to build a program or a game of something that's constantly running into problems [...], it's hard for it to run as smoothly as you'd hope. (Leander, Sept. 22, 2021)

When there were no technical problems, Stall Catchers was always available, usable, and accessible, and, as mentioned above, there was always data to analyze, even if it was only an already-completed dataset. From Leander's perspective, this contrasted with the scientific research behind it, which was subject to different "expectations" and "constantly running into problems." The different temporalities of scientific research, which does not unfold in a steady or predictable rhythm, did not always align with the temporalities of the game, which is expected to function in a predictable and stable way at all times. But these temporalities were also interdependent, since long gaps in which no new datasets were available would cause some participants to lose interest in analyzing videos, or even to contribute on a smaller basis. The frustration of data gaps was also shared by team member Paul from the Human Computation Institute, who expressed: "[F]or me and for the users, it's really important that they don't waste their time so there's constant flow of data, but the scientists, they have their own stuff going on, and things get delayed and they start too late and then there used to be really huge data gaps, and that was really frustrating" (Oct. 14, 2020).

On December 16, 2020, Michelucci sent a message on an internal Slack channel, addressing an ongoing data gap within Stall Catchers: "most of our active catchers have dropped off because we completed the last dataset, and seem to be checking daily to see when the new data arrives." Data gaps changed the game's flow and temporalities. At the same time, some laboratory members, such as Leander (Sept. 22, 2021), found the requests for new data from Stall Catchers to be stressful. This was because they had to shift their focus to preparing new data for Stall Catchers, regardless of whether there was an immediate need for analysis. The laboratory's PI Schaffer described this as "an unanticipated thing for us" (Dec. 07, 2021). Stall Catchers had not always been faster in analyzing data than the laboratory could provide. In contrast, Schaffer, explained,

> early on when we first started Stall Catchers up and going, we were desperate for more throughput from Stall Catchers because we were generating data and had this huge backlog. We were generating data at a faster rate than they could analyze and had a huge backlog of data. But as the number of players has grown and as Pietro [Michelucci] has developed more sophisticated methods of agglomerating answers, the capacity has grown quite a bit. (Schaffer, Dec. 07, 2021)

When the dynamic reversed from the laboratory waiting for their data to be analyzed to Stall Catchers being too fast for the laboratory to keep up with data provision, "the cart was in front of the horse." (Schaffer, Dec. 07, 2021)

> [T]here was some sense in the lab of [...] like it's our job to get data to Stall Catchers. And I never thought about it like that. And we're honest, we would tell people [...] there isn't new data to analyze right now. So, the game is just not going right now, and Pietro [Michelucci] was always much more concerned about that. I think from a player management or participant management kind of perspective [...]. I didn't mind there being a little bit of a tension there, but I did want to shift the perspective of the lab to Stall Catchers as a very valuable tool that we use. And we treat it with respect, just like we do every other tool. But it's not like this thing that we have to feed with data. And I think we're over that now. (Schaffer, Dec. 07, 2021)

Although, according to the PI, this period had already passed by the time of our interview in late 2021, some laboratory members still expressed the pressure they experienced in my interviews with them. It becomes clear from the quote above that the laboratory and the Human Computation Institute sometimes had different priorities. For the biomedical laboratory, Stall Catchers was a "tool" for their research and not the primary focus of their efforts. In analyzing the researchers' perspective, I noticed a hierarchy between the work Stall Catchers does and the work done in the laboratory, which I will elaborate on below.

Another expectation that some researchers found stressful and felt was directed toward them was the pressure to continually generate new research questions for each dataset submitted to Stall Catchers for analysis. According to researcher Jada, this was particularly experienced at the beginning of the collaboration with Stall Catchers:

> And at that time, the pressure was also up for everyone much more that kind of each dataset would have—it's still a little bit the case, which I don't like, but I do understand that sometimes it feels like each dataset has to be kind of a new question that we are answering in a way, but this is typically not how science works. It's typically, you run the same thing again and again […] and change something and see what turns out to be the best, and that was a little bit frustrating at the beginning because you need to get players on the one hand, you need to keep them going. So, you kind of make those questions, but in reality, what you need is running the same data several times with different parameters. (Jada, Oct. 27, 2021)

To motivate participants to continue contributing to Stall Catchers and to attract new participants, Jada felt a pressure to deviate from the way "science works," which requires running experiments repeatedly with small, controlled changes in their design. It was impossible to predict when a research question would be answered and how many experiments and changes in experimental conditions would be needed to arrive at an answer. For Jada, however, the CS game context required predictable processes and progress that were at odds with their scientific practice.

Hierarchies Between Play and Science

> Because we are all members of more than one community of practice and thus of many networks, at the moment of action we draw together repertoires mixed from different worlds. Among other things, we create metaphors—bridges between those different worlds. Power is about *whose* metaphor brings worlds together, and holds them there. (Star [1991] 2015, 284, emphasis i.o.)

In both case studies, I could observe a clear hierarchy between what was considered "play" and "science." This hierarchy was introduced and represented by different actors. It is, for example, generally inscribed in the funding logic of scientific research that endeavors related to game design or improving the play-related user experience are not typically considered legitimate uses of grant money, which poses challenges for online CS game designers and developers (Miller et al. 2023). In the words of Foldit team member José during our interview:

One of the other challenges is that our financial resources are driven by academic grants. So, we don't have money to make cool new backgrounds or add a story or something like that if we can't justify it with a scientific paper. So, everything that we work on in the game has to be directly connected to some scientific advance, which makes my job very hard because that basically cuts out all of our budget for game design. There isn't a budget for making the game feel better, making the UI nicer. Because that isn't some scientific advance. (Jan. 24, 2020)

Good game design and an enjoyable player experience are important for the success of any online game, including online CS games, which presents developers with a difficult problem to solve. However, the Foldit team also exercised control over which scientific tasks could be delegated to nonprofessional volunteers and which required professional training and expertise and, thus, could not be handed over to volunteer participants. Sometimes, they were approached by motivated participants wanting to help improve the design or fix bugs, specifically to help make up for the missing resources due to the funding problems described above. However, when team member Hugo told me about such offers in our conversation, he argued that even though they appreciated "those calls" (Jan. 28, 2020),

[w]e can't always take people up on it from a perspective of security and knowing how to conduct research and all those things. We have to make sure that we only parcel out certain parts of the game to players in those contexts where the player volunteers. I don't even know within the player community if it's known that other players [...] do this sort of thing and [...] when I say other players, I think we had a few volunteers, we only ever had one person actually doing anything after some very stringent screening. But I know we continue to get on and off those sorts of volunteers. (Hugo, Jan. 28, 2020)

The tasks to be performed by untrained participants were carefully selected and distinguished from other scientific practices that required training, specialized knowledge, or even security measures (see Chapter 6).

The hierarchy between science and play in Stall Catchers manifested in the game design and in the diverging temporalities. The top priority for the biomedical researchers with respect to Stall Catchers participants was simply that the scientific requirements be met to the same degree and standard as their previously established research process. Therefore, the goals of improving player experience and ensuring accuracy in scientific research were not always aligned. For example, the Human Computation Institute and the laboratory initially had the idea of letting Stall Catchers participants "see their progress if they finished a mouse or something like that. We talked about that [it] would have been really cool if you could see, OK, mouse, Fred and [...] Molly, and then we finish those things" (Nishimura, Dec. 7, 2021). This game feature, however, could potentially interfere with the scientific requirement of a "blinded" analysis in which participants would not know which mouse certain data belonged to. The PIs were concerned that "people would figure out that there was a trend in the data. So, in the end, we ended up doing it [...] the right way, but [...] we sacrificed [...] a little bit on the user experience where the whole dataset is blinded, which is, I think, the right way to do it" (Nishimura, Dec. 7,

2021). The "right" way to design the HC-based CS game in this example diverged from the most enjoyable way for participants. Here, from the perspective of the researchers (and, in this case, also from the perspective of the developers at the Human Computation Institute), the formula "play follows science" guided the development of Stall Catchers.

Similarly, regarding the different temporalities between the scientific research behind the game and the game itself, researcher Jada explained the importance of research setting the pace (fieldnote Oct. 28, 2022). When we discussed my observation that some researchers had expressed that they sometimes felt stressed by Stall Catchers' data requests, she argued that providing Stall Catchers with data always had to be justified. You could not "kill 20 mice" just to produce data, she said. Producing data had "real consequences" (Jada, Oct. 28, 2022). Therefore, Jada considered it important to prioritize scientific goals over user experience. This hierarchy between play experience and science was shared by virtually all participants who contributed because they wanted to help scientific research or had a personal connection to Alzheimer's disease. Most of the participants interviewed also valued the game/play and science parts differently, as in the words of Caitlin: "[S]o you can make games out of it too. But the main thing is to hopefully advance the research" (Caitlin, May 5, 2020). The hierarchies between play and science described here were, thus, not simply enforced by the scientists and developers on the participants but shared by most of the actors involved. They drove the formation of HC-based CS assemblages.

"Success" Has Different Meanings for Game and Science

Finally, as a last example of friction or misalignment between science and play in HC-based CS projects, I would like to discuss the different perceptions of the project's success, using Stall Catchers as an example. As an online CS project, Stall Catchers was perceived as very successful by all members of the laboratory I spoke with, and they were all very enthusiastic about it. Researcher Jada, for example, explained to me:

> [A]s a project, I think it's extremely successful [...]. I think it has been extremely successful to engage the people [and] has been successful especially also on the human computation side, from the institute, it really gets the people in and doing all this work. And also on how this whole thing has been growing. (Oct. 10, 2021)

The success of Stall Catchers and the importance of the participants' contributions were publicly acknowledged in blog posts and at live events. Participant contributions were also measured in "lab hours" to communicate how much time the participants saved the researchers by contributing to Stall Catchers. Prior to my field research visit to Ithaca, I had the impression that the project's success story was widely and unquestionably shared by all actors involved. However, when I spoke with researchers in the laboratory who had been working with Stall Catchers and asked them about the "success" of Stall Catchers without further defining the term, they distinguished between "success" in terms of the project in general and in terms of its scientific value: "[I]n terms of popularity obviously, it's doing really well. [...] I think it has the potential to be very successful," explained Leander (Sept. 22, 2021). When I asked why he was referring to the potential for success, he clarified further:

> I mean it's obviously successful as a game, it's growing, people are interested, so there's definitely something there. On our end, but I don't know if that's so much a flaw of Stall Catchers is like—what I was talking about earlier—where we didn't have the infrastructure in place to support it being successful. So, if it's successful, we need to figure out how to do that. So, I think it's […] almost there in terms of being successful from our end. (Leander, Sept. 22, 2021)

During my research at the laboratory, the challenge from the researchers' perspective was that the data results they received from Stall Catchers were unreliable due to problems in the data pipeline. The next chapter on researcher–technology relations will discuss the pipeline and its problems in detail. For now, it should suffice to note that because of the infrastructure problems, working with Stall Catchers was, at times, even more time-consuming for the researchers than manually analyzing the data themselves.

Similar to Leander, researcher Emily explained to me during a coffee break that the project's limited success on the laboratory side was not the participants' fault but rather a problem on their end. She also said that she sometimes preferred to analyze the data herself rather than sending it to Stall Catchers because of the problems they were facing with the data pipeline (fieldnote, Sept. 15, 2021). Despite the perception of Stall Catchers not being successful in terms of advancing and supporting the research as they had hoped, Leander believed that they were "almost there" (see above) because of the infrastructuring they had been focusing on over the last few months. Jada agreed that while the scientific side "was lagging behind […] I think it got way better by now. It's just tricky, that's all I'm saying, and it took many years. But I think it's—now looking back of course—, it was all worth it and it was really good" (Oct. 27, 2021). The hope was that the infrastructure, i.e., the data pipeline, would eventually solve the problem once it was "functioning."

Despite the frustrations, the researchers agreed that Stall Catchers was a successful game. As PI Schaffer argued, the "value of Stall Catchers" (Dec. 07, 2021) could not be reduced to scientific value. By "value," he was referring to "understanding culture and process and the fact that it's people who do science rather than any particular scientific fact" (Schaffer, Dec. 07, 2021).

> I think the other value of Stall Catchers […] is not just the quality of the data analysis or the data analysis that we get, but it's the opportunity to engage people who haven't been trained as professional scientists in the act of doing science. So, I do talk about the idea that […] it is authentic science that people are engaged with. Not every experiment works. Things go wrong. We're public about that. We try to—as much as possible as you can with this sort of amorphous cloud thing—but try to make this an opportunity for people who are interested to get a peek behind the curtain at how scientific decisions are made and how priorities are set and how it is slow and why it's slow and things like that. (Schaffer, Dec. 07, 2021)

While this value was important to and supported by the PIs, I, nevertheless observed some tension, or at least an inherent imbalance, between different perceptions in the laboratory. There remained frustrations with the project for the researchers working with the Stall Catchers data, to which I return to in Chapters 6 and 7.

Closely related to this aspect of divergent meanings of "success" is the balancing act that the researchers and the team of the Human Computation Institute had to perform to communicate the value and success of Stall Catchers, as well as its scientific goals and progress. The politics of communication, or the balancing act, was to motivate and encourage participants to contribute and to translate the scientific concepts and processes to the general public while, at the same time, not overpromising results or creating too much hope for rapid development of a treatment for Alzheimer's disease. Researcher Oliver Bracko expressed his concerns about this fine line of communication in an interview with *The Scientistt Podcast*:

> [As scientists] we really have [...] to watch our language because these people, they really play because they usually have a family member that had [...] died or has Alzheimer's disease or another form of dementia and I mean we as researchers often state or overstate our results and I think we really have, I learned my lesson to have to be much more cautious because things are really slow in science and the players [...] had to understand it and so did we that we cannot say it's basically this is the path to a new drug or this is the path and it is a path, there's a possibility that is a path to a new drug, but it's slow, and it's at least ten years from now. (Scientistt 2020, 5:46–6:27)

The researchers had to be very careful not to overpromise the results of their research, Bracko said. It was for this reason that the original name of Stall Catchers' umbrella project, "WeCureAlz" (Ramanauskaite 2016), was changed to "EyesOnAlz." In one of our conversations, Michelucci explained that while the idea behind the brand name "WeCureAlz" was to convey that they were "all in it together" to work toward the goal of curing Alzheimer's disease, they still changed the name so as not to mislead participants (fieldnote, Nov. 2, 2022).

The tensions described in this chapter, from the unpredictability of science and the need for rigidity and reliability of Stall Catchers' infrastructure to the hierarchies of science and play and the associated understandings of success in terms of the game and science, exist alongside the productive entanglements of play and science discussed above, requiring researchers, participants, and developers to find ways to accept and work around these tensions through trade-offs or to align their different interests actively. An example of such alignment practices is discussed in Chapter 7. However, these entanglements between science and play also led to the emergence of new play practices, to which I turn in the following.

Adaptations and Practices Beyond Design

In analyzing HC-based CS games as multiples (Mol 2002b) and assemblages of different human–technology relations, various participant–technology practices come to the fore that go beyond the play practices intended by design. The affordances of Stall Catchers and Foldit as games, for example, and the participants' active engagement with the platforms invited practices that went beyond the task of analyzing research data (in the case of Stall Catchers) and designing proteins (in Foldit). In the following, I present several examples of such practices initiated by participants in relation to technologies. While the next chapter focuses more on the nature of the relations from a processual perspective, I focus here on the practices that emerge within and from these relations. While this order of discussion seems to suggest that these practices precede the intraverting relations, neither this nor the reverse is fully the case. Instead, these practices are fundamental parts of the participant–technology relations and vice versa, as the evolution of the relations and the practices are interdependent.

These practices span a wide range of different activities. They include adapting the projects in ways that enhance participant–technological performance and interpreting the data presented in ways that go beyond protein folding or data from the brain vasculature of mice. They also involve building new games into the existing ones. Some practices even attack the designed functionality of the projects and their purposes. All of these practices reflect the creativity in HC-based CS games. This creativity arises from the system's affordances, existing relations, and intentional actions. It also stems from the ways in which participant–technology relations shape the projects alongside designers and developers. However, they are too often overlooked in existing definitions of HC.

Ameliorating the Participant–Technological Performance

While the intended workflow for data analysis in Stall Catchers is well-aligned with the platform's goal and is followed by most participants, especially by new and occasional users, some long-term and frequent participants have developed their own practices to improve their play and speed of analysis. Some participants, for instance, as they shared in the in-game chat in September 2022, have written scripts for their browser that allow them to use hotkeys that reduce the number of clicks required to annotate videos.[28] Others started using multiple browser tabs to avoid the long load times that sometimes slowed down their play. While Foldit explicitly allowed and supported participants to write their own scripts to automate protein folding steps, some participants went further, such as Aram, who additionally created "an environment in Autohotkey that allows me to monitor multiple Foldit clients and run them in parallel" (Feb. 28, 2021). Participants also often used multiple Foldit clients to work on several puzzles at the same time or to run different recipes. "Some do multiples on a single puzzle, on multiple designs they have," explained participant Brandon (Mar. 4, 2021). The "minimal" way to play Foldit is with one client running and one UI, on which one manually designs or folds a protein,

> but you can also use a recipe, like a script, that continues working on it. You can go eat or something, it will do it automatically. But you can also open a second window [so that] one is automatized, and you are working on the other. [...] So, this can also be done with another computer. I have old computers on which I have put Linux. (James, Feb. 11, 2021)

Participant Cleo described their daily routine of starting Foldit as follows: "I open up three Foldit clients for the three current puzzles. Then for each one I set a suitable recipe from my cookbook in motion" (Apr. 22, 2021.) Some participants kept multiple clients

---

28   Gray and Suri have observed similar tactics in their study on ghost workers (2019, 12).

running in the background 24 hours a day, seven days a week (Arthur, Feb. 12, 2021). Participant David even set up his Foldit infrastructure and environment so that he could

> access them from basically anywhere in the world via a secure connection. So, they are always, basically, they are always on, or on standby, if I don't need them right away, then I can also put them in sleep mode. But I've been on vacation [...], for example, and you're in the middle of nowhere, and then you can still connect to the servers via cell phone, and you can still do puzzles. (Mar. 4, 2021)

In this way, David enhanced his play and detached it from the need to be in physical proximity to the computers on which he was running Foldit to contribute. Taken together, these play practices in Foldit provide examples of how participants improve their play or contribution in a way that still takes place within the practices inscribed by design.

Similarly, in Stall Catchers, participant Alyson discovered a new way to interact with the flow of the platform's computer algorithms by testing different key combinations at a specific moment in the data analysis, which allowed her to move more quickly to the next video to annotate. In an email to the institute, Alyson explained:

> It required my hitting the refresh button before the [Next] button had a chance to come up. My mousing is too erratic to do that repeatedly, so I switched to playing with my left little finger on the keyboard [anonymized key] button and the index finger posed over the [anonymized key] so I could refresh immediately after annotating. I was amazed. I played that way for several minutes (over 100K points worth) and it seemed significantly faster.[29]

By seizing a timely moment (Mousavi Baygi, Introna, and Hultin 2021: 431), this tactic, as understand by de Certeau ([1980] 2013), altered the flow of play and created a new path for analyzing data in Stall Catchers.[30] With tactics such as those described above, participants creatively adopt the HC-based CS projects that go beyond what the developers or designers had in mind when implementing them. This nicely demonstrates the first principle Latour defined in the context of studying science in action: "the fate of facts and machines is in the hands of later users" (2014, 131).[31]

Reading Data Differently and Playing Games in Games

Another practice I observed in the Stall Catchers example was the visual reinterpretation of the research data displayed. This practice was occasionally discussed by participants in the in-game chat. Rather than simply treating the videos as scientific data, participants sometimes interpreted them as, for instance, images of the universe, supernovas, black holes (in-game chat, Feb. 2019), caterpillars or artworks by famous painters (in-

---

29   Quote from an email exchange with the Human Computation Institute on Apr. 16, 2021, with Alyson's permission to publish.
30   I describe this example in detail in Thanner and Vepřek (2023).
31   Of course, the practices presented here probably represent only a moderate sample of participant–technology tactics that users have discovered and developed. There might be many more that did not come to my knowledge in my ethnographic research.

game chat, Oct. 2019), sparking discussions among participants in the chat as they analyzed the videos. While this particular practice largely grew out of the Stall Catcher's community, the Foldit team actually initiated a somewhat similar practice. During their "Snowflake Challenge" in 2020, they invited participants to submit snowflake-shaped protein designs, of which the developers and team then rated the most beautiful creations (joshmiller 2021).

Despite Stall Catchers' game features described above, which are designed to make the repetitive task of analyzing black-and-white video sequences more fun, some participants came up with even more ways to stay entertained. Participant Caitlin, who had been contributing to Stall Catchers from its beginning, and as I quoted above, described in our interview that, in the early days, she had "much enjoyed [the] process of climbing [her] way back up to the top" (May 5, 2020) when her skill bar had dropped. But since she had now become one of the top players, her skill bar no longer dropped low, and she missed "the fun of climbing my way back up the way I used to" (Caitlin, May 5, 2020). To make Stall Catchers more enjoyable and fun again, she started playing "leapfrog" with other participants. In this in-game game, they played Stall Catchers by taking turns passing each other on the leaderboard. "I'll get one, then [they]'ll get one, and then I'll get one, and we will do this all the way up the board, and we'll banter back and forth in the chatbox" (Alyssa, May 14, 2020), described one of the leapfrog players. Sometimes, participants tried to match someone else's exact high score on the leaderboard (Caitlin, May 5, 2020). This practice of trying to match or beat another participants score was also used by a teacher to keep the students motivated:

> They do see, when I participate, [...] They see the teacher, they see my name: If I'm above them, just above them, then they want to pass me and then so we can motivate each other a bit. If I also usually, I nowadays collect quite a lot of points because my progress bar is quite high then, but I usually stop when I am just above them so they can catch up with me because otherwise, they go like, he is already much too far above me, I can't catch up, then they are demotivated. But if I'm just above them, then they can beat the teacher like, "hey, I'm above the teacher" and all that. (Ren, May 18, 2020)

While these examples did not undermine the official task of Stall Catchers but rather motivated participants to continue contributing and completing the task, in Foldit, some participants developed in-game games that created a parallel game space to the actual Foldit task. Building on the scripting feature allowing participants to write their recipes for automating the folding of the protein, different games were implemented using repurposed Foldit controls to, for instance, change the direction in the new in-game game (Friedrich, Mar. 9, 2021). One example of such a game is "The Game of Go minigame," developed by Foldit user zo3xiaJonWeinberg, an adapted version of the famous board game Go (2021a). Like other player-developed games in Foldit,[32] the game was shared with the

---

32 See, for example, the "action realtime minigame" and recipe "rate1star output anime Minecraft13pub.lua" (zo3xiaJonWeinberg 2021b).

Foldit public so that all other participants could download the recipe and play the game in their Foldit client.

Exploring Boundaries

"People may likely push on the edges of a mimetic world as part of exploration or even in an effort to hack it," notes computer scientist and video game designer Brenda Laurel in *Computers as Theatre* (2014, 131). Laurel defines players' attempts to try to find the limits and boundaries of a game as part of every game. As the following example shows, this also applies to HC-based CS games. In addition to the playful modifications of the engagement in Stall Catchers described above, which enhance the experience and introduce a new play mode without disrupting the game, there have also been attempts to hack, spam, or troll Stall Catchers.

Up until a certain point, for example, Stall Catchers participants could change their usernames. Eventually, this feature was exploited, as Michelucci recounted in one of our conversations. A small group of participants were playing Stall Catchers "all night long [...] and then pretty soon, they were the top three. They had the top three positions on the leaderboard. And then they coordinated their behavior, and they all switched their usernames to put it to create a three-line political message at the top of the leaderboard" (Michelucci, Jan. 14, 2021). Although this did not interfere with the core task in Stall Catchers, these users repurposed the Stall Catchers platform as a political bulletin board. As a result, the team had to "apologize to the community and sort of say we didn't intend for this to happen, we're [...] doing something about it" (Michelucci, Jan. 14, 2021). Fortunately, Michelucci explained, it was not "a very harmful thing that had happened" (Jan. 14, 2021) in terms of the scientific purpose and functionality of Stall Catchers. Moreover, in order to repurpose Stall Catchers, the participants who "hacked" Stall Catchers first had to make valuable contributions to the project—valuable in the sense of data analysis. Michelucci contacted the participants, and

> the first thing I did is I thanked them for all the research contributions they had made because by getting to the top of the leaderboard, they had annotated a lot of stalls. I looked carefully to see if they were using a bot to do it, and [...] it looked like they were actually doing it themselves. So, they were making a research contribution. There was value. I looked at their sensitivity. So, it was truly a research contribution. (Jan. 14, 2021)

After this incident, the team updated the platform to prevent participants from changing their usernames in the future. However, during my time at the institute in November 2021, due to an undetected and (re)introduced bug, some students participating in Stall Catchers discovered the possibility of changing their usernames again. The following excerpt from my fieldnote, written one day after the event, shows how this new incident unfolded:

> At around 4:40 pm yesterday, I noticed some strange chat behavior in the slack channel that forwards all chat messages sent on Stall Catchers. Some users were spamming the chat. I immediately went to Stall Catchers to see what was going on—ap-

parently, some schoolchildren had taken over the chat; some had copied the usernames of the supercatchers and had included a black star into their username, which, in Stall Catchers, stands for "supercatcher." They were also insulting users, not personally but by username. Some catchers had already asked them to stop spamming as they wanted to focus on analyzing research data.
I posted a message in the chat reminding everybody of the fact that spamming was not allowed in the chat.
I sent a message in the slack channel to report bugs. [Developer Samuel] had also noticed the spamming. Pietro [Michelucci] quickly jumped in and tried to stop the spamming.
He blocked some of the users from the chat ([Samuel] had identified the user IDs). Pietro [Michelucci] also renamed some of the users who had copied existing usernames, but shortly after that, some of them managed to change their names again. This should not be possible—there must have been some recent update to the code that overwrote the feature of not being able to change the username.
Pietro [Michelucci] also messaged the chat, and after that, a few students mentioned how great the game was, etc.
However, spamming did not really stop. At one point, a student wrote in the chat that a teacher had called the students off. After a while, the conversation (if you can call it a conversation) slowed down. (Nov. 17, 2021)

After the situation had calmed down, Michelucci apologized to the chat and to some of the participants who had been personally insulted, and developer Samuel deployed a new feature to prevent users from changing their usernames. This short note from my field diary shows how participants, in this case, schoolchildren, search for the game's boundaries and adopt the platform for their own purposes, which do not always align with the purpose intended by design. Situations like the one depicted require immediate attention from the team, regardless of what they are currently working on, because the spammers' messages also disrupted other Stall Catchers participants' game flow through the chat window, expanding with each new message and, thus, affecting the project's working in general. This incident exemplifies how the HC-based CS game Stall Catchers is continuously being negotiated by different actors and human–technology relations (in this case, participant–platform and researcher–code relations) in everyday life.

Emerging Spaces in Play/Science Entanglements and Frictions

These examples of different practices presented here show how gaming the projects is part of every game, as Laurel wrote, and how participants find new ways of engaging with the platform and software. By tinkering with it or modding, they adopt the platforms differently. At the same time, new possibilities for human–technology relations emerge beyond the designers' intentions and expectations. Here, the software's affordances or object potentials and the game frame (the "con-text" in Beck's terminology [1997, 342]) invited such different practices.

However, despite participants' rejection of the classification of the project or their contributions as play, I argue that the entanglement of play and science directly influences how the sociomaterial assemblage and its human–technology relations form. "Play" and "science" here refer to attributions I encountered in the field and were used to explain why, for example, Stall Catchers ran smoothly in contrast to the scientific research conducted in the laboratory, or why something did not work, and why different perspectives did or did not come together. Different goals and motivations unfold in parallel and in tension rather than together and in unison. It is in these interferences of play and science that the space for the HC-based CS games emerges in which participant–technology relations unfold and continuously intravert. In Deleuze and Guattari's words, these productive and tense entanglements can be understood as "movements of deterritorialization and processes of reterritorialization" that are "always connected, caught up in one another" (2013, 9).

While communication, game, and design scholar Mia Consalvo writes, following cultural historian Johan Huizinga, that play "occupies a time apart from normal life" (Huizinga [1938] 2016; cited in Consalvo 2007, 6), this chapter also illustrated the importance of the very embeddedness of GWAPs in the participants' everyday lives and the personal connection to the advancement of scientific research. In the case of Stall Catchers, Alzheimer's disease has always been part of the game, which is permeated with seriousness and leads some participants to refuse to call it a game. This was discussed in the first part of the chapter, which concentrated on the co-texts of the HC-based CS games studied with a focus on Stall Catchers. As I showed, most participants contributed because of a connection to Alzheimer's disease and not, as the team had initially imagined, to fill spare time while waiting in line, for example. Alzheimer's disease formed the overarching system of meaning "with which specific representations and perspectives, evaluations, and normative orientations of technical artifacts are established" (Beck 1997, 351). In this setting, calling Stall Catchers a game was perceived by some participants as devaluing their contribution, as they felt it did not accurately represent the seriousness of their endeavor.

From these play/science interferences in Foldit and Stall Catchers, specific values and normative claims were shared by participants, creating understandings of the *right* way to play and specific ethical subject positions. Participants with a connection to Alzheimer's disease, for example, positioned themselves as *assistants* or *helpers* to the researchers (e.g., Louise, May 07, 2020; Quinn, May 07, 2020; Jeshua, May 8, 2020; Elle, May 13, 2020) rather than as *gamers*.[33] Moreover, playing these games was considered more meaningful than playing other games. Foldit participant Aram explained that he had stopped playing other games partly because of time constraints but mainly because he would not "see much sense in it anymore because there is simply no real application behind it" (Feb. 28, 2021). This perspective of doing something meaningful was also part of Stall Catchers' designed narrative and was shared by participants. Using computer hackers as an example, anthropologist Gabriella Coleman describes such processes as

---

33   However, other participants stated that the actual motivation of participants to contribute did not matter because every contribution was meaningful. Stall Catchers participant Noemi, for example, explained that "every person participating [...] taking that step to assist is important whether they are just doing it to play a game or whether they are doing it because of personal reasons or just 'hey, yo,' it's all important" (May 14, 2020).

"ethical enculturation" (2013, 124). In such enculturation, the actors involved learn about "the tacit and explicit knowledge (including technical, moral, or procedural knowledge) needed to effectively interact with other project members as well as acquiring trust, learning appropriate social behavior, and establishing best practices" (Coleman 2013, 124). These values emerged in the relations between the programmed and intended play practices based on the assumptions, values, and visions of developers, designers, and researchers, as well as the participants' particular forms of engagement that fill Stall Catchers with multiple meanings. However, as should have become clear, the HC-based CS assemblages shaped by the interferences of play and science are never neutral but fragile and contain processes of deterritorialization. Similar to the social space described by Bourdieu (1985), they are contested and permeated by power relations, different interests, and, as I have shown in this chapter, different logics of play and science inscribed in them.

In the next chapter, I will shift my focus to concrete human–technology relations in HC-based CS projects. With the example of participant–software and researcher–technology relations, I analyze how they unfold in the setting described here and how they evolve and change, or intravert, over time.

6 Intraversions: Human–Technology Relations in Flux

HC systems, just like any sociotechnical system in everyday life, and despite the linear visions behind them and driving them, are constantly in a state of becoming, that is to say, they are "ontogenetic in nature" (Kitchin 2016, 18). Drawing on insights from his collaborative work with human geographer Martin Dodge (2011), geographer Rob Kitchin aptly describes this state of becoming for algorithms that are "teased into being: edited, revised, deleted and restarted, shared with others, passing through multiple iterations stretched out over time and space" (2016, 18). To some extent, they are "always uncertain, provisional and messy fragile accomplishments" (Gillespie 2014; Neyland 2015; both cited in Kitchin 2016, 18). In the previous chapter, I discussed Stall Catchers' (and Foldit's) multiple meanings, focusing on the participant's perspective with respect to the imaginations inscribed or intended by design, which materialize in and are often articulated through different practices and forms of engagement with and along sociotechnical entanglements. These engagements are both constrained and enabled by nonhuman actors, which afford certain practices and open up possibilities for action, thus, shaping the assemblages collectively with human actors. Together, the intentions of various actors, such as developers and participants, and the human–technology relations, which are sometimes aligned, and sometimes in tension, bring HC-based CS systems into being in continuous negotiations, leading to these systems never being closed or completed.

While this observation may apply to many sociotechnical systems, the intriguing point about HC systems is that they are not intended to be complete or finished in the first place. Instead, designers and developers understand them to be transitory. As such, HC systems do not simply implement known concepts and realizations of human–technology relations but experiment with ideas and new ways of combining humans and technology.

While HC researchers seem to agree that human–AI partnerships or other combinations are the future of AI research, the question of *how* humans, human intelligence, or creativity should be included in computational systems remains at the heart of current research. Humans are often viewed as "assistants" for AI systems (Kamar 2016a, 4071), and, in this sense, the role of humans or human intelligence is already defined in advance. Alternatively, in the field of AI in general, computers or software are often considered assistants to humans. Apple Inc.'s virtual "intelligent assistant" Siri (Apple Inc., n.d.) is one such example. However, as I argue below, this understanding of either the human or the computer as an assistant to the other does not sufficiently take into account how hybrid systems are constantly undergoing intraversions on their trajectories. Such a dynamic understanding of roles and relations was also proposed recently by computer scientist Zeynep Akata and colleagues in their "Research Agenda for Hybrid Intelligence" (2020):

> In HI settings, artificial and human agents work together in complex environments. Such environments are seldom static: team composition and tasks can change, interpersonal relations evolve, preferences can shift, and external conditions (for example, available resources and environment) can vary over time. Thus, competences cannot be fixed before deployment, and agents will have to adapt and learn during operation. As such, the ability of HI systems to adapt or learn is a prerequisite not only to perform well but to function at all. (Akata et al. 2020, 22)

As such hybrid systems tackle currently unsolvable (AI) problems, they cannot rely solely on existing systems but must create new human–technology relations. This means, for example, new combinations and interwoven systems of humans and software in which humans can take over certain computational tasks, ranging from simple classification to complex problem-solving that requires human-centered notions of creativity. From the previous chapter and related work on the construction of user representations by innovators and developers (Akrich 1995) and the co-construction of users and technology (Oudshoorn and Pinch 2005), we know that sociotechnical systems do not rely only on design and implementation. Rather, they rely just as much on practices of adoption and meaning-making by users themselves as they engage in and with the designed environment. Hence, human–technology relations in sociotechnical systems typically evolve continuously based on design choices and implementation details informed by user feedback and adoption practices that may go beyond the intended design.

Both examples, Stall Catchers and Foldit, can be understood as laboratories for human–technology relations. At the Human Computation Institute, HC-based CS projects function as laboratories for new human–technology relations in pursuit of particular visions, as discussed in Chapter 4, of how these relations should be for the "betterment of society" (Human Computation Institute, n.d.) in the future. For Foldit, the primary goal of combining humans and computers is to solve complex biomedical problems related to protein structure prediction and design. While the protein structure prediction problem has seen enormous progress in the last few years, especially with the development of the AlphaFold AI system, which I will return to below, there is currently no canonical solution to the protein design problem. Foldit can, thus, also be understood as a laboratory for finding new ways to solve these problems by combining humans and technology in novel ways. New computational developments are continuously being introduced into Foldit to "help players." In response to my question about the role of AI in Foldit (in early 2020), Foldit team member Gidon explained:

> [W]e always try to see it as humans and computers working together, and so certainly, I mean, if you consider AI to basically be more generally kind of algorithms, they have always been a part of Foldit like from the very beginning. [...]. So, I think

we are always trying to integrate the latest advances that we can get into the game to help players. And so […] it's sort of a collaboration, I think, between the human players and the AI optimization algorithms that are built into the game. And those are always advancing. (Jan. 31, 2020)

Human–technology relations in Stall Catchers and Foldit should, therefore, be seen as always preliminary and in constant flux, in a "dance" with each other (Pickering 2010 cited in Lange, Lenglet, and Seyfert 2019, 600), not only in their everyday appearance but also by design. A genealogical analysis of human–technology relations in Stall Catchers and Foldit reveals how these phenomena are shaped not only by their everyday becoming but also by the continuous pursuit of pushing the systems toward a goal, an abstract idea of ideal human–technology relations that has yet to be materialized. In the examples studied, this pursuit consists, for example, of the introduction of new tools and features, the attunement or restriction of different algorithmic flows (Mousavi Baygi, Introna, and Hultin 2021), and the opening up of new action spaces within the systems.

In this chapter, I apply the concept of intraversions to the analysis of the becoming and continuous changing of selected human–technology relations. The historical and processual study of human–technology relations takes into account instantaneous and gradual temporal developments and focuses on examples of participant–technology relations, followed by researcher–technology relations in the second part. The investigation of researcher–technology relations sheds light on other human–technology relations that are part of HC systems. As described in Chapter 4, these relations, although not often explicitly mentioned by HC advocates and designers, usually remain in the background. Of course, these relations also consist of participant–researcher–technology, developer–technology–participant relations, and other configurations and actors. For this chapter, however, I focus on these selected examples to trace the almost circular forward movements and shifts happening within these relations. Nevertheless, I include other actors who intervene and engage with the relations discussed. I argue that the ongoing changes are not merely incremental improvements to the systems and their (scientific) results but often represent intraversions of human–technology relations within the systems themselves. Inspired by Hutchins' work on distributed cognition, I show how these changes continuously transform and intravert the subject/object positions, the practices, and the nature of the tasks to be performed in these sociotechnical relations. Before concluding the chapter with a summary of key arguments, I use the example of ARTigo to discuss the dynamic interactions between human–technology relations within HC systems and the advances in AI that become visible with the concept of intraversions.

The motivation for this chapter also relates to how STS scholars Wiebe Bijker and John Law have described their interest and concern with technology: technologies or, in my case, human–technology relations, *"might have been otherwise"* (1992, 3, emphasis i.o.). Finding answers to why sociotechnical systems became what they are involves asking questions that concern all actors involved: the design and implementation decisions and their narratives, how human–technology relations actually unfold in the everyday, but also how participants, for example, "reshape their technologies" (Bijker and Law 1992, 3). Answers to such questions help us understand how they continue to evolve, since changes in human–technology relations always create new potentials and entanglements while, at the same time, constraining what is possible. In this sense, relations partly stabilize over a certain time. However, they always remain open to new intraversions and tweakings according to the tensions between existing human–technology relations, everyday becoming, and future-oriented visions of HI. In the following, I first turn to participant–technology relations in Foldit.

Never Obsolete: Intraversions of Participant-Technology Relations in Foldit

The story of Foldit, like that of Stall Catchers, goes back to a scientific problem that could not be solved satisfactorily according to scientific quality standards. In the late 1990s, the Baker Lab of the Institute for Protein Design at the University of Washington sought an automated solution to get closer to solving the difficulty of protein structure prediction by developing the Rosetta program (Zimmer 2017). Rosetta computes the probability of interactions between segments of amino acid chains. This process is based on energy levels, such that the structure with the least energy is the most likely to fold (Gonzalez 2007).

In the beginning, the program relied on randomly selecting and altering protein segments, trying many different protein structures until it found the one with the lowest energy. If a move resulted in a lower energy level, it was accepted, and the program continued to modify the protein structure until it found a more optimal arrangement. This brute force approach required a lot of computing power to be successful. The researchers had about 400 computers on which to run their calculations (Kim in Gonzalez 2007, 3:30). However, it became clear that to make real progress with the protein structure prediction problem, they would have to drastically increase their computing power, building on thousands of computers. Biochemist, computational biologist, and director of the Baker Lab David Baker explained in a 2006 interview with the *Team Picard Distributed Computing* team, a community of distributed computing participants: "[b]ut there was simply no way to scale up our in house computing facilities significantly" (Baker 2006). Inspired by VDC projects such as Folding@home, David Kim, a project scientist in the Department of Biochemistry at the University of Washington, started to modify and adapt the Rosetta program to connect to the BOINC (Berkeley Open Infrastructure for Network Computing) distributed computing platform (University of California, n.d.). BOINC allows individuals to donate their idle computing power to scientific research projects that require large amounts of computing power to complete their calculations. Rosetta@home, the distributed version of Rosetta, was launched in 2005 (Zimmer 2017). Kim explained that "now with BOINC, we have thousands of computers that we could run our jobs on located all around the globe, and it's really exciting to see how it developed" (in Gonzalez 2007, 3:33–3:42). Researchers gain access to the power of supercomputing to tackle their computationally intensive research by distributing the computational problem and inviting volunteers to provide their computing power in VDC (Holohan 2013, 28).

Baker described Rosetta@home in an explanatory video about his laboratory's protein folding research as follows:

> What we're doing with Rosetta@home is analogous to searching the surface of a large rocky planet for the lowest elevation point [...]. Imagine you have a team of human explorers working with you, and they're all exploring around the planet. If the team is small, it's quite likely that no explorer will actually find the lowest elevation point, in particular, if there are a lot of tall mountains that lead to explorers getting trapped in pretty good places on the planet. Now, instead, imagine that you have a very large team of explorers, and they each parachute down randomly on the surface of the planet and then start searching for the lowest elevation point. The more explorers you have, the more likely it is that at least one of them will find the lowest elevation point on the planet. Now, on Rosetta@home [we're] instead searching the energy landscape for a protein, trying to find the lowest energy structure for an amino acid sequence. The more computers there are doing these searches, the more likely it is that somebody will actually find it. (Baker in Gonzalez 2007, 3:43–4:35)

People could contribute to solving the scientific problem of protein structure prediction by downloading and installing the Rosetta software. Unlike the later Foldit project, participants would not interactively fold proteins themselves, or even personally invest any of their free time to contribute. Instead, Rosetta@home would simply run in the background, using the computational power of the participant's computer when it was idle. If they wanted, participants could observe Rosetta's moves through a screensaver that came with the software. Baker described this approach as a collaborative effort between professional scientists and the public that would also change the relationship between them:

> Because it's a whole new step forward in the relationship between scientists and the public. To solve the problem of protein structure prediction, it's quite clear that it's really not possible without the contributions of, of people from all over the world [...], like yourselves because it's such a big computing problem that there, it just cannot be done with any in-house resources. So we can only do it collaboratively as a collaboration between us and you and through this collaboration we can solve the problem which I really think couldn't be solved otherwise. (Baker in Gonzalez 2007, 6:00–6:32)

Baker's emphasis on the potential of Rosetta@home to change the relationship between scientists and the public must be considered in the context of how science was perceived by the public when VDC emerged as a new phenomenon "after a time when science's relationship with the public was at a low ebb. In the mid-1990s, Carl Sagan observed that the general public's attitude toward science was increasingly one of alienation and even hostility" (Holohan 2013, 27). Against this backdrop, VDC presented a promising approach to bridging this gap. Distributed computing—especially with the BOINC architecture—made computationally heavy science projects, which had previously been restricted to a selected group of professional researchers, accessible to a wider range of people (Holohan 2013, 27–28). Despite the enthusiasm for VDC and the growing number of projects in this space, Holohan declared in 2013 that "VDC is still in its infancy and the democratization of science enabled by the fewer barriers to knowledge production that the Internet offers has not yet crystallized into institutional paths" (Holohan 2013,

28). In addition, the user's contributions in the distributed computing setting remain somewhat passive. To consider the assignment of volunteers' roles in Rosetta@home, the active role of human participants—beyond downloading and installing the software—is reduced to providing computational power, to the very act of stepping away from the computer so that the idle cycles can be used for Rosetta@home. In this way, the following observation by Mackenzie, who refers to a similar distributed computing project called THINK created by researchers at the University of Oxford to analyze the interactions of specific molecules and proteins in cancer research (2006, 187–188), also applies to Rosetta@home's participants. "The relatively anonymous membership has no claim or control over the research. Their mental or intellectual effort has been deliberately figured out of the software, and their computers' execution of the 'virtual screening' processes largely disappears into a calculative background." (Mackenzie 2006, 188)

The observatory or passive role of participants makes Rosetta@home very different from Foldit, which, in fact, originated from Rosetta@home. This development may seem somewhat surprising at first. How did a project like Foldit, in which participants are the primary entity actively engaging with protein structures, evolve from a project that is all about harnessing scalable (machine) computational power? How was the decision-power in the individual protein folding steps transferred from the randomly proceeding program to human participants?

Foldit's Legend

The shared narrative (or "legend," as one Foldit team member described it) is that it grew out of Rosetta@Home participants actively requesting to play a different role in solving the protein structure prediction problem. Computer scientists and one of Foldit's creators, Adrien Treuille, described in an interview with Neil Savage for the *Communications of the ACM* magazine that participants "started noticing they could guess whether the computer was getting closer to or farther from the answer by watching the graphics" (Treuille in Savage 2012). Foldit developer Daniel further explained in our interview in early 2020 that participants observed what the program was doing, identified its move choices as questionable, and, thus, began to voice their feedback to the team (Jan. 24, 2020). Foldit researcher José described the feedback from volunteers in our conversation:

> [L]ots of them really liked this research, and [...] there were the screensavers that came with [the Rosetta@Home software] so they could watch what the computer was doing [...]. And there were a couple of requests to be able to interact with the computer. So that kind of, I think the legend goes, that blossomed into an idea to make an actual game where, instead of the computer running things idly, you could actually sit down at the computer, and you could try to direct how the simulation progressed. Well, not the simulation but how the computation progressed. (José, Jan. 22, 2020)

According to this "legend," the motivation of the participants eventually led to the development of a new CS game. "[T]hat's how Foldit was born," José recalled (Jan. 22, 2020). The collaboration between the Department of Biochemistry and the Center for Game Science at the University of Washington began working on the new project in 2007, and it was publicly launched in 2008. Gidon, who has been part of Foldit since its early days, described that when they started to develop Foldit from Rosetta's software,

> [they were] trying to [...] get some kind of human reasoning involved in [...] what was Rosetta@home's sort of purely computational kind of directed random approach to folding proteins, and maybe if we had people get involved and help direct some of that search essentially, then that might come out with something different or something better or something, like a different way of searching through this space of proteins than just the kind of computational approach. (Gidon, Jan. 31, 2020)

The hope, as Gidon described it, was to come up with better protein structure solutions and new ways of exploring the space of protein structures than was possible with the purely computational approach of Rosetta@home.

It should be noted that the reference to "human reasoning" in Gidon's quote echoes the imagination of the human in HC systems described earlier (Chapter 4), which reduces humans to thinking about something in a logical way and, hence, to their cognitive processes.

In the beginning, Foldit's development, driven mainly by a core team of three developers, was highly exploratory and included many experiments and iterations[1] because, as Gidon stated, the team could not simply build upon well-established processes to create such a CS project: "we worked on it for about a year trying lots of different things cause we didn't really know what was gonna work and what people were gonna be good at" (Jan. 31, 2020).

However, even after its official launch, Foldit was not complete or finished, but remained a laboratory for CS games and human–technology relations. Since its initial design, the Foldit team has continuously been improving and updating the design and features of the software so that the Foldit of today includes many more new means of interaction between participants and software than when it was launched in 2008. Even though the Foldit team has driven these main developments, in the following, I will show how all different human and nonhuman actors are involved in this process. The existing participant–technology relations have most prominently formed and influenced Foldit and its becoming in crucial ways.

In its earliest moments, Foldit consisted of a basic UI and controls for the participants to interact with the protein. These controls already included "automated algorithms where the player [could] let the computer take over and figure out some of the details" (Gidon, Jan. 31, 2020) and, over time, Foldit developers added more and more such tools to make the interactions more user-friendly and allow for more ways to manipulate proteins. These tools, as the team member with the username Zoran explained in a Foldit blog post, were designed according to "the way players tend to manipulate proteins, and according to the way expert biochemists would like to alter the configuration" (2009). The tools were, thus, presented to participants as an offer to facilitate protein

---

1   To test Foldit, they invited some of the Rosetta@home participants via the project's forums (Gidon, Jan. 31, 2020).

manipulation. Only a year after the launch, it was again participants who requested the introduction of automation tools that could help them in their approach: "players began requesting the addition of automation tools so that they could more easily carry out their strategies." (Cooper et al. 2011, 2) With the introduction of an editor and aforementioned *recipes*, effectively computer scripts for controlling Foldit itself, participants were given the opportunity not only to use the tools provided but also to create their own tools by writing and sharing recipes (Cooper et al. 2011, 2). The team explained in the announcement text that

> [i]n the spirit of allowing you to shape the course of scientific research, we've been planning to do something much more powerful: allow you to design, share, refine, discuss and rank new tools by combining the low level building blocks into more complicated operations through a simple visual interface. (Zoran 2009)

The goal of this new feature was that anyone, even participants with no programming experience, could build their own tools. Besides facilitating game play and puzzle solving for participants, the creation of tools or scripts also served to discover new approaches that would improve the computational performance of protein folding (Zoran 2009):

> It was also our intention to infer optimal strategies from the Foldit players and use them to improve fully automatic approaches. Rather than performing machine learning on gameplay traces of Foldit players, we decided that the players themselves would likely be much better at systematic abstraction of their strategies. (Cooper et al. 2011, 2)

Although Foldit's developers would continue to introduce new tools, participants could now build their own tools, automate the steps of their choice, and were, thus, equipped with new "powerful means of managing [...] complexity" (Cooper et al. 2011, 1).

First, so-called GUI (Graphical User Interface) recipes, now also referred to as "the original type of Foldit recipe" (Foldit Wiki 2017b), could be created. Participants could choose from existing Foldit tools, such as "Wiggle" and "Shake," and create new combinations and sequences of these tools in a "simple block-based visual programming interface" (Cooper et al. 2011, 2) without actually having to write code in the dedicated editor (see Figure 6).

While GUI recipes were relatively easy to create, they had drawbacks compared to text-based high-level programming languages, lacking several features often found in these programming languages. For example, it was not possible to create loops of specific steps, i.e., to repeat instructions until a condition was met, or to define code paths depending on conditional choices (Foldit Wiki 2017b). The GUI recipes also had a technical limitation that made it impossible to run them in the background when the client was minimized (Foldit Wiki 2017b), which, given the practice of many players to use multiple clients and run recipes 24/7 when not actively playing Foldit, could make their use quite cumbersome. They were discontinued in 2021. However, just a few months after the introduction of the GUI recipes and even before their end, the "Foldit Lua 1" interface was added, again at the request of participants who wanted to take advantage of the full potential of scripting languages and move beyond the limitations of GUI recipes (Cooper et al. 2011, 2). This interface allowed the creation of "script recipes" using the Lua programming language (Cooper et al. 2011, 2). The introduction of a "full" programming language enabled the creation of more advanced scripts with loops, if-then-else logic, and the ability to define and call functions. To this day, improving the editor and adding new capabilities and features remains a focus of the Foldit team. This also led to the introduction of a new editor, the "Lua v2" interface, which is the current version.

Figure 6: Screenshot of a Foldit GUI recipe

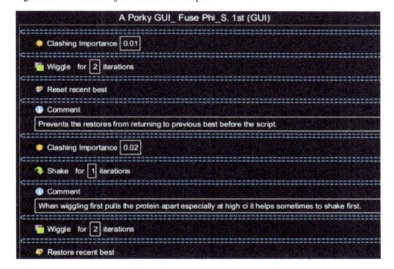

Source: LociOiling 2017

Participants in Foldit do not necessarily write recipes only for their own use, instead they often share them with other participants. In fact, many participants do not create their own recipes from scratch but refine existing recipes, which they can download and add to their "cookbook" (Cooper et al. 2011, 3). In these cookbooks, each participant can manage their collection of recipes in the Foldit client. Participant Lucas, for example, explained to me during our interview that he had an extensive collection of recipes in his cookbook. However, he argued, "I have a very short list that I use all the time and … I just tend to use the same ones over and over. And there are people who cycle through a much larger collection of recipes, and they can beat me; they can get more points" (Mar. 17, 2021). Participants can search for and navigate the recipe catalog on Foldit's website to explore new recipes (Foldit Wiki 2017b). This web-based catalog also allows participants to rate the recipes they use to assess their usefulness better.

With the introduction of the possibility to write, run, and share recipes, human–technology relations within Foldit intraverted from automated algorithms/tools as ready-made utensils used by Foldit participants to self-made means. Furthermore, with the introduction of the recipe editor, participants became developers[2] and tool creators, and even though participants could continue to play as before, their core activities within Foldit—folding and designing proteins— now extended beyond manually manipulating protein structures to programming scripts that, as participant Salma explained in her written response to my questionnaire, "automate actions that would get very tedious to do manually" (May 8, 2021).

Most participants in Foldit today rely heavily on recipes in their gameplay or are "carried" by them, as participant Sylke pointed out during our conversation in early 2021: "The algorithmic tools carry me; they are the basis of all my solutions. [...] I use recipes all the time" (Feb. 27, 2021).[3] With this development toward automating manual steps, recipes reintroduced what participant Friedrich described as "staring at what the algorithm does" (Mar. 9, 2021). Participants estimated that recipes would "work" on the puzzles about 90 percent of the time and that only ten percent of the time would consist of so-called "handfolding" or selecting and starting new recipes. In a paper on the use of recipes in Foldit that was written in collaboration with Foldit participants, Ponti et al. (2018) even calculated that, in the case of expert Foldit users, only two percent of the total time spent working on proteins can be attributed to human contributions, while computers contribute 98 percent. It is worth noting, though, that to become an expert in Foldit, human participants have to invest a lot of time not included in this calculation. Similarly, even if recipes require minimal editing, at a certain point, many hours have been invested in writing them, usually by different participants. James, who has been participating in Foldit for many years, described in our conversation that he had even written a recipe to automate the entire design of the protein (and not just the steps to fold it) (Feb. 11, 2021).

"Staring at what the algorithm does" is very similar to the pre-Foldit times when participants could only run the Rosetta@Home software on their computer and watch the computational steps via the screensaver. However, despite this similarity, there is a significant difference between observing Rosetta@Home and running scripts in Foldit. Individual participants in Foldit ultimately control the process and the individual steps of protein folding. Team member Daniel emphasized the player's choice of recipes as opposed to Rosetta@Home: "[L]ate game is running automated recipes and scripts to further refine and optimize. Which, at that point, is almost resembling Rosetta@Home in

---

2   "Developers" here refers to the development of scripts for automating steps in the Foldit game and not to the development of Foldit itself. This distinguishes participants as developers from the Foldit team developers.

3   At the same time, some participants still prefer manipulating proteins manually and advocate strongly for "hand folding." Brandon argued, for example, in his written response to my questions that he no longer uses recipes because they would still include bugs and have to be restarted upon interruption: "Early on (in my Foldit experience) I would use them a LOT, and found them to be seriously helpful to my work. The last 6–8 months, I've actually all but ceased their use, becoming entirely a 'hand folder.' This primarily came to be due to a bug that was resulting in Foldit randomly crashing, which is not conducive to a Recipe (algorithmic tool) producing results. Primarily because if they are interrupted, they have to start over (there is no 'resume'-like feature), but also because Foldit hasn't the ability to start itself back up and automatically restart what it was doing. (It's very much like a DVD player and watching a movie, where you're half-way through it, and the power goes out. You can approximately get back to where you were, but it won't resume back to the split-second that it left off before the power outage)" (Mar. 4, 2021).

terms of the computer doing a lot of the work. But the player is choosing what scripts and what algorithms and computations to run." (Jan. 22, 2020)

"Late game" refers to one of three stages that are distinguished in Foldit. "[E]arly game is getting the basic structure, the basic shape. Mid-game might be like moving some things around and refining that" (Jan. 22, 2020), Daniel explained. In the late game, the main puzzle-solving task that the human participant has to perform is what I call "orchestrating" recipes. It consists of selecting, starting, and stopping recipes, practices based on knowledge of which recipes are best applied at what time, with what parameters, and for how long they should run in each case. David, for example, a participant with many years of experience, described his approach as follows:

> [I]f I have to make a model, then I first make a rough draft, I then set it up on those servers and then let the servers run scripts to further develop it, and after a few hours, I look again: How's it going? Is it good? Is it not good? Do I need to run another script, [...] and that's how it goes all day. And in the evening, I prepare things to run overnight. So those PCs also run 24 hours a day, at least some of them. And for important puzzles that I consider important, such as corona puzzles, or in this case, the grip-binder, I also use more machines to do larger processing. (Mar. 4, 2021)

While David manually creates the first drafts of the protein, scripts or recipes then take over to refine those drafts. But David still stays in the loop by "directing the algorithm," as Foldit team member Gidon explained. "So, the player can sort of do whatever they want, including starting up some kind of algorithm and then stopping it basically. So, [...] I would say for the most part the player is sort of in control of the algorithms" (Jan. 31, 2020). As becomes clear, playing Foldit using recipes involves more than just running a script. Instead, it includes monitoring the process of each recipe and continuously analyzing and evaluating its progress and performance. It is crucial to understand the scripts and know which recipes to use and when to use them. Participants who contributed to my research agreed that "[s]electing which tool to apply in which situation is the art of it" (Salma, May 8, 2021) and requires "human creativity" (Aram, Feb. 28, 2021):

> The tools only help little if the initial design is not good. And here, human creativity is needed. Almost everything depends on it, and the tools are then only used for "finalizing" and refining. On the other hand, you will hardly get a top-score design if you do not use any additional tools. So here, there is a significant mutual dependence. (Aram, Feb. 28, 2021)

The relations between the players and the software had, thus, shifted or intraverted again. They now resembled a symbiosis; neither manual handfolding nor pure execution of recipes alone lead to a successful protein design. Rather, it is precisely the symbiosis, as described by player Aika (Apr. 10, 2021), that leads to solving a protein puzzle which can be tested in the lab. David, therefore, described the relations between humans and algorithms as "mutually enhancing" (Mar. 4, 2021). With this intraversion, instead of one subject controlling or merely using algorithmic or human input, algorithmic tools and participants played equal roles and were partners in solving a puzzle.

Some participants even took this symbiosis a step further by increasing their computing power with additional servers, running multiple clients to parallelize different approaches to the same puzzle, or working on different puzzles at the same time. Arthur, for example, described "actively" playing around for about two to three hours in the evening, but "the five clients, of course, automatically run 24/7 a week in the background with different programs" (Feb. 12, 2021). David's additional servers, which he acquired specifically for Foldit, were always ready for a Foldit job (Mar. 4, 2021). Multiple servers and clients allowed participants to go beyond working on one puzzle at a time. Participant–technology relations are extended over a larger network of computers and clients that exceed previous relations.

In this way, human participants and computational entities, in fact, become indistinguishable when viewed from the outside. For example, if recipes are running while the human participant is doing something else, the corresponding user in Foldit will be recognized as active in the game statistics.

In addition to participants who actively seek to expand and amplify the human–technology relations of which they are part, some prefer to fold proteins manually, using the algorithmic tools simply as *tools*. Foldit's task is divided into the manual manipulation of proteins and the writing and orchestration of recipes, thus delineating two main practices for approaching proteins. Interview partners commonly distinguished between a biomedical approach to solving the puzzles, which relies on a biomedical understanding of proteins, their elements, and how they fold, and a computational approach, which, instead, relies more on writing and orchestrating scripts and optimizing toward points. Not all Foldit participants follow one approach exclusively. Some have adopted a combination of both. Moreover, since the participants interviewed for my research were highly engaged in Foldit compared to the majority of registered users, it is reasonable to assume that others probably adopted other approaches, such as an "intuitive" method of folding. Or, as I did in my first attempts to fold proteins, they may primarily rely on trial and error, experimenting with the tools and scripts provided without adhering to a specific strategy. Although the two approaches described, the biomedical and the computational, are very different, they can both lead to successful protein designs. Participant Lucas described these different approaches as different options while clearly identifying himself with the biomedical approach:

> [S]ome people, their interaction is primarily choosing what script to run and being familiar with how the scripts behave and which is the one they want to use next. And they might drive that entirely based on their score. My interaction is, I'm interacting with the protein, and I'm trying to shape the protein based on what I believe it should be like, and I'm using the tools only as tools to do that. (Mar. 17, 2021)

It was important to Lucas that the tools remain *tools* that facilitate their manual handfolding so that he can interact with the protein rather than with the scripts. Similarly, when I asked what tools they would use most often, Brandon explained:

> My brain! Then my arm/hand and my mouse... :) But in terms of Foldit's internal tools... Wiggle, Shake, Bands.... Honestly, as a 'hand folder'—someone who spends

the majority of their time on a puzzle doing things by hand instead of using Recipes—it's probably easier for me to list what I don't often use! (Mar. 4, 2021)

Not only were their brain and hands in an extension of the mouse the first "tools" they mentioned, but Brandon also distinguished between "internal tools," referring to those tools that Foldit developers had implemented from the beginning, and recipes, which they generally did not use often. They also considered themselves to be a "hand folder," someone who mostly folds the protein with manual moves and only used certain recipes, which Brandon considered to be "'utility' tools" (Brandon, Mar. 4, 2021).

Following the distinctions between hand folders vs. participants relying heavily on recipes or biomedically vs. computationally oriented participants, the introduction of the ability to create their own tools with recipes created another distinction among participants: tool makers vs. tool users. While tool makers were typically tool users, the latter built on recipes provided by tool makers. Since recipes can be complex programs, their exact functionality was not always accessible to all tool users: "I'm a recipe user and I'm not a recipe maker. So [...] as much as the description would convey to me. [...] [T]here comes a time when [I] [...] say all right, this is a black box, but I know it works" (Aika, Apr. 10, 2021). Therefore, participant–technology relations also differ depending on the individual player's approach to solving puzzles and their practices. The role played by participants, simple algorithmic tools such as "Shake," and recipes within the relation varies: While in some cases the subject/object positions seem to be clearly separated, for example, between participants *using* algorithmic tools, in other cases, recipes and participants are equal partners forming a symbiosis.

Along Foldit's continuous development, various intraversions of its participant–technology relations can be seen in its movement from fully automated algorithms designed to solve the protein prediction problem to a hybrid system in which computational tools assist humans in manually manipulating proteins, followed by a shift on the participants' part to rely on automated "tools" (recipes) to manipulate the protein better, to today's sprawling interplay of participant-developed algorithmic recipes and individualized styles of applying the former to achieve ever more performant gameplay. Ultimately, these emerging relations even lay the foundation for entirely new human–AI relations to form beyond Foldit's initial scope in the form of the highly advanced AI system AlphaFold.

Introducing the Artificial Intelligence System AlphaFold in Foldit

DeepMind, a British-American company and Alphabet subsidiary, introduced the AlphaFold2 model in 2020, which significantly outperformed the previous state of the art. AlphaFold2 is an AI system for predicting protein folding, which far exceeded expectations at the *Critical Assessment of Structure Prediction* (CASP) conference, a scientific competition in which participating groups attempt to either experimentally assess or compute structures for specific amino acid sequences (University of California, Davis, n.d.). "This will change medicine. It will change research. It will change bioengineering. It will change everything," said Andrei Lupas, an evolutionary biologist at the Max Planck Institute for Developmental Biology in Tübingen, Germany, in an interview with

Robert F. Service from *Science* (Service 2020). While DeepMind had already participated in CASP in 2018, the protein folding problem was, in fact, declared to be solved in 2020 with AlphaFold2. AlphaFold2 builds on DL and a so-called "attention algorithm," imitating how humans might approach a jigsaw puzzle (Service 2020). What is particularly interesting in the context of Foldit and its continuous intraversions is that AlphaFold2 itself was not only partially trained on data that originated from Foldit[4] but, according to the DeepMind co-founder Demis Hassabis, its approach was directly inspired by Foldit and its game-like approach:[5]

> [T]here was this game called Foldit, where some people had created a puzzle game out of proteins. Gamers played it – and what they were doing was actually trying to turn the protein into a particular shape. It turned out that through playing this game ... they actually discovered a couple of very important structures for real proteins. Firstly, it was a fascinating use of games in science – and games is another one of my interests. But secondly, it kind of suggested to me that somehow these gamers had trained their intuition and their pattern-matching capabilities so that somehow they were able to do what brute-force computer systems couldn't at the time – and actually come up with the right shapes. That made me think that AI could maybe try to mimic that intuitive capability that those gamers were demonstrating. (Hassabis 2020)

Furthermore, as indicated in this quote, the Foldit participants' "intuitive capabilities" and their development of "intuition and pattern-matching capabilities" were seen as fundamental to AI development. By mirroring the strategies used by human participants, it was considered feasible to train AI systems in protein folding.

Looking back at 2020 and summarizing the best news about Foldit from each month, Foldit member Dev Josh lists AlphaFold2's win at CASP, mentioning that Foldit was cited as an inspiration (2021a). In what can be understood as another intraversion in Foldit, an integration with AlphaFold was then introduced in Foldit itself so that participants could contribute solutions they had jointly created with algorithmic tools to the AI program. The AlphaFold feature was introduced as a button that, when clicked on, opened an interface where participants could upload their protein solutions to Foldit's server, where the AlphaFold algorithm would run and compute its prediction for the uploaded solution (Bkoep 2021a). Participant David explained the AlphaFold process after submitting his solution to the AI algorithm as follows: "[S]ee what you [AlphaFold] come up with in an hour or so. And then you'll get a result back, and you can see if it's the same shape

---

4   AlphaFold was trained on solved protein structures that were stored in the Protein Data Bank, to some of which Foldit participants have contributed.

5   AlphaFold is not the only AI system that has attracted attention and advanced protein structure prediction. Influenced by the successes of AlphaFold, the Institute for Protein Design at the University of Washington first developed the transform-retrained Rosetta (TrRosetta) algorithm for protein structure prediction (Yang et al. 2020; Baker Lab 2021) and later, an even more accurate model called RoseTTAFold, which was "[i]ntrigued by the DeepMind results" (Baek et al. 2021, 871). In 2022, Baker, along with Hassabis and DeepMind's senior staff research scientist John Jumper received the Wile Prize in Biomedical Sciences "for procedures to predict highly accurate three-dimensional structures of protein molecules from their amino-acid sequences" (Rose 2022).

you had in mind. If it's not, then apparently there's still something in there [the protein structure] that's not right" (Mar. 4, 2021). The ascribed role of the AI model is that of an assistant that should check the uploaded solutions and "help out Foldit players," as a team member with the username bkoep already anticipated in 2019 in an online chat with participants (2019). AlphaFold, thus, functioned as a control body and enhancement guide, once again transforming participant–AI relations.

Rather than making the game and the participant–AI interplay or the human in the loop redundant—a fear often expressed in response to AGI narratives, as was described in Chapter 4—this intraversion actually allowed new relations to emerge. With the possibility of using AI models for most cases of protein structure prediction, the role of human participants could move on from their original task and work on other problems.[6] As bkoep clarified in a forum post on RoseTTAfold, a model inspired by AlphaFold and developed by the Institute for Protein Design at the University of Washington (Baek et al. 2021, 871):

> These deep neural networks are very, very good at protein structure prediction. There will always be cases where the network predictions fall short or do not tell the story about a protein, and human predictions may still be useful in some of those cases. But, by and large, these neural networks seem to be the better option for raw protein structure prediction. We think that humans have more to contribute to other problems, like protein design or model-building with experimental data. (Bkoep 2021b)

This shows how the target and the very purpose of the system itself keep transforming. While Foldit started out as a protein structure prediction program, it now focused on protein design, which was even more difficult to tackle but became possible to address by the continuous developments, changes, and intraversions of the participant–technology relations described. Tracing human–technology relations in Foldit and unraveling their intraversions over time shows how the roles of individual elements of the system cannot be ascribed and fixed once and for all. Not only do the participants themselves contest meanings and their assigned roles, as described in Chapter 5, but their roles also change with intraversions due to new technological developments or seized potentials. Likewise, the roles of automated tools, scripts, and AI system are also subject to intraversions in Foldit. In fact, the intraversions discussed in Foldit's participant–technology relations describe only a small part of the ongoing evolution of these relations. In September 2022, a new version of Foldit was introduced that included new AI-based tools alongside the existing "classic" algorithmic tools. For instance, "Neural Net Mutate," which was much faster than its classic version, was now added to the older "Classic Mutate" tool (Haydon 2022, 0:43). This is another example of how intraversions continue to unfold within Foldit's human–technology relations.

I would now like to turn from intraverting human–technology relations in solving the problem of protein structure prediction to human–technology relations in Stall Catch-

---

6   Additionally, as participant Salma pointed out when explaining why she thinks humans will still be needed in the future even as AI's capabilities continue to advance, AI models like AlphaFold demand extensive computing power, whereas "[c]itizen scientists are pretty low-cost!" (May 8, 2021).

ers, where the focus is on advancing Alzheimer's disease research. Since Stall Catchers was the primary fieldsite of my research, it will be discussed in more detail regarding the overall project to be able to unravel not only participant–technology but also researcher–technology relations in the second part of this chapter.

"First Use No Humans:" Intraversions of Participant–Technology Relations in Stall Catchers

"We've always had a division of labor between machines and humans in Stall Catchers" (Michelucci and Egle [Seplute] 2020), write the director and the CS coordinator of the Human Computation Institute about an ML competition organized in collaboration with the data science crowdsourcing platform DrivenData. Although the biomedical engineering laboratory's initial attempts to automate the analysis of Alzheimer's disease research data using ML techniques failed due to a lack of AI model accuracy, which then led to the development of the CS game Stall Catchers, human–machine relations have played an important role in Stall Catchers from the very beginning. Participants' annotations, for example, were computationally reviewed before being included in the final crowd answers, and ML had been part of the data preparation steps in Stall Catchers from the beginning. Various developer–technology and researcher–technology relations had to be formed and aligned to enable the analysis of the Stall Catchers research data in the first place. The researcher–technology relations building the data pipeline create the data around and on which the participant–technology relations operate and, hence, set the stage for and frame these other relations.

In the early years of the Stall Catchers project, the ML algorithms within human–technology relations were considered "preprocessors" to facilitate the tasks to be performed by human participants. The role of human participants, by contrast, was to take over the task of annotating the presented data, which in technical terms can be understood as an image-recognition problem, of which the AI was not yet capable. Here, as in the HC imagination of the "human in the loop" discussed in Chapter 4, humans were indeed meant to be "assistants" to algorithms in the sociotechnical system. But the relations between participants and technology were mutually supportive because even though the role of AI in Stall Catchers was limited to preparing data—or rather, taking over a few steps in the very complex process of preparing data—participants would not have been able to annotate data without the preceding preparation by a network of humans, computers, and ML models. Or, to put it differently, even though computational models could not take over the analysis task, they still controlled and evaluated human input—similar to the example of CAPTCHA described in Chapter 4. The annotations of individual participants were not considered reliable because participants were imagined as nonexpert humans in the loop. To ensure the quality of the crowdsourced annotation results and to build trust in the sociotechnical system, customized algorithms were implemented in the game to evaluate individual participant's analysis skills.

The algorithmic assessment of each participant's "skill level" began with the tutorial once a participant had registered with the platform. Here, participants were presented with videos to which the correct annotation answer was already known in order to evaluate their "sensitivity" for identifying stalls. But even after the tutorial was completed and the participants were analyzing "real" research videos to which the correct annotation results were not known, the algorithmic evaluation of their skill level continued with so-called "calibration movies," which were regularly presented to participants between "real" research videos. These "calibration movies," to which the expert answers were known and which, for participants, were indistinguishable from research videos, served to continue the evaluation of an individual participant as they played Stall Catchers. The blue bar to the right of the video frame in the game interface indicates the participant's skill level, which is calculated from the number of videos they correctly classified. These calculations also consider the difficulty of a given video, that is, how hard professional researchers rated it based on the detectability of the stalls. In this way, each participant's skill level was not only tracked "backstage" by the algorithmic system but also revealed to the participant, who could monitor the *value* of their contribution.

Crucially, and building on the idea of the "wisdom of the crowd" (discussed in Chapter 4), algorithms also calculated so-called "crowd answers," the final output of Stall Catchers that was ultimately sent back to the laboratory. This was done by combining the individual answers and weighting them according to the skill level of each participant. Despite the fact that humans perform the core task of analysis, their impact on the results is still governed or curated by automated algorithms. However, the algorithmic control was not perceived negatively by participants. John, a Stall Catchers participant, described it more positively:

> So, you are the little machines doing the work and trying, and the algorithm makes sure that if you're having a bad day and you wanna beat somebody and you're just zipping through files, that's still okay. And if you're taking it serious cause you realize that's not what you should do, that's good too. And it's all gonna work out, and [...] things are okay, and we just need you to spend some time for us if you'd like. (May 7, 2020)

John described the algorithms as safeguards to ensure that participants only provide information that is beneficial and not "harmful" to the platform, and, consequently, to Alzheimer's disease research. John even referred to the participants as machines rather than the algorithms, which he anthropomorphized. John's response shows that an interesting participant–technology interplay emerges in Stall Catchers, which goes beyond the HC imagination of humans and algorithms working in computational systems as mere assistants to each other. Instead, while participants take on the analysis task, they and their annotations are continuously evaluated and rated by algorithms due to their unreliability in consistently delivering the same quality. Valuable annotations, therefore, emerge only in the interplay of participant–technology relations.

These relations also open up new spaces and potentials for entanglements that the designers had not intentionally implemented. This became possible due to the possibility of engaging with the software differently and the participants' creative practices. Following what Mackenzie describes for software, I argue that these participant–software relations unfold within the situational context and the different practices and intentions of actors (2006, 6). The example of participants bypassing the programmed participant–platform procedures described in the previous chapter shows the everyday becoming of these relations, which, despite endless efforts to design and implement them in a specific way, are always open to new and unintended formations and practices. These tactics, which Michel de Certeau describes as undermining pre- and inscribed directives ([1980] 2013), allow participants to enter into different relations with technology and contribute to the formation of Stall Catchers as a sociotechnical assemblage in the everyday.

However, participants and their practices are not the only actors directly influencing their engagements with the Stall Catchers platform. Instead, servers can fail, maintenance work can crash the platform, and expired certificates can make Stall Catchers unavailable. The latter, in particular, was a disruption I experienced several times during my fieldwork and collaboration with the institute. Most of the time, participants would send an email to the team reporting the unavailability of Stall Catchers. However, even though the problem was known, updating the expired certificates sometimes still led to complications. As discussed in Chapter 4, such disruptions required the full attention of the Stall Catchers team and kept them from working on their primary tasks or focusing on meetings (fieldnote, Aug. 19, 2021). These breakdowns were as much a part of the everyday unfolding of Stall Catchers as the careful design and implementation of human–technology relations (Jackson 2014). During one of these disruptions, the entire platform came to a halt, with participant–technology relations unable to form and analyze research videos. This example, similar to the data outages described previously (Chapter 5), shows how participant–technology relations in Stall Catchers must be actively ensured for Stall Catchers to "always [be] there," as participant Akin explained in our conversation (May 11, 2020).

Thus, these participant–technology relations—even if they did not always unfold due to subversive user practices that the team had not anticipated or disruptive certificate issues—had to be ensured at all times. At the same time, they were not even meant to be complete and remain as they were but were considered a work-in-progress and, as such, preliminary in nature. This is already indicated by the title "The machines are coming! (but the humans are staying)" (Michelucci and Egle [Seplute] 2020) of the 2020 blog post cited above. In fact, changes to the originally designed human–technology relations in Stall Catchers were an inherent part of the vision to develop hybrid thinking systems (Bowser et al. 2017). It was, thus, already part of the HC imaginations and, as I show below, considered a moral obligation for the institute that participant–technology relations would intravert.

"The Machines Are Coming:" Artificial Intelligence Bots in Stall Catchers

The Human Computation Institute follows the normative oath "First use no humans," discussed in Chapter 4, which means that if a computational solution to an analysis problem exists, it is not justifiable to ask humans to do it. While there was no computational solution to the task of annotating Stall Catchers data in 2014 when the platform's development began, this may no longer be the case. Therefore, the institute has been pursuing different efforts to explore new and modify existing human–AI relations.

During my collaboration with the institute, it, together with computer scientists and human–AI collaboration researchers Kori Inkpen and colleagues, conducted an experiment on human–AI partnerships in 2020, in which an AI system played the role of an assistant to human participants (Inkpen et al. 2023). Visualized as a robot icon in the UI, it pointed to the annotation label the AI system predicted to be correct. In this experiment, the roles across the human–technology relations were intraverted; it was no longer the human who was considered to take an assisting role but the AI system, whose recommendation the human participant was allowed but not required to follow. The study also sought to investigate under what circumstances participants would trust the AI the most. This intraversion happened in a sandbox version of Stall Catchers, i.e., an experimental environment that was closed off from the actual Stall Catchers platform. While this study presents an interesting development shaping Stall Catchers' possible future participant–technology relations, to which I return in Chapter 7, another example of intraversions in Stall Catchers did not occur in a sandboxed experimental setting but rather during a Catchathon and, thus, directly impacted the participant–technology relations.

Around the same time as the collaborative study with Microsoft Research scientists, the data science competition platform DrivenData in collaboration with the Human Computation Institute and MathWorks, launched an ML competition called "the Clog Loss Challenge" to try again to automate the data analysis problem in Stall Catchers. This time, the automation was based on hundreds of thousands of human annotations that had been collected over the past few years and could serve as training data for AI models. The task was to build "machine learning models that could classify blood vessels in 3D image stacks as stalled or flowing" (Lipstein 2020). More than 1,300 solutions were submitted by over 900 participants in this challenge (Lipstein 2020). In a blog post published during the course of the competition, the Human Computation Institute expressed its hope that ML models could be used to create new human–AI relations:

> We could then use such models to label all the 'low-hanging fruit' – the easier vessels, and save the more difficult ones for our human catchers. In this arrangement, even if the new AI systems can only reliably label 20% of the images, that's still 20% less time that volunteers would need to spend on a dataset, and a 20% speed up in the overall analysis time. (Michelucci and Egle [Seplute] 2020)

Once again, computational systems were considered effectively as preprocessors for humans, picking the "easier" videos from the datasets before the remaining videos were presented to the participants. Other ideas included building ensembles or teams of AI systems that together could achieve the accuracy required for analyzed data (just as human annotations were combined into crowd answers) or introducing standalone AI users that would play Stall Catchers alongside humans (Michelucci and Egle [Seplute] 2020).

Even though the models resulting from the Clog Loss Challenge in 2020 were significantly better than those from the first attempts in 2014, the institute's blog post after the end of the challenge states, "machines are still falling quite a bit short of our high quality requirements in Stall Catchers" (Michelucci and Egle [Seplute] 2020). However, this did not mean that the models could not be included in Stall Catchers in "useful ways to speed up [the Stall Catchers] search for an Alzheimer's treatment" (Michelucci and Egle [Seplute] 2020). The Human Computation Institute, thus, invited the winners of the challenge to work with the institute to program an AI bot to be introduced to Stall Catchers and to "play" it alongside human participants.

For the first Stall Catchers bot experiment, one of the winners joined the institute's team to build an AI bot based on their ML model. The team started building a "bot-wrapper," an API through which an ML model could interact with the Stall Catchers platform. The development took place during a time when I was actively collaborating with the institute and remotely attending many of the institute's meetings, such as the regular development meetings. The focus of one of these meetings in April 2021 was what the nonhuman actors in Stall Catchers could and should look like. The team quickly agreed that AI bots should be treated as similarly as possible to human participants, in part because the goal was not just to automate Stall Catchers but also to explore how "nonhuman agents" would participate in a "community with humans" in Stall Catchers (fieldnote Apr. 6, 2021). The approach was also deemed pragmatic because of certain dependencies in the Stall Catchers source code, such as the requirement that all players have a profile (fieldnote Apr. 6, 2021). Therefore, AI bots would be equipped with user profiles and points, and assigned a skill level. Once it was agreed that bots and humans would be treated almost equally in Stall Catchers, the discussion turned to how to introduce bots to human participants. It was not clear how participants would react to their new fellow AI players, since introducing AI bots to participate in a CS game alongside human participants would be a novelty. Hence, the goal was to "remain as neutral as possible with the bot and to hear from the Stall Catchers participants how they perceive the bot," explained Michelucci (fieldnote Apr. 6, 2021).

The first test bot named "Kaos" (Human Computation Institute 2021, 41:16), which had no actual underlying model to predict whether a vessel was flowing or stalled but would randomly select annotation answers, was introduced as a baseline. The second bot released on Stall Catchers was GAIA. Named by its creator Laura Onac, one of the winners of the ML challenge, GAIA is named after the second goddess in Greek mythology, the personification of Earth, who followed the god Chaos. Onac explained in an interview published on the Human Computation Institute's blog: "[s]ince [this bot] was the first one, we gave it the name of a Greek primordial deity—the great mother of all creation" (Vaicaityte 2021a). Additionally, GAIA could also be read as an acronym for "Gateway for Artificially Intelligent Agents," referring to the history of AI bots in Stall Catchers now starting (fieldnote Apr. 12, 2021).

In terms of Stall Catchers as a laboratory for new human–AI relations, the introduction of AI bots once again modified its participant–technology relations. And, as before, these relations did not fully unfold in the ways the team had imagined, designed, and implemented. For example, despite the team's prior discussions about the potential reactions of Stall Catchers participants to AI bots, they refrained from clearly defining the AI bot as a partner in the announcements to the participants, even though they aimed to encourage the formation of human–AI bot teams. The introduction of these artificial participants in Stall Catchers led not only to such teams but also to competitive human–AI bot relations, as I will show in the following. These would pose new challenges for the design of human–AI combinations in Stall Catchers.

GAIA's Debut

GAIA made its debut in Stall Catchers during the 24-hour Catchathon event on April 28, 2021, described in the introduction of this work. The goal of the Catchathon was to reanalyze a specific dataset that Stall Catchers participants had previously analyzed. This time, however, not only human participants would annotate the data, but GAIA would play alongside humans. The goal, as described by the Human Computation Institute in a blog post informing participants of the upcoming event, was to explore how well human participants and the AI bot would work together and to see "if we can do it just as well with our bot friend GAIA, as we did on our own!" (Egle [Seplute] 2021b).

During the final hour of the competition, when all participants were invited to a Zoom hangout with the institute, the team introduced GAIA as a new fellow participant who,[7] just like humans, was not perfect. When a participant asked if they could assume that all of the bot's answers would be correct, Michelucci responded by pointing out GAIA's human-like imperfection:

> Well, if it were a perfect learner, then we might expect that. But you know what it's like when you have, when you have many different teachers telling you different things, then you have to decide who you're supposed to believe, and this bot has had 35,000 teachers, and so somehow it has to integrate all of that—you know by 35,000 teachers I mean everyone who's ever played Stall Catchers. So, in principle, I definitely see what you mean, on the other hand, when you start to kind of, you know, how do they say that the devil's in the details, right?! (Michelucci in Human Computation Institute 2021, 15:02-25:40)

Here, while invoking one of the well-known metaphors of AI, that AI systems "learn" (and even extending it by referring to Stall Catchers participants as "teachers," humanizing GAIA as a student), this narrative simultaneously departs from common AI narratives or "myths," such as the understanding of AI systems as "coherent object[s]" (Bruun Jensen 2010, 21) that are neutral, acultural, and, therefore, infallible (Carlson and Vepřek 2022). Instead, GAIA's imperfection is acknowledged and turned into a supportive argument for the "wisdom-of-the-crowd" approach underlying HC development, according to which the combination of diverse answers from different information-processing approaches is considered particularly valuable.

Despite, or perhaps because of, the anthropomorphizing of GAIA with its imperfections, the bot's appearance on the Stall Catchers leaderboard still evoked competitive feelings, an outcome the team had not intended. This perception of GAIA as a competitor can be described as a "tacit consequence of an explicit design decision" (Forsythe [1996] 2001e, 99), namely, not actively defining or announcing the intended human–bot relations but choosing to include GAIA in the game and on the Stall Catchers leaderboards, and having it accumulate points. Once again, the play/science entanglements and how they shape the participant–technology relations in Stall Catchers become apparent. Mike

---

7   Because of its name and the reference to the goddess, GAIA was often referred to as "she" and somehow personified.

Capraro, a long-time active participant in Stall Catchers (known as a "supercatcher"), joined the live Zoom meeting during the Catchathon's final hour and described his perception of the bot: "I woke up, and that pesky bot was ahead of me by almost 2,000,000 points," he explained with a smile (in Human Computation Institute 2021, 29:19–29:24). "I'm pretty impressed it was, she was tooling along at about skill level 96 to 108, but then around 11:30, she got all the way up to 100 percent she was getting 116" (Capraro in Human Computation Institute 2021, 29:32–29:46). Participants can instantly gain between one and 116 points, with the latter being the maximum, in Stall Catchers scoring system for annotating a video. Capraro explained that he had been closely observing GAIA's gameplay over the course of the Catchathon. As mentioned in the introduction of this work, I got the impression from attending and observing the hangout myself that Michelucci's response to Capraro's experience with GAIA carried almost a bit of relief as he summarized his preliminary analysis of the bot: "GAIA is fast but not quite as skillful as [the best human participants]" (in Human Computation Institute 2021, 32:28–32:31). In the end, some human participants had even succeeded in beating GAIA, who came in fourth in the final ranking (see Figure 7).

Figure 7: Final leaderboards of the April 2021 Catchathon

	Catchers: Score			Team: Score			Catchers: Research vessels			Team: Research vessels	
1	starider	16805616	1	Tracker	16605616	1	starider	12716	1	Tracker	12716
2	caprarom★	6441304	2	I See Stalls	7708764	2	caprarom★	5025	2	I See Stalls	6071
3	christiane	3697823	3	krissi	3697823	3	Bot GAIA	3223	3	Raider Team	4621
4	Bot GAIA	3628976	4	Bots	3628976	4	christiane	2879	4	Bots	3223
5	sean4046	2295325	5	Raider Team	2525706	5	sean4046	1939	5	zion science 8L	3207
6	KarisFraMauro	1408186	6	Alz Together Now	2492427	6	Carol_aka_Mema★	1046	6	krissi	2879
7	Carol_aka_Mema★	1267460	7	zion science 8L	1983552	7	KarisFraMauro	1042	7	Alz Together Now	2300
8	Brogan	630715	8	Canada	1431975	8	Arie1234	851	8	PTS Falcons	1609
9	ababbie	583571	9	Cookie	773171	9	Sean_Ettner	819	9	UniqueMappers	1377
10	EYEWIRE.ORG	547361	10	PTS Falcons	766970	10	Zinnykal	665	10	Canada	1078

Source: ©Human Computation Institute 2021 (Egle [Seplute] 2021c)

This first experiment involving "AI participants" in Stall Catchers had produced a competitive relation between humans and AI bots. As a next step after this event, Michelucci explained in the final hangout that the team would now thoroughly analyze the data collected during the Catchathon, including the performance of all participants and the annotations of the dataset, "to see if the bot is good enough to help the research" (in Human Computation Institute 2021, 39:12–39:15). If so, GAIA would become a permanent part of Stall Catchers, though the details of GAIA's participation would still have to be worked out:

[W]e're going to do this in consultation with others, with all our Stall Catchers players and community, to make sure everyone is comfortable with it. Whether we keep the bot visible on the leaderboard or whether it becomes something that's working in the background, we need to figure out where everyone's comfort level is. But the benefit of having a bot is that, again, it can play 24/7. If it's doing a good job, we could potentially start introducing other bots! (Michelucci in Human Computation Institute 2021, 39:21–39:53)

After the Catchathon, I reached out to some of the participants via the Human Computation Institute's forum to learn about their experiences with the bot. Most of the few participants who shared their impressions agreed that it was fun to compete with GAIA and described this first encounter as "relatively non-threatening" because GAIA was slow and inaccurate enough to be beaten by the best human participants:

While her skill level was generally pretty good, she was not on par with our best catchers. That combination of speed and skill, I felt, made her a relatively non-threatening first encounter for those of us unfamiliar with bots as collaborators. I'm looking forward to seeing the next iteration, and how we can leverage AI to improve stall-catcher productivity. (Caprarom 2021)

A Stall Catchers participant with the username Christiane added that she hoped to get the chance to "work with her even more from time to time" (Christiane 2021). While Capraro had described GAIA as a "pesky bot" during the competition, Christiane could imagine working together with GAIA beyond the end of the event. Participant starider still perceived GAIA as a direct competitor but, at the same time, described the bot as an encouragement to focus even more during the Catchathon:

Gaia was [...] my biggest concern during the catchathon. The fact, she was capable of undertaking the task without having to take any rest, was the biggest advantage she had over us. That was enough pressure and encouragement to keep me going. If not for the initial hitches that prevented her from performing continuously during the event, I'm certain she would have beaten me. Did notice that her sensitivity level wasn't 100%. She should be at her best for the next event. It was fun and challenging, competing against her. (Starider 2021)

These impressions suggest that AI bots in Stall Catchers might not only accelerate the analysis of the data (if proven not to harm the scientific data quality) but also impact the human participants' own play practices. While GAIA was perceived as an annoying competitor during game-play due to its advantage of not having to take breaks, it was seen as a valuable resource in furthering the goal of Stall Catchers, which was speeding up Alzheimer's disease research in the laboratory.

Three Bots in Stall Catchers

The Human Computation Institute decided to run another bot experiment with the support of a grant to continue CS bots research, building on the experience and lessons learned from the first AI bot experiment in Stall Catchers. This second experiment was scheduled for October 2021, just a few months after the introduction of GAIA, and emerged from a lack of new research data and not necessarily from a thorough analysis of the last Catchathon's results, as Michelucci had announced in the final hour of the April event. Additionally, the institute was expecting more participants than usual on the platform in October due to a collaboration with the company Microsoft, which had chosen Stall Catchers as one of its featured events in its annual "Giving Campaign."[8] Michelucci shared his idea for another bot study with his colleagues in an internal Slack channel:

> The Cornell Lab, which provides new biomedical datasets for analysis by Stall Catchers will not have new data ready for a while, and the current dataset is fully analyzed. We also have new players coming to Stall Catchers as early as tomorrow, and I'd like to have a new dataset running. For these reasons, I would like to run a new bot study using a previously analyzed dataset. We would not run it as a challenge for a particular time period, as with the previous event, but as normal ongoing data analysis. (Michelucci, Aug. 23, 2021)

Only a few weeks remained to prepare the "bot study."[9] Starting from October 6, 2021, three bots, including GAIA, joined Stall Catchers as participants, just like their human counterparts. The other two bots, named clsc2 and ZFTurbo, were based on the other two winning ML models from the 2020 Clog Loss Challenge and were built by the model creators using the same bot wrapper as GAIA. Similar to GAIA, clsc2 and ZFTurbo were named by their creators (Vaicaityte 2021b; 2021d). This time, the bots were active not just during a special event with a preset duration but for an indefinite period of time until the hybrid human/AI bot crowd had finished annotating a particular dataset. While the first experiment with GAIA aimed to understand the impact of an AI bot on the annotated data in general, the research question of this study was to investigate "how well [different] bots can work with humans and other bots to analyze Stall Catchers data" (Vaicaityte 2021a). Since each bot was based on a different ML model, they not only had different needs in terms of hardware and configuration, for example, but they also differed in their performance. These AI bots would be considered just like other human participants, so the team did not see it as a disadvantage that each had its own shortcomings and biases. The team articulated their research interest and the advantage of having differing AI bots in Stall Catchers in an interview setting in a blog post that preceded the new bot study:

> P. [Michelucci]: […] If all the bots always gave the exact same answer for the same vessel movies in Stall Catchers, then there wouldn't be any value in having more than

---

8   During this campaign, Microsoft donated money to Stall Catchers for every hour employees contributed to Stall Catchers (Vaicaityte 2021c).

9   The process of developing the second experiment can be described as following more of an "engineering ethos," as Forsythe described such practice-oriented approaches ([1993] 2001c, 44) rather than a theoretical, thoughtfully designed approach. The little time available to prepare the study meant that there was no time to discuss further the human–AI bot relation as had been announced in the final hour of the first experiment. Instead, the focus was on developing the technical side.

one bot playing. But what we discovered is that the 50 bots created in the ClogLoss machine learning challenge are all fundamentally different in their design, how they are taught, and how they decide on their answers.

L. [Onac]: Every bot is different and uses a different machine learning algorithm, each with their own strengths and biases. [...] And even then no single bot is accurate enough so that we don't need the help of humans anymore. We are currently studying the performance of hybrid crowds, with both humans and bots. (Vaicaityte 2021a)

In this excerpt from the interview conducted by another institute member, Michelucci and Onac argue for the value of combining bots with different models and, hence, with different strengths and weaknesses. A few days after the publication of the interview, the bot study was launched on Stall Catchers. It took only a little time for AI bots, which now had a small bot icon next to their usernames, to consistently outperform the human participants on the leaderboard. Although some participants were motivating each other to beat the bots and congratulating their fellow human participants when they passed a bot on the leaderboard,[10] the bots not only had the advantage of not having to take a break but this time, they were even designed to be faster than individual humans. The goal was, in fact, for the three bots together to annotate data at roughly the same speed as all the human participants combined, which almost inevitably led to the bots topping the leaderboard. Observing the developments on the platform and leaderboards and noticing human participants' resentments toward the bots expressed in the in-game chat, the team tried to address this issue behind the scenes of the ongoing study, eventually informing participants about their motivation for enabling the bots' increased speed and inviting them to share their perspectives on the bot engagement.

However, even this explicit knowledge about the intended, programmed dominance of the AI bots did not lead all participants to accept their place on the leaderboard. Two human participants managed to regain the top two spots, possibly in part by "redeeming points," a Stall Catchers feature allowing participants to redeem accumulated points (fieldnote, Oct. 24, 2021). These points are accumulated in the following way: Participants initially receive only a few points for annotating each "research movie," for which there is neither an expert answer nor a computed crowd answer, since it is not yet clear whether their answer is actually correct. Once enough participants have annotated the video, an actual "crowd answer" is calculated. At this point, if a participant has previously annotated that video correctly, they receive additional points, which are added to their "redeem account." Participants can redeem these points when the redeem button turns green, indicating that points have been accumulated. If no points have been accumulated, the button is blue and states "No points yet. Check back!"

By employing this tactic, which, at the time, could not be performed by AI bots in the game, participants were able, at least for a short time, to pass the bots and regain their

---

10   I refrain from directly quoting participants' responses, as they sent these messages to their fellow human participants while contributing to Stall Catchers. The messages were, thus, not intended to be analyzed. I discuss this question of including chat data that was not primarily created for the purpose of my ethnographic research in Chapter 3.

lead in Stall Catchers. A participant had chosen the username ALZ_BOT_X, further blurring the programmed differences between human participants and bots. By the end of the month and near the end of the AI bots' involvement in Stall Catchers, the participant with the username starider had managed to earn more points than all the AI bots combined. This impressed the other human participants, who expressed their appreciation and pride for starider in the in-game chat and even surprised Michelucci.

Stall Catchers' participant–technology relations, as illustrated by this example of AI bots in Stall Catchers, intraverted. Initially, software prepared the videos for the participants to analyze, and participants took over the annotation task. At the same time, their competence was evaluated by computer algorithms. The latter weighted individual participant answers and combined them into the final crowd answers. Despite the team's careful design, participants came up with other practices and ways of engaging within these relations, for example, by exploring key combinations that could introduce shortcuts into the game (Chapter 5) and, therefore, changing them as well. Just as in the case study of Foldit, existing human–technology relations opened up the potential to train ML models based on the data annotated by participants over several years and to introduce AI bots into Stall Catchers, thus, transforming the participant–technology interplay. With the introduction of AI bots, even though the first two settings were experimental, the Human Computation Institute aimed to explore human/AI bot teams and how they could contribute to Alzheimer's disease research together. Participants and AI bots were treated almost equally by the system, allowing bots to earn points and climb the leaderboard. These new intraverted participant–technology relations increased the overall performance or speed of Stall Catchers, but, at the same time, they risked destabilizing it, at least slightly. Although participants appreciated the additional help of the AI bots in analyzing data to advance Alzheimer's disease research, the participant–AI bot relations were also competitive. Here, the play–science tensions described in Chapter 5 became apparent once again, as AI bots were highly appreciated for the scientific purpose of Stall Catchers. However, during gameplay itself, the introduction of AI bots as participants in Stall Catchers was perceived as unfair competition. This also illustrates the dynamic nature of participant–technology relations, which depend on context and perspective as Coleman writes for the example of hackers: "Hacker technical practices never enact a singular subject-object relation, but instead one that shifts depending on the context and activity. There are times when hackers work with computers, and in other cases they work on them" (Coleman 2013, 99). Similarly, participants sometimes worked with the AI bots toward the goal of accelerating Alzheimer's disease research, and, in other cases, they worked against and competed with them. In Don Ihde's terms, the AI bots became the "quasi-other" in Stall Catchers (1990, 107).

While the Human Computation Institute's team had deliberately designed GAIA to run on a smaller server and more slowly in the first experiment to compensate for the advantage of not having to sleep and eat, the three AI bots in the second study were programmed to work together to annotate roughly the same number of research videos as all the human participants combined. This resulted in the bots annotating much faster than individual participants. When the team noticed that participants in the second experiment perceived this as unfair competition, they responded by explaining the scientific purpose of the bots' speed to reassure them. In the end, some participants still found ways to jump ahead of the AI bots on the leaderboard, at least for a short time, by redeeming points.

This example is one moment in the continuous intraversions of participant–technology relations. After the end of the second bot study, the institute analyzed the results of the challenge to see how well different human–AI combinations had performed and which ones were most promising. This also raised the question of how, for example, decision-making power should be distributed across these sociotechnical systems in the future to facilitate human–AI bot teams and not necessarily competitors. Ideas included having separate leaderboards for bots and human participants. In the beginning, the introduction of AI bots into regular Stall Catchers play was intended and anticipated to transform the human tasks into more challenging or interesting ones, as the AI would take over the easier videos, leaving participants with the "more interesting" ones. The institute explained in a blog post providing an update on human participants and AI models in Stall Catchers: "If and when these new bot catchers join the game, regular catchers will see less boring (easy) vessels and get a higher percentage of stalls to look at – the more interesting ones! And the dataset progress bar will hopefully move twice as fast!" (Egle [Seplute] 2020b). In line with the Human Computation Institute's oath not to use humans when machines can perform a task, AI bots *should* indeed take over this part of the analysis if their accuracy meets scientifically required data quality standards. The ethics of automation are guiding Stall Catchers' developments here. This could interfere with the meanings that Stall Catchers has for some participants as a practice and means of coping with everyday life marked by Alzheimer's disease (see Chapter 5) or as a pastime, which could then be sidelined in this regard. However, rather than simply replacing human participants, the institute's researchers ultimately aimed to change the existing tasks in Stall Catchers, including adding new ones, and to develop entirely new HC systems for humans to tackle meaningful problems that AI cannot yet solve. The Human Computation Institute's blog post "Stalls, machines and humans—an update" hinted at such potential new projects: "[W]e have new projects on the horizon with very different and quite interesting data where we will need to produce a similarly huge dataset to help teach machines. These new projects will address other disease research" (Egle [Seplute] 2020b). In this way, as in the case of Foldit, not only the tasks but also the purpose of the system itself are constantly changing with the intraversions. Stall Catchers must be adapted to stay at the edge of computational AI capabilities and to continue to legitimize HC's imaginaries.

Reconfigurations of Participant–Technology Relations

Tracing the human–technology relations in Stall Catchers and Foldit over the historical development and everyday unfolding of the projects has shown how these are neither completely predetermined by the systems' designs nor static in nature but continuously evolve and transform within and along the development of HC systems. In the example of Foldit, these intraversions unfolded from purely computational attempts to tackle the protein structure prediction problem, to participants manually manipulating protein structures with the help of algorithmic tools and, eventually, to symbiotic relations between human participants and automated recipes, as well as the AI system AlphaFold. In Stall Catchers, relations intraverted from humans as assistants to the computational system, itself a "preprocessor" to humans and evaluator of their answers, to participants as teachers of ML models, to human–AI pairings working in both collaborative and competitive relations.

As discussed when introducing the intraversions concept, Hutchins' *Cognition in the Wild* analyzed naval navigation as distributed cognition, tracing how the introduction of new tools not only facilitated the cognitive processes involved in navigation but also presented users with different problems to solve which required different sets of abilities and their organization (1995a, 154). In a similar way, participants' tasks, practices, and forms of engagement within the HC-based CS systems were transformed, and their relations reconfigured not only by the addition of each new feature or automated tool but also through new potentials arising from within the human–technology relations themselves, by participants seizing timely moments (Mousavi Baygi, Introna, and Hultin 2021) and the software's affordances or object potentials (Beck 1997, 244). As such, changing practices and relations within the systems also led to renegotiations of responsibilities across the sociotechnical systems. And, as shown earlier, with these intraversions, even the purposes of the systems themselves are in continuous motion.

Tracing human–technology relations in HC, thus, reveals how these relations are in constant intraversion. These processual forward movements and shifts emerge not only from developers, designers, and future visions but also from the existing human–technology relations and practices. Participants' requests to developers and play practices aligned with the flow of algorithms, opened up new possibilities that both enable and constrain *how* human–technology relations unfold in the future.

As I argued in Chapter 4, researchers in the field of HC typically refer to *participants* when talking about human-in-the-loop computing. However, HC systems are not constituted solely by participant–technology relations, even if these are the focus of HC endeavors. Rather, developer–technology relations, the team–technology relations, or the researcher–technology relations, to name the most prominent human–technology relations besides participant–technology relations, are also an integral part of the formative relations of HC-based CS game assemblages. Like participant–technology relations, they also change and intravert as projects evolve, often in mutually reinforcing ways. In what follows, I turn to the study of intraverting researcher–technology relations in Stall Catchers, illustrating this dynamic with the example of the researchers' infrastructuring of and working with the data pipeline.

Extending the Loop: Intraversions of Researcher–Technology Relations

The biomedical engineering Schaffer–Nishimura Lab studies the role of stalls in capillaries of genetically engineered mice as part of their Alzheimer's disease research. As discussed earlier, they investigate how these stalls occur and how blood flow can be restored by exploring different treatments. The researchers use multiphoton microscopy to image the brains of living mice. By taking images of successive layers in a specific brain region and combining the individual images into so-called TIFF stacks,[11] simply referred to as image stacks, it becomes possible to manually scroll through these images layer by layer through time, and thereby through the depth of the brain, creating a somewhat fluid view of the 3D structure. This representation allows researchers to analyze blood flow and identify stalled vessels more easily. These created image stacks form what is referred to in the laboratory as "raw"[12] data, which were initially analyzed manually by researchers in the laboratory prior to the introduction of Stall Catchers (and, as shown in Chapter 7, sometimes still are today).[13]

It should be noted that from the very beginning of the imaging process and the creation of raw image data, *"the materiality of the process gets deleted"* (Law 2004, 20, emphasis i.o.). Similar to the rats Latour studied (Latour and Woolgar [1979] 1986), the following transformation steps are not manipulations of the materiality of something like rats or mice themselves but of representations produced by scientists in relation with technologies (Latour [1988] 1993; cited in Law 2004, 20). New representations of previous representations are created with each step in the data pipeline.

Figure 8: Data transformation in summary performed in the laboratory's data pipeline

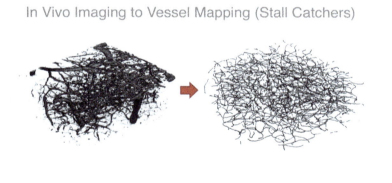

Source: ©Schaffer–Nishimura Lab 2021

---

11   TIFF stacks are single files of the Tag Image File Format (TIFF) that contain multiple raster graphics images. A raster graphics image displays a two-dimensional image as a grid of pixels.
12   "Raw data," as Bowker (2008) has aptly put it, is an oxymoron. Data, in this example Alzheimer's disease research data, are produced as part of knowledge production and, as such, "need to be imagined *as* data to exist and function as such, and the imagination of data entails an interpretive base" (Gitelman and Jackson 2013, 3, emphasis i.o.). In this chapter, however, I use the term to refer to its usage in the field, specifically the data produced during imaging. This data is subsequently processed and presented on the Stall Catchers platform, where it contributes to Alzheimer's disease research in the laboratory.
13   "Manual analysis" in the example studied already referred to human–technology relations in that researchers analyzed imaging data with the help of software.

In order for Stall Catchers to contribute to the analysis of the Schaffer–Nishimura Lab's research data, the data (representations) had to be transformed into short video sequences analyzable by participants without prior training. This required the creation of an infrastructure, more precisely, a data pipeline, through which the data follows a "chain of translation" (Latour 1999): from the form of image stacks, the data travels through the data pipeline of the laboratory where it is cleaned, normalized, and a vessel map is created (see Figure 8; fieldnote, Jul. 27, 2021).

After a laboratory member has submitted the data to the Human Computation Institute and further translation has taken place, the data is finally ingested into Stall Catchers as short video sequences with a highlighted vessel segment, where it is presented to participants.[14] Infrastructuring, to build and maintain the pipeline, therefore, played a major role for the laboratory from the very beginning of the project. Since the complex process to be achieved by this pipeline, consisting of several individual steps of data cleaning, transformation, and preparation, could not be fully automated by the programmers, biomedical researchers had to intervene at different points to complete a specific step manually, or to correct errors made by an ML-model, for example. In fact, human–technology relations play an important role in many steps of the pipeline. Hence, with the start of the Stall Catchers project, the very processes and practices of Alzheimer's disease research in the laboratory changed, as did the corresponding human–technology relations. It both unsettled existing, established practices and led to the introduction of new tasks, challenges, and ways of engaging with research data. The data pipeline was seen as a means to speed up the analysis of Alzheimer's disease research data and free up researchers' time for other tasks. As biomedical researcher and laboratory member Leander explained in one of our conversations during my first research visit in Ithaca:

> I don't know if it's just in research or this lab, but you kind of go for something, and it's not quite ready yet, but it's like, oh, let's try this. And then it's like, oh, it's working, keep going! And then it's like: But we don't have … [these] steps to support it. […] it's like things are working well, and then when you try to grow it, you don't necessarily have the infrastructure in place to actually support it. […] It was like, oh, this idea is working really well. It's a good idea. [LHV: But who's going to] yeah, who's going to do and […] do we have the stuff in place to actually make a pipeline that's not going to end up … costing more time. Because the goal is for the citizens to be helping the science, not for the scientists who feel it's like […] an extra thing that it's, […], I don't know. But it's getting there. (Sept. 22, 2021)

To get there, as I observed during my field research in 2021 and 2022, the data pipeline turned from a means to an end in itself (Vepřek 2022a). Researchers had to shift their

---

14 The data that Stall Catchers participants analyze on the platform are the preprocessed, normalized images presented as short video sequences. Other processing steps performed at the laboratory serve to create a vessel map of individual vessel segments, which ultimately facilitates the preprocessing that takes place at the Human Computation Institute. Here, on the institute's cloud servers, the orange outlines highlighting individual vessel segments are computationally drawn and the corresponding videos, focusing on such circled vessel segments, are created (step six in Figure 9).

attention away from the goal-oriented work of generating new scientific results to the "functional" work of infrastructuring. In this process, the data pipeline, which was supposed to be a means to enable and facilitate the research in the background, itself became the focus of everyday working practices. The development of the fully automated infrastructure necessary to allow Stall Catchers participants to analyze data had not been completed with the launch of the platform. Instead, when I joined the laboratory in August 2021 to learn about the laboratory's perspective on Stall Catchers, this infrastructuring, not the analysis of the crowd's output data, was the focus of the laboratory's work around to the CS game. Infrastructuring, here, was accompanied by the hope to "get there" (fieldnote Aug. 24, 2021; Leander, Sept. 22, 2021). However, the pipeline and its steps were not introduced once and for all but instead continuously developed, modified, and improved to facilitate researchers' practices. Ultimately, the goal was to achieve a fully automated pipeline. Since infrastructures must not only be built but also maintained as the design and corresponding requirements of the downstream/overarching system change, they are never complete but in an ongoing state of becoming.

Along with these developments and in infrastructuring, researcher–technology relations unfold and develop. For example, the manual tasks humans have to perform together with software are constantly changing. To show how these intraversions emerge, the next subchapter aims to walk through the individual steps of the data pipeline and its human–technology relations at the laboratory, focusing on selected moments.[15] The following descriptions are not only about human–technology relations but also about how data is created, translated, and new data are generated as representations of existing data, which is itself an imperfect representations of "real" events and information. The steps to be discussed can be summarized as follows (see Figure 9): first, the generation of research data in the laboratory in the imaging process, followed by manual and automated preprocessing of the data to prepare it for the ML model DeepVess. DeepVess then processes the data before several automated post-processing steps are performed on the ML model's result. Next, the data is curated by researchers with software tools before being sent to Stall Catchers. Finally, Stall Catchers' data annotations are analyzed in the laboratory to close the laboratory–Stall Catchers platform–laboratory loop. The ethnographic description is followed by an analytical section focusing explicitly on the intraversions observed.

In order to focus on the HC system Stall Catchers and its human–technology relations, my analysis starts from the practice of *imaging*,[16] with an ethnographic note on

---

15 It is not intended to represent a comprehensive technical overview but aims to provide in-depth, microperspectival insights into some steps which have proven to be particularly interesting for this analysis, while only roughly touching on others.

16 It is in this practice that the human–technology relations associated with the digital research data, its transformations and representations within the Stall Catchers project become visible. For this reason, some essential tasks of the Alzheimer's disease research I observed at the laboratory will be left out here, such as the preparations preceding the imaging sessions, including craniotomies, preparing materials, injecting dyes, setting up lasers, or maintaining microscopes. Similarly, the end of the project or research process could be traced to the publications in scientific journals or even beyond. However, to answer my research question, I follow the digital imaging

"getting lost in the brain" in the context of how research data are generated at the Schaffer–Nishimura Lab. As will become clear, data generation in the *in vivo* imaging process of mouse brains with multiphoton fluorescence microscopy itself unfolds in various human–technology-mice relations.

Figure 9: Simplified stages of the dataflow. Green refers to stages performed at the laboratory, light blue refers to the processing stage at the Human Computation Institute, and blue refers to the Stall Catchers platform

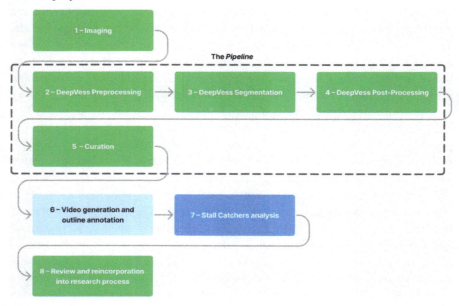

Source: ©Veprek 2023

Getting Lost in the Brain

During the first imaging session (step one in Figure 9) I observed at the laboratory in August 2021, researcher Benjamin explained that "it can be very easy to get lost in the brain." Sitting in front of two large computer screens and using the software Scanimage,[17] he was searching for a good spot in the mouse's brain to image (fieldnote, Aug. 18, 2021). A "good spot," Benjamin explained to me, was a region of the brain with many capillaries—which they analyze for stalls in their Alzheimer's disease research—and only a few larger vessels, which were not of interest for this particular analysis. "I don't know

---

data through the pipeline to the Stall Catchers platform and back, stopping at the analysis of the Stall Catchers results by the scientists in the laboratory.

17 "ScanImage is a software package for controlling multiphoton and laser scanning microscopes" (MBF Bioscience, n.d.). It can be customized toward specific research and microscope requirements.

where we are," he muttered (Benjamin, fieldnote, Aug. 18, 2021). Sometimes, it took several attempts to find such a spot in one of the brain hemispheres displayed on the right computer screen.

Meanwhile, the subject of inspection, the mouse, was out of sight behind a curtain to prevent light interference with the detection optics. Breathing regularly, as indicated by the breathing monitor on the right computer screen, the mouse lay beneath the objective, the optical component that gathers light from the mouse's brain, on a stage that Benjamin was controlling from his desk. Although out of sight, the mouse was still present, or represented, via the breathing monitor and the microscopic representations of its brain on the computer screen. Benjamin moved and adjusted the stage to a desired position with the computer program controls and the cursor traveling through the brain on the screen. On the left computer screen were two images representing the right and left hemispheres of the mouse's brain. These images had a magnification factor of four, much lower than the objective's magnification factor of 25 on the right screen. The right screen was divided into five frames, one of which was the breathing monitor. The other frames included the microscopic representation of the mouse brain, each displaying one of three different "channels." Each channel visualized different dyes used to color objects, vessels, and specific features of the brain. The remaining frame showed all channels merged together.

After a while, Benjamin found a good spot. He drew a rectangle to mark the region he wanted to image in the hemisphere on the left screen before returning to the magnified image on the right screen. Satisfied with the location, he moved the focus to the top of the brain to start the actual imaging process. After checking the settings, such as the number of slices to be imaged and adjusting the power of the laser to get a good view of the vessels, the actual data acquisition process began, imaging from the top down deeper into the brain. The laser "raster-scanned" the mouse's brain, moving in one direction, then turning, and moving back in the other direction to cover the entire region. In this process, it captured the emitted fluorescence and converted it first into electrical and then digital signals. Benjamin closely observed the process on his screen, where the completed frames incremented quickly. As the laser moved deeper, he occasionally adjusted the settings to maintain the best view and, hence, the subsequent image quality. Multitasking, he frequently checked the breathing monitor while saving the rectangle to find it later for the next imaging session with this mouse, which would take place in a few weeks. Depending on the experiment, mice were imaged several times over a certain period of time, usually several weeks. This first session would be the baseline imaging and, therefore, included finding the brain regions that would be used again in subsequent sessions.

The process ended once the laser had reached the set final depth and completed the final scan. The high-resolution images generated were stored as one image stack on the computer before Benjamin moved on to the next imaging spot. A total of six image stacks were created per mouse and imaging session. Depending on the experiment and research study, several mice were imaged—some receiving a specific treatment to be tested and others serving as control mice—in approximately three sessions, generating 72 image stacks (if four mice were imaged) per experiment. On average, a dataset sent to Stall Catchers contained around 170–190 image stacks.

Although the imaging session also included measuring blood flow and placing the mice back in their cages, here, I continue to follow the data created in human–technology–mouse relations as "raw" image stacks, the first step in the data pipeline. These 3D high-resolution images of the vasculature, including the shape and size of the blood vessels, are represented as connected voxels[18] in the digital representation and bundled into an image stack of around 500 images. In Latour's terminology, these high-resolution images are inscriptions produced by instruments and lie behind scientific texts (e.g., in Latour 1987, 64–70). They represent both time and space, or depth, and, in fact, differ in many ways from the short video sequences with highlighted vessels that are ultimately presented to Stall Catchers participants in their "virtual microscope." These raw stacks, for example, ideally include many capillaries but also other larger vessels, liposomes, or dura, making it difficult for the untrained eye to analyze individual vessels for stalls. The data then had to be cleaned, processed, and transformed in order to delegate the analysis to Stall Catchers and the crowd of participants and to create these short videos focusing on a single capillary to be analyzed; a complex process undertaken both by the laboratory, which handled the image data preparation, and by the Human Computation Institute, which then created the actual videos and calculated and drew the vessel outlines around them.

With this description of the origins of the raw research data that are ultimately analyzed in Stall Catchers in mind, I will now turn to their subsequent transformations or translations (Callon 1984; Latour 1999), which themselves involve new data generations, happening at the laboratory in the data pipeline.[19]

Processing and Translating Data

At the time of my research, the core process of the entire pipeline was centered around the ML model DeepVess, which was included to perform vessel segmentation. In the latter, individual blood vessels are isolated from tissue and other structures in the surrounding image region. Multiple pre- and post-processing steps had to be introduced because this model could not be run on arbitrary data; it performed differently depending on the quality of an input dataset. Both pre- and post-processing here refer to the data pipeline at the laboratory itself, as well as the processing that takes place before the data is sent to the Human Computation Institute. In the following, the analysis follows two scales of chronological developments, with the primary focus on the data and its transformation steps in terms of their order within the data pipeline and the secondary focus on the longer-term changes made to the data pipeline itself. While I begin with a focus on the state of the pipeline in 2021, I subsequently describe the changes that were made over the course of my research until late 2022.

---

18 Voxels can be thought of as volumetric (3D) pixels, depicting values on a grid in 3D space.

19 The term *data pipeline*, also used in the laboratory, creates the fiction of an *object through which data is sent*. While I use this term to describe the human–technology relations and data translations and generation steps, I understand the *pipeline* as an incomplete process that, along with its researcher–technology relations, is continuously changing and becoming.

The imaging data had to be preprocessed for DeepVess to perform as expected (step two in Figure 9). The raw images, for example, contain elements and parts of the brain vasculature that are not considered to contribute to the understanding of Alzheimer's disease and are, therefore, seen as "noise" in the research data. Dura mater,[20] for instance, could be visible in the images at the top of the image stack, taken from the upper layers of the brain, but DeepVess could not handle dura, and, additionally, it was not relevant to the scientists' research interest. Therefore, dura had to be manually removed from the images. In this process, researchers identified the image regions containing dura and drew the boundaries around these image regions. In this context, cleaning meant that the information in a given image region was "masked" by the researcher, i.e., overwritten with zero-value (black) pixels, thereby, making it invisible to both human vision and ML models.

Stalls only occurred in the capillaries (the small vessels). Therefore, large vessels were also not needed for the Stall Catchers analysis and would be a confounding variable in the overall analysis. For a long time, researchers had to manually mask these large vessels in the images before running them through DeepVess because the ML model's performance was negatively affected by them: "[T]he problem is not the capillaries. It does a great job with the capillaries. The problem is [...] the larger vessels," explained researcher Leander (Sept. 22, 2021) in one of our conversations. DeepVess was often confused by large vessels and, as a result, misidentified parts of them as vessels. While masking dura was quite efficient and fast, masking all individual large vessels was (and remained) quite a time-consuming task:

> [Y]ou'll have a large vessel going down. And the best way we had at that time was to go through frame by frame and mask out the vessel. And you can't just do it straight because the vessel moves, you have to [, for] every frame, move the little thing. And there's usually multiple, and it's taking me hours. (Leander, Sept. 22, 2021)

Therefore, to improve their data cleaning and preparation work, the researchers modified or improved the data pipeline by developing an automasking algorithm to take over their task.

After these initial data cleaning steps, laboratory members actively pushed the data to the next automated data preprocessing step, which referred to "basically [...] just normalization with the possibility of motion correction," as laboratory member Charles explained (fieldnote, Aug. 17, 2021). Normalization,[21] in this case, refers to adjusting the intensities of the image by mapping the values to the range between zero and one. A threshold value of 0.8 was set for a mask, meaning that any value above 0.8 is considered a logical one and included in the mask, and anything less is not included (Charles,

---

20   Dura mater describes the outermost layer of the membrane surrounding the brain and spinal cord, which provides protection for the central nervous system.
21   Following the feminist science and technology sociologist Hannah Fitsch and media studies scholar Kathrin Friedrich, who draw from literary scholar Jürgen Link's understanding of the "normal" (e.g., Link 2014), "[n]ormalization [...] describes the processual steps required to constitute normality" (Fitsch and Friedrich 2018, 3; *cf.* Bowker and Star 2008).

Sept. 14, 2021). As Charles pointed out, this process raises the question of what is considered a capillary in the laboratory's research. It was not considered perfect but "operating fairly well" (Sept. 14, 2021) regarding the researchers' goal of identifying capillaries. This step was performed using the MATLAB programming language and environment (The MathWorks, Inc., n.d.), which provides various tools for image processing. This frame-by-frame image normalization took approximately two minutes per image stack and could be parallelized to process four stacks simultaneously (fieldnotes, Jul. 27, 2021; Aug. 17, 2021).

The "possibility of motion correction" referenced above refers to the fact that mice are active actors in Alzheimer's disease research whose actions and movements do not always conform to the scientific, or more specifically, the imaging process. If a mouse moved during the imaging session, the image quality was impacted and could include smears or waves caused by movement (fieldnote, Aug. 17, 2021). These movements could be minimal, as Michelucci explained to me during one of our interview sessions: "The mice are breathing. Their heart is beating, these introduce motion artifacts into the image stacks" (Jan. 21, 2021). These motion artifacts are described in the paper detailing the DeepVess algorithm as "one of the major challenges for 3D segmentation of *in vivo* MPM [multiphoton microscopy] images" (Haft-Javaherian et al. 2019, 5, emphasis i.o.). The laboratory implemented a motion correction algorithm based on previous work on an image registration tool (Thirion 1998; Vercauteren et al. 2009; cited in Haft-Javaherian et al. 2019, 5). However, at the time of my field research, this algorithm was not necessarily applied to every dataset as it could also worsen data quality, sometimes to the extent that further data analysis was no longer feasible (fieldnote, Aug. 17, 2021). This could mean, for example, that vessel outlines would not line up correctly due to cumulative errors introduced by motion correction (fieldnote, Jul. 27, 2021). The decision to apply motion correction was, therefore, based on the researcher's assessment of how they thought DeepVess would handle a specific dataset. In these researcher–technology relations, more of the decision-making power was on the side of the researchers, who determined the algorithm's role and the course of action. At the end of the preprocessing program, a so-called H5 file was generated, a data format that stores data hierarchically as multidimensional arrays. This file constituted the input to DeepVess.

Machine Learning for Data Segmentation

Having described the preprocessing steps performed before and *for* DeepVess, I now turn to the ML model itself (step three in Figure 9). The model was developed and introduced to automate the segmentation of vessels, which is important for the later analysis of the vessels for stalls. Stall Catchers participants should be presented with the complete segment of an individual vessel to determine if it is stalled (fieldnote, Aug. 16, 2021). Prior to using DeepVess, researchers manually traced and marked each vessel in the image stacks using the open-source image processing software ImageJ (National Institute of Health, n.d.). Manual segmentation of a 3D image stack took around 20–30 hours per stack. The authors[22] state in the article introducing DeepVess that manual segmentation, thus, was not feasible since it "slows down the progress of biomedical research and constrains the use of imaging in clinical practice" (Haft-Javaherian et al. 2019, 12). Researcher Emily explained that manually tracing each vessel without the help of DeepVess would be "a colossal task" (Sept. 8, 2021). Adding up the number of images to be analyzed and vessels to be traced illustrates this problem: In each experiment, several mice, for instance three mice, had to be imaged in three phases of imaging with six image stacks per session and mouse, adding up to 18 image stacks per mouse and, in this example, 54 image stacks in total. If each image stack contains around 500 images, this would result in 27,000 images to analyze, each containing multiple vessels. Hence, DeepVess' lead developer Haft-Javaherian's "aim throughout his tenure with EyesOnALZ [the name of the former overarching project of Stall Catchers] [was] to replace himself with machine automation," states a blog post by the Human Computation Institute (Egle [Seplute] 2019).

Haft-Javaherian developed a convolutional neural network (CNN) to automate the segmentation of 3D vessels. CNNs are a form of deep neural networks, sometimes also called DL. Deep learning is a specific form of ML (Goodfellow, Bengio, and Courville 2016, 95). To give a slightly simplified account of what neural networks are, they combine different mathematical operations whose composition is described by a directed acyclic graph (Goodfellow, Bengio, and Courville 2016, 163), often simply a linear chain of multiplications and subsequent application of so-called "activation functions." The different components in the chain are also called "layers," while the "length" of the chain describes the depth of the model (Goodfellow, Bengio, and Courville 2016, 163). When there are two or more layers, the network is typically referred to as "deep." The CNNs are a particular type of "neural network for processing data that has a known grid-like topology" (Goodfellow, Bengio, and Courville 2016, 326), especially when there are "spatial invariants" in the data, i.e., "local" patterns that can occur independently of their "global" location. They are, thus, particularly successful in computer vision and image data processing (LeCun, Kavukcuoglu, and Farabet 2010; Albawi, Mohammed, and Al-Zawi 2017) and owe their name to the mathematical linear operation called "convolution," which is performed on matrices (Goodfellow, Bengio, and Courville 2016, 321; Albawi, Mohammed, and Al-Zawi 2017, 1). Mathematically speaking, the input and output of the individual layers and the overall model are essentially sets of matrices (or, generalized to higher dimensions, "tensors") or, in terms of code, simply "arrays." The outputs of intermediate layers in such a network are also called "feature maps" (LeCun, Kavukcuoglu, and Farabet 2010, 1). In the example of the 3D multiphoton microscopy imaging data, both inputs and outputs would be 3D tensors, each represented as a three-times nested array. The laboratory implemented their models using Tensorflow (Martín Abadi et al. 2015), a popular open-source software library for AI and ML applications for the Python programming language (Haft-Javaherian et al. 2019).[23]

The paper introducing DeepVess describes it as a system consisting of preprocessing steps (as described above), the actual segmentation with the model, and post-processing

---

22 The paper was authored by laboratory members and researchers of the Meinig School of Biomedical Engineering at Cornell University, Mohammad Haft-Javaherian et al.
23 The precise architecture of the laboratory's CNN is described in Haft-Javaherian et al. (2019).

steps (Haft-Javaherian et al. 2019, 3). However, during my field research at the laboratory, researchers referred to DeepVess as the ML model performing the segmentation and discussed its pre- and post-processing separately. Therefore, I follow the laboratory's usage, referring to the actual model itself as DeepVess.

DeepVess proved to achieve better accuracy computationally on the task of vessel segmentation than any of the "current state-of-the-art" (Haft-Javaherian et al. 2019, 2) computational approaches. Moreover, it took only ten minutes to calculate the segmentation of the images compared to 30 hours of manual work (Haft-Javaherian et al. 2019, 12). Once the model was trained, researchers could simply run it without thinking about DeepVess' internal steps. Laboratory member Charles explained that he had "never actually looked at [the DeepVess source code] file" (Sept. 14, 2021). When he eventually did during one of our meetings, he "realized, oh, it's only 276 lines long. It's not very long for such a program" (Charles, Sept. 14, 2021).

To actually process the data and perform the vessel segmentation, a researcher would invoke it via a simple command in a terminal. Once this process was complete, a researcher would take the resulting output file and similarly initiate the post-processing on it (fieldnote, Aug. 17, 2021). However, although this process flow was very clear and straightforward, it did not always go as designed, typically due to software errors interrupting the segmentation of individual samples. These errors were often related to the fact that DeepVess eventually had to be updated (fieldnote, Aug. 17, 2021). DeepVess had been introduced at the beginning of the Stall Catchers project and its maintenance involved, for example, updating versions of software libraries on which it depends, such as TensorFlow, from time to time. At the time of my field research, however, this maintenance was lacking and the code, or the libraries it used, were no longer up to date. Part of the problem was that none of the current laboratory members were deeply familiar with DeepVess and its codebase (fieldnote, Aug. 10, 2021) and, as researcher Leander explained to me, they had difficulties "decipher[ing]" (Sept. 22, 2021) it after the developer left. "So, it's been hard for us to […] improve it. We also don't have someone who's as knowledgeable, I think, in machine learning" (Leaner, Sept. 22, 2021). During my field research, DeepVess was sometimes referred to as a "black box" or "magic" by researcher Anna (fieldnotes, Jul. 27, 2021; Aug. 10, 2021). This is a common narrative used to convey praise for the ineffable capabilities of technology, or ML in this specific case, while simultaneously emphasizing its inscrutable and unknowable nature (e.g., Elish and boyd 2018, 63).

To get around the error messages that kept appearing, a workaround was found that allowed data to be pushed through DeepVess without interruption. Overall, Isabel summarized, DeepVess worked well "as long as there's stuff in the frame" (fieldnote Aug. 10, 2021). As it turned out, however, there was not always "stuff in the frame," and DeepVess did not always perform equally well, so the resulting segmentation was not always of the quality desired. The results could "contain some segmentation artifact such as holes inside the vessels, rough boundaries, or isolated small objects" (Haft-Javaherian et al. 2019, 8) and "[i]mages from deeper within the brain tissue […] suffer[ed] from more segmentation errors" (Haft-Javaherian et al. 2019, 14). Additional automated post-processing steps were introduced to mitigate such errors (see Haft-Javaherian et al. 2019, 8).

Automated Post-Processing

The post-processing (step four in Figure 9), as I learned during my field research at the laboratory, was significantly more complex than the preprocessing steps used to prepare the data for DeepVess (fieldnote, Aug. 17, 2021). The process, again implemented in MAT-LAB, consists of three main steps, which I briefly describe below. Like the segmentation process, post-processing is triggered manually by a laboratory member. The first step, Isabel described,

> [is] just the cleaning up [...] of the segmentation. So, what [...] we get out of DeepVess [...] got lots of noise in it. So, there's, first a cleaning app and that's just smoothing the surfaces [...]. We do that twice just to make it smoother, I guess. Fill in any single holes and then get rid of any single voxels that are left isolated. (Sept. 14, 2021)

Here, images are "re-filled" and boundaries "cleared" in order to recreate the shapes of the vessel segments as they appear, or would have appeared, in mouse brains, even though this digital representation never recovers its original but is practically a new image drawn on top of the already processed representation. Here, it becomes clearly visibly—not only in the post-processing but, in fact, from the beginning of the imaging process—that the object of analysis, in pursuing the aim of revealing more about its origin, is further alienated from its "original" form with each additional step.

In the next step, these new data representations are augmented with what the researchers call "skeletons." This means that the foreground regions of the stack are condensed into a "skeletal remnant that largely preserves the extent and connectivity of the original region while throwing away most of the original foreground pixels" (Fisher et al. 2003).[24] The resulting structures are thin, single-voxel-wide paths creating a 3D skeleton that traces the structure of the underlying vessels (fieldnote, Jul. 27, 2021). Due to the skeletonizations, the images will also include "background noise" (Charles, Sept. 14, 2021) because "any little scattered bits will end up with a little tiny spot on them or a little line, lots of little bits of mess everywhere" (Charles, Sept. 14, 2021). Therefore, additionally, Anna explained to me in our Zoom meeting as we went through the code line by line on the shared screen, walking me through the individual algorithmic steps:

> [T]o remove any background noise and non-vessels, group the voxels together into connected components, [...] that's what this [...] function means. It basically says [...] well this is for anything that is just off on its own and if it's off on its own, then just get rid of it. [...] So, it just counts the number of voxels involved in each connected component. And [...] there is a whole bunch of little things on their own and it just pulls them out. (Sept. 14, 2021)

Once the background noise has been reduced, the program shifts its focus to the actual skeleton, which has "lots of noise, there's gonna be gaps in it and stuff as well" (Anna, Sept. 14, 2021). Anna and I were now looking at an example image of DeepVess' output on

---

24   This process is applied at the level of voxels, not pixels, because the stacks are 3D structures.

the computer screen: "[y]ou can see down here it looks connected and it ends up being broken at some point" (Anna, Sept. 14, 2021). The next step, therefore, is to connect the vessels across these gaps, which may actually contain stalls (fieldnote Jul. 27, 2021), by expanding each voxel in the mask using a sphere of a specified radius. Anna explained that they "just inflate everything and that will cause things to then merge across gaps. So, if there [...] is a stall somewhere and that ends up with a break, this will end up dilating across and they'll, the spheres on either side of the stall will end up colliding and they'll join together again." (Sept. 14, 2021)

This step of connecting the inferred vessel structures was followed by another iteration of the previous steps of smoothing, filling holes, and skeletonizing again to improve the result (Anna, Sept. 14, 2021). The individual vessel segments in the new skeleton were then reviewed, focusing on their length. If the number of voxels connected in a segment was too low, the segment was removed (fieldnote Jul. 27, 2021). In other words, connected voxels are only considered valid vessel segments if they reach a minimum length.

In the next automated post-DeepVess processing step, segments that were wrongly separated but *actually* formed one vessel should be reconnected. Here, the laboratory built upon and extended an approach described in previous research (Schager and Brown 2020), and the endpoints of vessel segments formed the focus of the operation. Suppose two endpoints of different segments are within a certain distance and "point at each other" at a certain angle. In this case, they should be connected because they were probably separated by mistake. At this point, segmenting and mapping the vessels in the network becomes a mere geometric, mathematical problem to be solved. This reconstruction step was added later and not included in the initial automated post-processing. Similarly, the computer code for most of the other steps had changed from the original implementation. New versions of MATLAB, for example, were released over time, including new functions that could be used to replace or optimize those that the DeepVess developer had previously programmed from scratch. As Isabel explained, "a lot of the functionality is now built into MATLAB so ... it simplified things a great deal" (Sept. 14, 2021).

After connecting separated segments, the previously mentioned step of masking large vessels, which in the past was manually performed by researchers before sending the data through DeepVess, was now performed automatically.[25] Similar to the preceding steps, the problem is translated into a geometric problem in which the vessel's radius is compared to a "logical sphere" (Isabel, Sept. 14, 2021) of a certain size (fieldnote Jul. 27, 2021). According to the researchers' understanding, if the vessel's radius is smaller than that of the sphere, it classifies as a capillary. If, however, it is larger than the sphere's radius, it exceeds the size of capillaries and can, hence, be removed from the images by deleting its centerline/skeleton (Isabel, Sept. 14, 2021). Finally, a last round of skeletonization follows "just to make sure that there's nothing funny happening" (Charles, Sept. 14, 2021), junctions are removed, and the previous step of removing segments not

---

25 Therefore, the data were normalized regarding the masked volumes and compared to the "original image"—with everything that is to be masked replaced with a logical zero, and everything within a sphere are logical ones (Isabel, Sept. 14, 2021). Masking dura, by contrast, was still performed by researchers but at an even later stage in the data pipeline (see below).

meeting the required voxel count is repeated.[26] This yields the final skeleton, "which is the XYZ [coordinates] of all of the connected voxels for each segment and the number of objects in that is the number of segments [...] in the skeleton" (Isabel, Sept. 14, 2021), from which the output of the automated post-processing is created.

In the first years of the Stall Catchers project, this post-processing was the final step in the data transformation process before a laboratory member would take the resulting data and sent it to the Human Computation Institute. However, while DeepVess took over the overall task of manually creating a vessel map, it did not always work as intended, introducing errors that the post-processing algorithms could not smooth out or repair:

> [DeepVess] obviously fails because [...] our imaging data is very heterogeneous and sometimes we have very nice images, sometimes we have images that have a ton of noise, a ton of autofluorescence, and so DeepVess will pretend or will ... DeepVess doesn't do a very good job in excluding things that are not vessels. So, it will detect things [that] are not vessels and will say, "Oh this is a vessel," but it's not; it's like a random thing that is autofluorescing. So, and then if we don't correct that, it will go into Stall Catchers, and then the users will be like, "Oh, this is not a vessel, this is like a random thing," and then they will flag the movie. [...] [W]e can eventually see which movies are bad [= flagged], and there are a lot of bad movies. But then, at the end, because DeepVess doesn't really recognize a lot of the capillaries, [...] the result that we get back from Stall Catchers ... we get very few stalls. And we get so few that it's just not possible to work with that data. (Emily, Sept. 8, 2021)

In our conversation during my first research stay at the laboratory, Emily described not only the failings of DeepVess but also their reverberations through the Stall Catchers project and how they led to the researchers' inability to "work with that data." Stall Catchers allowed participants to "flag" videos if they could not analyze them due to ambiguous vessel segments or unclear, or "grainy" images (e.g., Eliza, May 18, 2020; Asher, May 20, 2020; Daan, May 26, 2020). However, researchers needed to know the correct number of vessel segments in an image stack to report on the impact of stalls and learn from the experiments. If DeepVess did not recognize a certain amount of the capillaries in the images, the number of detected vessels and stalls were incorrect. Therefore, despite the extensive automated post-processing put in place to correct many of the errors introduced by DeepVess, it still did not produce the required quality. While some vessel segments were drawn where they should not have been, others were separated or not detected for "convoluted reasons" (Isabel, Sept. 14, 2021; fieldnote Aug. 24, 2021). Although the post-processing steps described were "supposed to compensate for potential breaks in the vessel[s]" (Leander, Aug. 16, 2021), this did not "always work," as Leander noted:

---

26 "Junctions" or "branch points" refer to voxels with more than two neighbors that are logical ones (Charles, Sept. 14, 2021). These points get subtracted from the skeleton to receive the individual segments. By, once again, connecting the endpoints of the then isolated segments, "a long list of all of the connected voxels and [...] every single connected segment becomes its own [...] object in a structure" (Charles, Sept. 14, 2021).

> [W]hen we send [the data] to Stall Catchers, we want it to include the whole segment even if there's a stall because that's how they find whether or not there is a stall. And if it's a pretty big stall, [DeepVess] would make two vessels, which is not what you want. You want it to send the whole thing so … part of the DeepVess post-processing [is that] it will […] expand the segmentation so that it overlaps across any potential stalls and then it comes back as one segment. But that doesn't always work. (Leander, Aug. 16, 2021)

Data quality varies due to experimental conditions, the movement of mice during imaging, DeepVess, and the algorithmic processing's performance, along with the interference of other "things" in the images that float around and can be misidentified as parts of vessels by DeepVess because it does not distinguish between vessels and other bright spots (fieldnote Aug. 10, 2021). As Emily explained (see above), the quality of images differed from experiment to experiment and from imaging session to imaging session. Moreover, at the time of my field research, DeepVess had been trained several years earlier. "The neural network […] was trained on data that's older now, and things have changed a little bit," explained Leander (Sept. 22, 2021). In fact, at the time of my second research visit, the model was performing better on bad data than on data considered good quality due to the data on which it was trained (fieldnote, Oct. 25, 2022). Since then, not only has the computer code been continuously updated but so have the experimental settings, such as the microscopes, which influence the imaging data generation and, hence, the data run through DeepVess (fieldnote Oct. 18, 2022). In addition, the laboratory's PI Schaffer stated that four to five years are an "eternity in AI" and that there are already better libraries and models that could be used (fieldnote, Oct. 18, 2022). Retraining the model with current data was, therefore, discussed as an important measure but had not yet been done at the time of my research.

Together, these contingencies, changes, and data fluctuations led to varying data results at the end of the automated data processing. In the first years of the Stall Catchers project, however, the laboratory's focus had been on validating the CS approach to ensure that the crowdsourced analysis would meet the required accuracy. Only after that did they concentrate more on the data pipeline and improving the quality of the data sent to Stall Catchers. Until then, even when Stall Catchers participants accurately classified the data presented, these might have included broken or incorrectly identified vessel segments, or the image quality was sometimes not good enough to classify an outlined vessel. Following the principle of "garbage in, garbage out" (fieldnote, Oct. 22, 2022), the classified data returned to the laboratory from Stall Catchers could then also be difficult for the researchers to interpret because if DeepVess incorrectly separated vessel segments, Stall Catchers participants would sometimes separately classify videos of segments that actually belonged together (fieldnote, Aug. 10, 2021). Reflecting on the Stall Catchers developments from the laboratory's perspective, Anna saw bad data quality as the main problem from the beginning (fieldnote, Oct. 25, 2022).

"[W]e really hadn't anticipated how much the preprocessing […] was going to be necessary, […] and that was just something that we couldn't have predicted from […] the images and the data," explained PI Nishimura (Dec. 07, 2021) in our interview. In fact, at the beginning of the project using HC to speed up the analysis of the laboratory's Alzheimer's disease research data, the plan had been to create two CS projects: Stall Catchers, to analyze the vessels, and another project to identify the vessel segments. The platform for the second project could be based on the 3D puzzle game and CS project Eyewire (Seung Lab, Princeton University, n.d.), in which participants help map neurons in the brain. This second project was never realized, however, because the laboratory and Human Computation Institute decided that "the blood vessels were pretty straightforward enough that we thought we could just automate that from the beginning," Schaffer explained (Dec. 07, 2021) in our interview. The assumption had been that the vessel segmentation could be fully automated: "But, honestly, I mean, you see, we're still struggling with getting […] the final details about the automation right and it still takes quite a bit of us manually intervening […] in order for us to have confidence in the outcomes" (Schaffer, Dec. 07, 2021). As it turned out, automating the vessel segmentation with DeepVess did not just make the researchers' work easier but also introduced new tasks and problems to be solved manually.

An additional process was introduced to improve the quality of the data sent to Stall Catchers. This process was called "data curation" in which the researchers' role was to "see if DeepVess did a good job and edit what it did not do well" (Sean, fieldnote, Nov. 4, 2022). This change also brought about new researcher–technology relations. My field research in Ithaca coincided with a crucial time regarding the data pipeline, as the researchers were in the midst of implementing this new data curation process. To have a solid, smoothly functioning infrastructure to rely on, starting in early 2021, they had been focusing on improving the data pipeline by introducing new steps and tools to facilitate the researchers' data preparation tasks and improve the quality of the data sent to the crowd. They were testing and improving a new customized editing tool for researchers to review the automated processing results (fieldnote, Nov. 4, 2022). The new editing tool "to clean up [DeepVess'] results" (fieldnote, Aug. 10, 2021) was also implemented in MATLAB. In a meeting of the laboratory in July 2021 about the data pipeline, Charles described the manual curation as an "opportunity to ensure stall count from Stall Catchers [was] directly applicable to experimental results" (fieldnote Jul. 27, 2021). With this new step, researchers now had to clean the images before sending them to the crowd for analysis.

This move introduced a new intraversion within researcher–technology relations in the data pipeline. The supportive work around the ML model that researchers had to do became tedious and temporarily even more time-consuming (at least, that was the experience of some researchers) than manually annotating stalled vessels from the start.

Data Curation: Manually Intervening and Editing

After the results of the automated processing described on the previous pages were saved in a laboratory-wide shared repository for processed data, researchers could access the data to edit it in the new curation tool, which allowed them to review the results of the automated processing steps and decide whether to omit, accept, or adjust them (step five in Figure 9). The curation tool was implemented as a MATLAB program, but unlike previous tools that were effectively just scripts run via a terminal, this tool featured a simple UI. This UI included a large image frame for the data to be edited and a collection of menus, settings, view options, and controls for scrolling through the image stack. Data was displayed in the tool as black images with white vessels and a few white spots scattered across the frame. The vessel mask depicting the individual vessel segments, i.e., the segmentation created by DeepVess, was layered over the white vessels in red. After loading an image stack into the program, researchers could choose how many images they wanted to view at a time. Multiple consecutive image slices from the stack were then accordingly projected into a single condensed image, with the top slice being considered the image currently displayed, which allowed them to go through the image stack faster. Vessels in the currently displayed image were colored light red, while vessels in the other projected image slices were colored dark red. Some white lines were not colored in red. Researcher Leander explained that DeepVess had not identified these white lines as vessel segments. Sometimes, the resolution of the vessels was too low, or they were not bright enough, making them difficult to identify not only for the AI model but also for humans analyzing the images later. In this case, Leander continued, it was better to omit vessels because "you don't want to include vessels that cannot be analyzed" (fieldnote, Aug. 16, 2021). Most of the time, Leander "agreed with" DeepVess' results (fieldnote, Aug. 16, 2021).

After having "cut" the stacks to remove low-quality images and dura, researchers had to perform two main tasks in the editing tool. These were to reconnect vessel segments that had been incorrectly separated and to remove segments that DeepVess had wrongly identified as capillaries. While Sean usually took several passes through the image stacks, going through the projected images from the top of the image stack to the button and checking for broken or redundant segments and segments to connect separately (fieldnote, Nov. 14, 2022),[27] he emphasized that others would have their own routines for editing the data.

These two main steps of connecting vessel segments and removing falsely identified vessels allowed the researchers to improve the vessel map created by DeepVess. However, it did not allow them to retroactively include vessels that DeepVess had missed. Even though Leander considered it to be better to omit vessels that had not been identified by the ML model and could not be analyzed anyway, some vessels that had been missed could be analyzed, and he would have preferred to be able to add them manually after the fact. Some researchers, such as James, therefore, found the curation process occasionally frustrating because he would see vessels DeepVess had missed but was unable to correct the data, and had "to live with knowing that the data is not perfect" (fieldnote, Nov. 16, 2022). In addition to this discontent, editing also consumed a lot of time. In 2021, it took Leander about three hours to curate a single image stack (fieldnote Aug. 16, 2021). Although researchers agreed that this editing was necessary to get the total number of vessels in an image stack, which was subsequently needed to know what percentage of

---

27 In the first iteration, he focused on removing noisy segments which DeepVess often incorrectly classified as vessels, or large vessels that were overlooked in the automated preprocessing (fieldnote, Nov. 14, 2022). Once he arrived at the bottom of the image stack, he scrolled up again, checking for additional segments to be removed again before starting the second pass in which he focused on connecting segments (fieldnote, Nov. 14, 2022). Similar to the automated process of post-processing, researchers would focus on the endpoints of the vessel segments to connect broken vessels, this time manually (fieldnote, Aug. 16, 2021).

all vessels were stalled in their experiments, it was still perceived as tedious and sometimes described as a "misery" (fieldnote Aug. 16, 2021). "I even feel like this is more tedious than [manually counting stalls] cause I've been just manually counting stalls, but this is three hours per stack versus maybe ... a little bit over an hour for manual," explained a researcher in a meeting related to the Stall Catchers pipeline (fieldnote Aug. 16, 2021). It is worth noting that this comparison of the work they now had to do in the curation process versus manually analyzing stalls did not include the additional 20–30 hours it would take to also segment a stack manually.

At that time, in late summer 2021, the goal was to improve the curation process so that it would take no more than an hour per image stack. The goal was to do "anything [...] to make it faster" (fieldnote, Aug. 24, 2021). Laboratory members worked together to improve the tool by providing and implementing feedback. By the time I left Ithaca at the end of October 2021, researchers were confident that this goal would soon be achieved, with the process improved to the point where it would no longer be perceived as painful. Once the curation process and, hence, the data pipeline were improved, Stall Catchers could be successful not only as a game but also "successful from [the researchers'] end" (Leander, Sept. 22, 2021) (*cf.* Chapter 5). The problem, according to Leander, was that they "didn't have [...] the infrastructure in place to support it being successful. So, if it's successful, [...] we need to figure out [...] how to do that. So, I think it's like almost there in terms of being successful from our end" (Sept. 22, 2021). Once "there"—which, in an August laboratory meeting, was defined as one hour of manual work to prepare the dataset—they could start focusing on other things again (fieldnote Aug. 24, 2021) and the data pipeline could fade into the background of Alzheimer's disease research processes.

However, when I returned about a year later, in fall 2022, the laboratory was still working on improving the data pipeline. It was not yet finished, Nishimura explained (fieldnote, Oct. 21, 2022). There had only recently been a new round of discussions about problems in the curation tool, and the understanding that they would be "there" soon had shifted. The data pipeline was now seen more as an "ongoing process," as Sean described it (fieldnote, Nov. 14, 2022), or as a "work-in-progress" (Emily, Oct. 18, 2022). Working on the data pipeline was understood as a "long-term investment" since the pipeline could be used for different data and different aspects of the data beyond the Stall Catchers project (fieldnote Oct. 21, 2022).

Compared to the previous year, however, the PIs stated that they were now in a much better situation (fieldnote, Oct. 18, 2022). The work on the data pipeline and particularly the curation process in 2021 was retrospectively described as very frustrating. Compared to 2021, when the process had been "very sloppy" (Emily, Oct. 18, 2022), Emily agreed that they were in an "okay spot." The vessel maps they now had were "much cleaner" (Oct. 18, 2022). The latest improvements included a new automated masking procedure and new features in the curation tool, such as the option to add vessels that DeepVess had missed.[28]

---

28 Furthermore, they even included a full re-implementation of the tool from scratch. The initial tool had been built as an image viewer with many features not required for the curation process, creating a researcher experience that was not smooth (fieldnote, Oct. 25, 2022). The new curation

However, when I asked James how they experienced the data pipeline now, he said that it was generally the "same old, trying to curate data" (fieldnote, Nov. 16, 2022). While all laboratory members seemed to agree that the changes to the data pipeline, including the curation tool, were improvements, the overall problem of efficiently and effectively preparing data for Stall Catchers was not yet considered solved. The pipeline was now producing data of sufficient quality, but curating was now (again) taking up too much of the researchers' time (fieldnote, Oct. 18, 2022). Indeed, some researchers still found the whole process related to Stall Catchers too time-consuming and saw no advantage over manually annotating the raw research images. The PIs had to invest a lot of effort in convincing researchers to "use" Stall Catchers (fieldnote, Oct. 25, 2022) and send data to the platform instead of manually annotating it themselves. This effort included improving the data curation tool and trying to standardize curation practices across the laboratory, which could vary widely between different researchers, especially in terms of the time it took. The new goal was to consistently reduce curation time to 15–20 minutes (Sean, fieldnote Nov. 4, 2022), since each research project would require several hundred stacks. They, for example, had to stop drawing lines again to speed up curation because it was too time-consuming. Drawing lines, however, was something that researchers described as very helpful and had greatly improved their curation experience. James, who had previously been dissatisfied with not being able to draw lines, described that, with this new feature, he got "the sense that I actually do it accurately" (fieldnote, Nov. 16, 2022). Now, with the need to speed up the curation process and, thus, skip drawing lines, he faced the risk of losing that sense of accuracy.

The shift in focus from achieving the required data quality through curation to minimizing the time required for curation was also accompanied, or even driven, by a new understanding of the ultimate goal of the processing beyond preparing data for Stall Catchers: While the researchers' initial goal was to create a "holistic perfect vessel network," they now focused on further improving certain segments within specific image slices that DeepVess had actually correctly identified (fieldnote, Nov. 16, 2022). This was related to the new insight that they only needed a certain total number of vessels for their research rather than correctly identifying every vessel in each individual image stack. Therefore, one of the previous first steps in the data pipeline, masking out dura, was changed completely and moved to the curation step after all the automated processing. Researchers would no longer manually mask out dura before sending the data through DeepVess. Instead, entire image slices were excluded from the final editing step, reducing the size of the final image stack and, thereby, the amount of data sent to Stall Catchers by more than half (fieldnotes Nov. 4 and 16, 2022). While the top was excluded to reduce visible dura, the image slices at the bottom were excluded because image quality tended to deteriorate with depth due to the weakened signal, sometimes causing DeepVess to incorrectly identify vessel segments where there were none. Sean described this new practice as "cutting aggressively." "[T]he more you cut, the less work you have, the faster you can go" (fieldnote, Nov. 16, 2022). "Cutting aggressively" was a relief to the researchers.

---

tool was cleaner and seemed to be working better, though it still took too much time to curate an image stack (fieldnote, Oct. 25, 2022).

James, for example, explained that he no longer felt like he was doing "pointless work" but could spend his time on more "productive" work (fieldnote, Nov. 16, 2022).

Speeding up the curation step and improving the researchers' experience with the tool was not the laboratory's only focus regarding the Stall Catchers data pipeline in fall 2022. Laboratory members were also dealing with a new problem that had only recently been identified. As researchers and Nishimura explained, they had not known until recently that they "hadn't closed the loop" (fieldnote, Oct. 21, 2022) of Stall Catchers. In one of their last internal meetings about Stall Catchers, they had realized that only a few researchers had been pulling data from the platform. Many researchers at the laboratory had been "dumping data" (fieldnote, Oct. 21, 2022) into Stall Catchers but had not studied the results coming back from the CS game. Therefore, these researchers did not really know what the Stall Catchers results looked like and did not include them in their research. A new goal was, thus, to close this loop by focusing the infrastructuring work on the phase after Stall Catchers participants had analyzed the data.[29] Yet another software tool was introduced to close this loop, describing another step in the continuous infrastructuring practices related to Stall Catchers and its researcher–technology relations.

Closing the Loop

Once a dataset was fully analyzed in Stall Catchers, a Human Computation Institute member notified researchers that they could download the results from the researcher interface[30] on the Stall Catchers platform. The results in the downloaded CSV files were structured with one vessel per line and columns for different information. While the file included more than ten columns, Sean focused on only three columns, one of which was the link to the corresponding video to validate the results. Even though this had not been intended in the design of Stall Catchers (as I discuss in more detail in Chapter 7), researchers went through the list of annotated vessels to review the generated annotations and to manually validate them by analyzing the video. The vessels, i.e., the lines of the file, were ordered according to the confidence level of the crowd answers for a vessel being stalled. High confidence reflected the agreement between different Stall Catchers participants that a vessel was stalled. Sean explained that he usually created an extra column for his answer and then reviewed each stall annotation with a confidence level between 1.0 and 0.5 (fieldnote, Nov. 4, 2022). The challenge of "closing the loop" was not only to get researchers to look at the data results and learn to trust them (see Chapter 7) but also to make this step more efficient and convenient for researchers.

The problem with this review approach was that the researchers were looking at the same data as the participants, i.e., the short videos of blood vessels that had been created from the original 3D imaging data through the multiple steps of data transformation described above. To answer their research questions, however, the researchers had to trans-

---

29   Closing the loop, PI Nishimura suggested, would also help build the researchers' trust in the Stall Catchers project as they could then engage with the results (fieldnote Oct. 22, 2022; see Chapter 7).

30   In addition to the user game interface, there is an additional interface for researchers to access their own projects and related datasets.

late the videos back or compare them to the original imaging data (and, ultimately, to the individual mice). This task was very cumbersome, as researcher Jada explained:

> I check the video and then I go back to the stack, which is a little bit difficult [...] [T]hen I have a quick look whether I identify those [stalls] as well in the original video. [...] [I]t's just [...] for my sanity; I just would like to see these in the full stack. [...] I'm kind of extremely used to it and see it in the context and there are also a lot of regions, edges, and stuff where stalls occur where it gets difficult for the program to really see those. (Oct. 27, 2021)

While the abstractions and transformations of the data were necessary both to count the vessels in a stack and to allow Stall Catchers participants to analyze research data, they also made it more laborious for the researchers to continue working with the results.

The laboratory also implemented yet another tool for the manual post-processing of the data returned from Stall Catchers to facilitate this process and further improve the researchers' experience with the project. This tool enabled researchers to open an unprocessed image stack and select the vessel they wanted to look at from the list of all vessels (fieldnote, Oct. 25, 2022). The tool then drew an outline of the requested vessel, similar to what the Stall Catchers participants had seen but on top of the original raw imaging data. This way, instead of watching the Stall Catchers videos and then trying to find the vessel in the original image stack manually, they could directly examine the vessel and its environment. Bringing the latter into view was another objective of the new tool, since the data sent to Stall Catchers only included one channel and, thus, not all the information, such as other cell types, was visualized in other channels. It was now crucial for the researchers to understand not only *if* a vessel is stalled but also *why* it was stalled.[31] Therefore, it was important to know what is around a vessel and to learn about the behavior of these other cells in relation to stalled or flowing vessels. The new tool allowed the researchers to see all channels.

At the time of my field research in 2022, this tool had just been implemented and was not yet being used by researchers (fieldnotes, Oct. 25 and Nov. 4, 2022). This new researcher–technology configuration carried the hope of closing the Stall Catchers loop and facilitating their everyday research practices, thus, solving some of the remaining problems with the project. However, how this new tool will shape and impact the data pipeline and practices surrounding Stall Catchers remains open at the time of writing.

What seemed to be clear, though, was that closing the loop would not be the end of the infrastructuring related to the Stall Catchers project. Improving DeepVess, for instance, had already been identified as a potential next undertaking, as the model was seen to require retraining or perhaps even replacement. The ML model was considered to be "lagging behind the rest of" (Anna, Oct. 14, 2021) the pipeline. Researchers Sean and Leander expressed their hope that "it would be great if we did not have to curate" (Sean, fieldnote, Nov. 4, 2022) any longer and if, instead, "the AI" were to be improved. "[I]f we can get it to the point where we don't have [to make] all of these little connections ... that'd be really nice, so I don't know. The more I think that we can jump onto [improving the]

---

31   Charles, April 2023, private correspondence.

DeepVess part as opposed to adding on tons of post-processing" (Leander, Aug. 16, 2021). A new data pipeline focus with DeepVess had already been identified for future work.

The example of closing the loop shows how various intraverting researcher–technology relations are entangled in the data pipeline and how new potentials and tasks emerge within the same relations and enable new configurations in other interwoven ones. Closing the loop only emerged as a problem once data preparation had been improved to a certain point, and once that was solved, DeepVess could become the next target. Researcher–DeepVess relations will intravert yet again with the work on the ML model.

Reconfiguring Data, the Pipeline, and Researcher–Technology Relations

While the previous sections described in detail the implementation and steps of the Stall Catchers infrastructure on the laboratory's side, in this section, I summarize and analyze them with a focus on intraversions. I concentrate on two main points: first, how the human–technology relations keep changing with each reorientation and modification, focusing on the development of the infrastructure. This includes how the idea of the data pipeline as a means shifted to the idea of it as an investment alongside and because of the intraverting human–technology relations, what this tells us about the imagination of infrastructures, and how this impacts the researchers' working practices. Second, I show how the data pipeline and the work on it can be understood as part of an HC system itself.

The detailed descriptions of the improvements, modifications, and related reorientations revealed the complex intraverting sociotechnical interplay between humans and technology in which automated steps were interwoven with manual human tasks and vice versa. Agency and tasks within the relations were not fixed but shifted and redistributed along with these developments. For example, they shifted from the ML model originally developed to support and accelerate the researchers' work to the need for researchers to support DeepVess and other automated algorithmic steps by cleaning up their results. Similarly, the subgoals or tasks associated with infrastructuring kept changing, such as, as described above, the initial subgoal of creating a perfect vessel network, which was changed to a required number of correctly mapped vessels. DeepVess and the researcher–tool relations initially complemented each other toward the same goal of identifying the complete vessel network. In contrast, later, researcher–DeepVess–tool relations focused only on those parts of the DeepVess output considered "good enough" to achieve better efficiency and throughput, i.e., freeing up time and resources for the researchers. Retraining DeepVess to achieve better automated performance would be another step in this direction, which will go hand in hand with reconfigurations of researcher–DeepVess relations. Along with the development of the Stall Catchers pipeline, similar to what Mackenzie observed with the example of machine learners, both human and nonhuman actors were assigned new positions. "These positions are sometimes hierarchical and sometimes dispersed," Mackenzie writes (2017, 186), but the subject position in this hybrid relation is always mobile (2017, 186). Much like the introduction of new tools in the example of naval navigation described by Hutchins (1995a), the introduction of the ML model created new tasks for the researchers, such as data curation and image masking. While pushing the data from one automated step to the next was considered "pretty easy" (Isabel, Oct. 14, 2021), and the researchers' role here was simply to guide the data through the pipeline, taking it from the computer and handing it back (Isabel, Oct. 14, 2021), new tasks, such as curation, now became a bottleneck in the data pipeline. Instead of relying on the ML tool to take over the researchers' task, researchers had to focus on assisting the tool and evaluating its performance.

It should have become clear that the data pipeline, which was originally conceived as a means to enable data analysis via Stall Catchers and to speed up Alzheimer's disease research at the laboratory, itself became a central focus of work at the laboratory, which was not completed even five or six years after the project was introduced. In fact, during the time of my field research, researchers were particularly focused on working on the data pipeline itself. This was not always perceived as satisfactory by laboratory members, as for some of them, their actual research goals fell out of sight while working on the data preprocessing. There are many translation steps from raw pixels depicting vessels in the mouse brain, to knowing the total number of stalled vessels in the dataset, to deriving insight into the effects of reduced blood flow in Alzheimer's disease in mice. In contrast to working on the data pipeline, when researchers were still manually annotating stalls—the tedious activity that the CS project was designed to replace—they, at least, had a stronger sense of directly working toward their research goal; the stalls were the center of interest, and though tedious, their work was directly concerned with analyzing them. This sense of working toward their research goal was lost when they were working *on* the infrastructure rather than *using* it; this work was effectively one level of abstraction removed from their actual research. Nevertheless, the focus on the infrastructure was also seen as a future investment that would eventually facilitate the research once the data pipeline was running smoothly. In 2021, James explained: "I guess it's even like [a] long-term investment in the future [...] if this will be faster than manually, [...] but right now it's hard cause it's a lot" (fieldnote, Aug. 16, 2021).

However, during the time of my research, even as the data pipeline, data quality, and the researchers' experience improved gradually, the idea of the data pipeline as a means that could slowly disappear into the background turned into an understanding that the pipeline was an "ongoing" project that would never be finished. With the solution of one problem and a better understanding of the current processes, new problems arose that had to be tackled, such as the loop that had to be closed or the curation of data with the introduction of the ML model DeepVess. By fall 2022, the laboratory was closer to its goal of obtaining correct results, i.e., not only correctly classified vessels from Stall Catchers but also the correct number of vessels in an image stack. However, it seemed that the closer they got and the more they worked on the infrastructures and improving the processes, the more they identified gaps and problems they had not thought of before (fieldnote, Nov. 4, 2022). "[N]ew instrumentation gives new perceptions," argues Don Ihde (1990, 56, emphasis i.o.). Viewed through the lens of intraversions, each improvement and change in the pipeline opened up a space for new scaling, revealing questions or problems that had previously been out of sight. In fact, infrastructuring, in the example of the data pipeline for Stall Catchers, involved several intraversions in the research–technology relations, most notably the introduction of an ML model to do the segmentation necessitating manual preprocessing steps, which were then refactored to semiautomated postprocessing steps, again creating the need for manual curation.

The observation of the incomplete pipeline is not to say that the efforts to improve the infrastructure did not make a difference but to emphasize that infrastructure is unruly (Wynne 1988). One laboratory member explained that the processes related to the Stall Catchers project would probably never be "standardized" compared to other established collaborative processes because of the "variability within the data:"

> [Y]ou have to spend some time, and you also have to make sure for your science that it's [...] sort of correct because it is a little bit of a black box. But I don't think it's negative. We also have many other black boxes. If I send something out to some facility, they are [sending] me some excel sheet back with data that I also don't have the real insight to, but I think it's more standardized and this is just not as standardized and probably will never be. It's just because of variability within the data. (Jada, Oct. 27, 2021)

In addition to the variability within the data mentioned by Jada, Charles explained that there will always be some "nonlinear pacing" (Oct. 14, 2021) in animal experiments, which cannot be predicted and prevents the process from being completely standardized.

The continuous changes and modifications to the pipeline's steps that went hand in hand with the shifting tasks researchers had to perform were difficult for new laboratory members who were still being introduced to Alzheimer's disease research with Stall Catchers. As Benjamin explained, for example, every time he tried to learn how to curate data, the task or program had changed or a new tool had been introduced that he had to get used to again, resulting in him not being eager to try it and preferring to analyze data manually instead (fieldnote, Oct. 26, 2022).

Even as the researchers' understanding of the pipeline changed from a means that would be completed one day to an ongoing process, they were confident that Stall Catchers would eventually save researchers time (fieldnote, Nov. 4, 2022). At the end of my field research in November 2022, Sean stated they were getting closer to the point where Stall Catchers would make the analysis process easier than manually counting stalls (fieldnote, Nov. 4, 2022).

Focusing on the development of the pipeline further illustrates how digital infrastructures in the example studied, and particularly trained computational models, carry inertia. They often lag behind spontaneous changes in processes and practices, such as variations in experimental settings. While it is comparatively easy to exchange a microscope objective, adjust the laser trajectory, or alter the treatment of mice with a direct impact on research, it takes a long time to retrain an ML model to account for these changes or to develop a custom data annotation tool for researchers to use.[32] Not only does code need to be changed, but the adjustment of downstream systems and the lengthy evaluation and feedback cycles involved further increase the effort for such changes. This is

---

32 This is not necessarily generalizable to all ML models but specific to the situation described. Some ML models, for example, are designed to continuously get retrained and updated. However, this depends on having the necessary meta-infrastructure in place, i.e., the tools, processes, and systems required to support and maintain ML models over time, as well as enough personnel resources to support ongoing model development.

not always feasible within the constraints under which researchers work, and as the series of pipeline modifications at the laboratory shows, instead of tackling such problems at their deepest level, other workarounds are often introduced. The decision to improve the output of DeepVess via complex post-processing instead of improving the model itself, i.e., retraining it, is an example of such an improvised, temporary solution, which is also related to the fact that the laboratory lacked the necessary ML expertise after the researcher who had originally implemented the model had left. Similarly, earlier on in the pipeline, manual masking of the dura and other noisy remnants of the imaging process could also be seen as such a workaround.

The data pipeline, understood initially as a means to enable outsourcing the image analysis to the Stall Catchers platform, had become a major focus and investment. This discrepancy in the meaning of the pipeline led to tensions in the laboratory, since, even in 2022, it was still considered a "side thing" (fieldnote, Oct. 26, 2022) by some researchers rather than a core part of their Alzheimer's disease research process. Curation was often practiced in the late evenings when researchers had time to do it (fieldnote, Oct. 26, 2022). In addition to this attitude, the fact that the curation program did not work well on all laptops due to the large amount of computing power it consumed played an important role here. Benjamin explained that he preferred to curate data every now and then while in the laboratory waiting for an experiment to be completed (fieldnote, Oct. 26, 2022). But since the program was too slow on his computer, this was not an option. Sean, whose laptop managed to run the program as long as he pulled the data from a solid-state drive, tried to find time to curate data whenever possible (fieldnote, Nov. 4, 2022). Data curation was something researchers typically incorporated into their daily working practices with lower priority, organizing it around their other tasks. Changing this understanding of data curation, therefore, was one of the main aims of the laboratory's PIs during the time of my research. Stall Catchers were to become a central part of the research and not just a side thing, explained laboratory member Isabel, who tried to get other researchers to "use it" (fieldnote, Oct. 25, 2022). According to Isabel, what was needed was a "paradigm shift" in their thinking (fieldnote, Oct. 25, 2022).

Looking at Stall Catchers, including the data pipeline as a whole, the pipeline can be understood as a part of a "higher-level" HC system, even though it was not designed and considered as such by the Human Computation Institute and laboratory. When I suggested this idea of a "higher-level" HC system and the pipeline as a part of it, Michelucci explained that he had not done so because the human–computer and human–AI handshakes in the pipeline are not automated but manual, meaning that the individual steps are not fully streamlined (fieldnote Nov. 4, 2022). To move from one step in the pipeline to the next, humans must actively intervene, for example, by moving the data from one folder to another or by starting a successive program. However, I here aim to show how this is, nonetheless, an insightful perspective for understanding how different human–technology relations intravert in their entanglements and how the overall system evolves.

Viewing the laboratory as one actor in this higher-level system and Stall Catchers as another, the laboratory's role shifted from fully (but slowly) performing all of the analysis to delegating a key part of the analysis to Stall Catchers. The laboratory's own focus shifted to providing better data outputs for it to use, while introducing a new manual validation step to reincorporate Stall Catchers' data. At the end of my field research, this system still faced considerable friction and the efficiency gains on the side of the laboratory were not yet as substantial as researchers had hoped. However, it was already on a path of continuous change and intraversion that both actors expected would eventually lead to significantly outperforming the system's initial configuration. The intraversions occurring at the scale of this overarching system are slower than those occurring within either system viewed in isolation. Given this and the project's short existence, it is still too early to give a full account of how human–technology relations and the actors' roles within this system are intraverting over time. However, during my field research, I could already observe some tendencies and imaginations of potential future modifications in these relations, tasks, and responsibilities. These included ideas for changes to the task on the Stall Catchers platform itself (fieldnote Oct. 26, 2022) as well as involving Stall Catchers participants in earlier steps of Alzheimer's disease research conducted at the laboratory and improving the ML model with input from the game platform. Even though neither the laboratory nor the Human Computation Institute seemed to actively consider or describe this higher-level HC system as such—only Stall Catchers itself was usually described as an HC system[33]—there was still an awareness among researchers that Stall Catchers' capabilities and infrastructure could be further integrated into their work.

The researchers had discussed the idea of using the Stall Catchers participants' identification of poor-quality videos by "flagging" them as input for improving DeepVess based on an HC approach (fieldnote, Aug. 10, 2021). Another idea that was discussed during my field research with the common goals of both reducing the laboratory's own workload and involving Stall Catchers participants in more steps of Alzheimer's disease research was to invite some of the more experienced Stall Catchers participants to take on some more complex tasks (fieldnote, Oct. 21, 2022), as data curation turned out to be more time-consuming and laborious for researchers than anticipated. The idea was also related to the fact that more and more participants were analyzing data on the Stall Catchers platform, increasing the data throughput to such an extent that the researchers in the laboratory sometimes could not keep up with generating and preparing new data for Stall Catchers. Improvements in the data pipeline even further reinforced this, as better data quality reduced the number of segments sent to Stall Catchers. However, this meant fewer videos for participants to analyze, which could exacerbate the problem of lack of data (fieldnote, Aug. 10, 2021). Participants asking for more data was a new dynamic for the researchers. Initially, the laboratory had a huge backlog of data to send to Stall Catchers and researchers had to wait for participants to analyze it. One concrete idea to address this asymmetry was to allow participants to work closer to the raw data in the curation steps (fieldnote, Oct. 21, 2022). According to researcher Sora, this could benefit both biomedical researchers and Stall Catchers participants, as it would contribute to the research process and potentially be more interesting for participants (fieldnote, Nov. 1, 2022). In some ways, this would not only lead to new transformations of the infrastructure and related human–technology relations but also change the relationship between researchers and participants, since the latter would

---

33   This is related to the human-in-the-loop understanding behind HC that I discussed in Chapter 4.

then be included in the scientific process of vessel mapping and cleaning (fieldnote, Nov. 1, 2022). When asked how this would be different from the current involvement of Stall Catchers participants, a laboratory member explained that participants would then be introduced to and confronted with the variability and uncertainty of science, which were "pretty much removed" in Stall Catchers (fieldnote, Oct. 21, 2022). If participants were included in data curation, they would have to make their own judgments, since not everything in science can be decided with 100 percent certainty (fieldnote, Oct. 21, 2022) and, hence, would have to be trained as researchers first. This illustrates the boundary work biomedical engineers do to distinguish between their research and the analysis step outsourced to CS participants. Although this idea had not been realized at the time of my research and was also controversially discussed in the laboratory, it points to possible future intraversions within the human–human and human–technology relations in Stall Catchers.

Looking back at the first five years of the Stall Catchers project and the development of the interplay between human participants and computer algorithms, the idea of introducing new types of tasks to be solved via participant–technology/computational tools relations was also considered a very complex and long process: "I've learned […] that it's difficult to make a single task pretty consistent. So, the challenges of making multiple tasks would be a whole other thing. Well, I'm still hopeful," said Nishimura (Dec. 7, 2021).

The question of whether and how this intraversion will unfold remains open at this time. However, two key points I would like to make are already evident in the stages the system has gone through so far. On the one hand, HC-based CS systems are imagined differently by their actors, with their own specific aims and needs in mind. In this, the overarching structure of the system may not always appear to or be seen in the same way by each actor. On the other hand, such systems are never understood as complete or closed systems. This is not only because, like many systems, they exhibit problems in their development and maintenance. Rather, they are specifically thought of as being continuously open to change in the hope of surpassing their current function, which manifests itself in intraversions of human–technology relations in the HC-based CS systems. As Schaffer put it: "[I]t's not *done* […]. I mean, […] there'll be a new wrinkle in the data. There'll be some new problem that comes up because we're not going to keep doing just this exact same thing over and over" (Dec. 7, 2021). This continuous evolution of HC systems is particularly apparent in the analysis of the dynamics of the ARTigo example.

Moving Forward in Concert

Shifting our focus from the microanalytical perspective on HC systems in their everyday situatedness to the broader evolution of HC-based assemblages over extended periods (i.e., years), a pattern emerges. This pattern describes the dynamic and nontrivial interactions between human–technology relations within HC systems and advances in AI computational capabilities. In the following exploratory remarks, using ARTigo as an example, I show how these also undergo a continuous transformation pattern, evolving in parallel with the intraverting human–technology relations. They influence and create each other in concert as they move forward.

The main reason for my focus on ARTigo at this point is that the development of ARTigo's life cycle presents a revealing example of the changing relations between HC and AI research despite its less extensive treatment in previous chapters. When I started my ethnographic fieldwork in the last weeks of 2019, ARTigo had been an active project for over ten years. The project idea emerged in 2007 from computer scientist Bry, who was inspired by Von Ahn's ESP game (Von Ahn 2005),[34] which "was the first system where work activity was seamlessly integrated with gameplay to solve Artificial Intelligence problems" (Von Ahn 2005, 70; see also Law 2011; Lazar, Feng, and Hochheiser 2017). In the two-player ESP game, players (without seeing each other or knowing the other player) had to describe images with words but only gained points if their words or tags matched those of the other player. Bry brought this idea into the area of art history. Computer scientist and ARTigo team member Ben described in our conversation in early 2020 that "[f]or art history, [...] the idea of doing it this way and making it this big and also combining it with the several games, that is to say, making it a platform, was a very good idea. No one had done that before" (Feb. 24, 2020). In an article by digital humanist and data scientist Stefanie Schneider and art historian Hubertus Kohle, who are both part of the ARTigo project, the authors describe ARTigo and its twofold goal. They characterize it as "an internet platform in which digital reproductions of artworks are presented to an audience with unknown qualifications, who then annotate these artworks in a playful and competitive way" (Schneider and Kohle 2017, 82). Furthermore, they define it as "a semantic search engine which can master large image sets based on these crowdsourced annotations (*tags*) without having to rely on the expensive *manpower* of specialists—or even on artificial intelligence from the field of computer vision" (Schneider and Kohle 2017, 82, emphasis i.o.). There was a need for such a semantic search engine in the field of art, since several million digital reproductions of works of art existed in electronic repositories, but there was no way to retrieve them based on specific criteria (Schefels, n.d.). Computational search, at that time, was "still very limited" (Schneider and Kohle 2017, 82).

In addition to these official goals, for Ben, ARTigo was also an opportunity to "bring art history [...] into the computer age and to prove with ARTigo that computer science can also do research in art history, that is, with data science, data analysis" (Feb. 24, 2020). While participants contributed text-based tags, an unsupervised ML method[35] processed these contributions to build and constantly improve the semantic search engine (Bogner et al. 2017, 53). The advantage of a semantic search engine is that it can be used to "search for [art]works whose identity cannot be determined by identifying the author and title, which are available as metadata in traditional image archives" (Schneider and Kohle 2017, 82).

While the ARTigo project was funded by the German Research Foundation from 2010 until 2013, the team continued to work on it and maintain it beyond the end of the fund-

---

34   The game is named after "extrasensory perception" (ESP), which is also known as the "sixth sense." Even though Von Ahn does not explicitly state so in his dissertation, it becomes clear from the context to what he is referring.
35   More precisely, the team developed a "Higher-Order Latent Semantic Analysis" (Wieser et al. 2013; Wieser 2014).

ing period. Up to 2016, over nine million tags had been gathered from "on average 150 persons a day playing on ARTigo" (Bry and Schefels 2016, 5), and by 2023, Schneider and colleagues report on 10,679,711 annotations (Schneider, Kristen, and Vollmer 2023, 4) by several tens of thousands of participants (Kohle 2018).

The ARTigo platform included several games to collect different tags. As Ben explained:

> There are different games for different tag groups. So, the ARTigo game, the classic one, that's very general, it collects tags very broadly, while the other games, they keep refining that. They then build on the dataset and then also again the other way around, other [games] build on that dataset then again and this way, it keeps refining. And this way, you then get a better description for the images. (Feb. 24, 2020)

The different tags and their corresponding games can generally be divided into "simple descriptions" collected by the ARTigo game, which resembles the ESP Game (Bogner et al. 2017, 53; screenshots of the initial ARTigo UI can be found at Citizen Science Games 2019), and "more specific descriptions [that] are collected by 'diversification games' using simple descriptions collected by description games" (Bogner et al. 2017, 53), such as the ARTigo game.

Finally, "integration games" collected "annotation clusters" based on tags collected by all different games (Bogner et al. 2017, 53). Together, these different games and tag forms or annotations—all of which were text-based—allowed the "ARTigo gaming ecosystem as a whole [to] perform[] better than each of its games alone" (Bogner et al. 2017, 56). ARTigo, like Foldit, is explicitly built on the idea that nonprofessionally trained participants will provide different tags than professional art historians (Kohle, Dec. 4, 2019). The value of annotations by nonprofessionally trained people lies in the fact that while individual descriptions may refer to objects or colors in images of artworks, when combined, "the tags demonstrate a wisdom not present in any single word, but only in the collection" (Kohle 2016, 3). By saving only those tags contributed by at least two participants, ARTigo aims to exclude "deliberately false input" (Kohle 2016, 2).

In late 2019 and early 2020, I interviewed art history researchers and computer scientists involved in the project to learn about their experiences with the project and how they would describe ARTigo's journey so far. Team member Finn explained:

> Overall, you have to see that it is somewhat of a flagship project for all citizen science projects. It was one of the earliest projects to experiment a bit. And also one of the projects that, I think, had the most far-reaching consequences because art history […] has no datasets with actual annotations available. Currently, obviously, it has decreased a little bit. Because I also believe that this system of actually assigning text annotations for images, that is no longer popular. (Dec. 16, 2019)

The team universally agreed that ARTigo was very successful as an early HC-based CS game, especially since it was launched before ML, computer vision, and deep neural networks became "such a cool research topic again" (Emilia, Nov. 8, 2019). The technologies that exist today to automate tagging the semantic content of artworks were not available at the time, computer scientist Emilia explained in our conversation (Nov. 8, 2019).

By 2019, ARTigo seemed to have passed its peak. Due to a lack of funding, it was difficult to sustain such a large project. While various research studies (including bachelor's and master's theses)[36] had been conducted on and around ARTigo during its peak period, Emilia explained,

> not much is happening there at the moment. [...] It's just kind of a bit of an old project at the moment that kind of just keeps going [...]. But currently, in terms of the underlying technology, it is just [...] more or less end of lifetime. So, it also constantly crashes. So, if you try to access the ARTigo page, then it can be more often that 50[3][37] or so, so that the server is not working because somehow the system is not working. And I think the [...] daily business now is basically maintenance and restarting the server when it crashes again. (Nov. 8, 2019)

Thus, at the end of 2019 and the beginning of 2020, the team's main tasks were keeping ARTigo alive and maintaining the platform by occasionally restarting the server and answering press inquiries.[38] Finn sighed and explained that it "happens relatively often that I have to restart [the server] three times a day" (Dec. 16, 2019). At the same time, the platform was somewhat outdated.[39] Bry argued that "[i]f you want a platform to continue to remain popular, you need to work on the platform constantly. A game platform cannot stay the way it was for ten years. And that's hard work" (Feb. 3, 2020). For ARTigo to remain popular, Kohle agreed, it would have to be reimplemented from scratch with different games and less text-based approaches (Dec. 4, 2019). Finn suggested that participants today prefer to drag or click on things rather than type (Dec. 16, 2019). Such a reimplementation could also, according to Kohle,

> address a weakness of the application related to its logic. In particular, in cases where we also want to use the annotations to train computational neural networks, such that the computers will eventually be able to recognize the objects depicted in the works themselves, the information is too imprecise: If an image is tagged with the term "dog," then a dog appears somewhere in the image, but it remains unclear which part exactly contains the dog, and can only be deduced by human intelligence. Another version of the game could be to assign terms to certain areas of the image, for example, to drag the term "dog" with the mouse to the image representing the dog. (2018, 2–3)

---

36 E.g., Schemainda (2014); Taenzel (2017); Greth (2019).
37 Emilia mentioned the HTTP response status code 505, which refers to "version not supported." It is, thus, likely that she was referring to the response status code 503, which refers to "service unavailable."
38 Due to ARTigo's collaboration with museums, in which some version of ARTigo was featured in exhibitions, for example, the project still received public interest and media coverage from time to time (Finn, Dec. 16, 2019).
39 The sustainability problem of such CS projects and platforms was not only raised by ARTigo team members but was also a primary motivation for Michelucci to build the Civium ecosystem (see Chapter 4).

While researchers from the ARTigo team had already worked with clustering algorithms in 2017 to divide artworks into groups based on the crowd's annotations (Schneider and Kohle 2017), the resulting clusters could later be used to train CNNs. These could eventually allow the automated classification of images of artworks without prior manual annotation (Schneider and Kohle 2017, 88). However, I would like to remain a bit longer with the state of ARTigo at the beginning of my field research.[40] At that time, compared to other software and game interfaces or UIs in general, ARTigo no longer met the users' expectations (Ben, Feb. 24, 2020).[41]

From a user perspective, not much seemed to be happening with the ARTigo platform for the next year and a half during my main field research period (2020–2021). Yet, I knew from my interviews with ARTigo team members that they were working with computer science students (Greth 2019) to reimplement the entire platform. I frequently checked the website to see if the project was still running or if a new platform had been launched. At some point in 2021, the platform was only accessible from the Munich Scientific Network provided by the data and supercomputing center Leibniz Supercomputing Centre connecting academic institutions in Munich, including the LMU and the Technical University Munich.

To my surprise, when I checked the status of ARTigo in November 2022 (as I continued to regularly do throughout my research), an entirely new iteration of the platform appeared (*cf.* Schneider, Kristen, and Vollmer 2023). Instead of the eight mini-games available on the previous platform, players could now choose between two game modes on the new platform, which was released in November 2022. The first resembled the old games in that it asked participants to annotate images of artworks in a text-based manner. However, in addition to asking for descriptive tags, the game also asked participants to explain, "What feelings does it [the digital reproduction of the artwork] trigger in you?" (Ludwig-Maximilians-Universität n.d.a). The second game mode invited participants to annotate image regions based on the word tags provided. The latter presented participants with a completely new task. Instead of creating text-based tags, they now had to draw outlines around the object region(s) corresponding to one of the provided tags (Figure 10).

---

40   Despite the fact that the ARTigo platform itself was not much used at the end of 2019, annotations created on the platform were still used in other projects. For example, they were included as training data in the new research project "iART" funded by the German Research Foundation (Kohle, Dec. 4, 2019). The project, funded from 2018 to 2023, was a collaboration between the Chair of Medieval and Modern Art History at LMU Munich, the Visual Analytics Research Group at the Leibniz Information Centre For Science and Technology University Library, and the research group Intelligent Systems and Machine Learning at the Heinz Nixdorf Institute, Paderborn University, and included two researchers of the ARTigo project, Hubertus Kohle and Stefanie Schneider (Deutsche Forschungsgemeinschaft n.d.).

41   ARTigo participant Helena, who had responded to my public call for participation on ARTigo's Twitter account, expressed her wish for a more "modern" website (Jan. 27, 2020).

Figure 10: ARTigo Play mode "annotate image regions based on tags"

Source: Screenshot taken by LHV on Feb. 27, 2023 (https://www.artigo.org/de/game)

This development of the ARTigo platform can be compared to the evolution of CAPTCHA described in Chapter 4, which brings us back to the very first HC. As the algorithms for optical character recognition of written text improved, the task was changed to identify objects in images, such as traffic lights or taxis, to improve the AI algorithms for image recognition. The introduction of this new game mode in ARTigo provides an interesting example for my analysis for at least three reasons. First, it corresponded to the ideas presented by team members in our conversations in late 2019 and early 2020 described above. According to these, participants today were less interested in text-based tasks and more interested in tangible modes of interaction. In the mobile version, participants could circle the areas with their fingers. This new form of engagement was understood to be more entertaining in times when smartphones and touch screens had become standard for many people. As a GWAP, ARTigo depends on volunteer engagement and, therefore, needs to be (and remain) entertaining.[42]

Second, new modalities of information beyond test-based tags are collected with the new game mode. This new information then allows researchers to improve computational analysis tools and AI models and create new models, such as a CNN, which was already described as a future idea in the quote from Kohle (2018) above. In this way, it allowed AI research to advance. When I contacted the ARTigo team in March 2023 to learn about the goals behind introducing this new game mode, they confirmed this observation with their email reply that the new game mode served the purpose of "increasing precision. So far, [the label] dog could be used during annotation, but it was unclear which area of the image was being referred to. However, this would be beneficial to teach the computer to recognize a dog independently."[43]

---

42   Additionally, the new ARTigo platform included an input frame that was inspired by messaging apps, in which a chat, or chatbot, was simulated to create something that resembled a dialogue between the chatbot and participant (Schneider, Kristen, and Vollmer 2023, 3).

43   E-mail exchange between LHV and the ARTigo team from March 26, 2023.

Finally, with the advancements in computer vision research in recent years, the state of the art in AI research in 2020, shortly before the launch of the new ARTigo platform, could not be compared to 2007, when the idea of ARTigo was conceived. ARTigo, according to PI Kohle, is based on a normative understanding similar to the Human Computation Institute's Hippocratic oath: "First use no humans" (Michelucci and Egle [Seplute] 2020) (see Chapter 4). In an interview with the science blog *Bürger Künste Wissenschaft*, PI Kohle argued that "[w]hat is important is that people perform real tasks and do not do things that a computer could do, just so they are involved somehow" (Kohle 2019). Following this understanding, it could be argued that the new games were necessary to ensure that participants continued to perform tasks that computational systems could not, in fact, solve. Even if this was not the primary motivation for the new games, this new development could be understood as yet another iteration in a series of intraversions. Instead of asking participants for descriptions of the images, the software now presents them with information previously entered by humans to confirm the tag and specify the image region(s) corresponding to it.

ARTigo's early games built upon each other, forming an "ecosystem" in which participants filled in where the capabilities of the computational parts of the system fell short. Subsequently, two things happened: significant amounts of useful data were collected and AI research, particularly in computer vision, made significant advances that likely made it possible to perform at least some of the human tasks in ARTigo using computational methods. This combination of achieving the system's initial purpose (at least to a large extent) and the advances in the capabilities of computer vision models led to the emergence of a new iteration of the HC system by opening the space for the system to fulfill a new, elevated purpose. ARTigo's evolution here resembles that of the visual database project ImageNet described by Gray and Suri, who discuss its evolution as an illustrative example of the paradox of the last mile of automation: "Humans trained an AI; only to have the AI ultimately take over the task entirely. Researchers could then open up even harder problems. For example, after the ImageNet challenge finished, researchers turned their attention to finding *where* an object is in an image or video" (Gray and Suri 2019, 8, emphasis i.o.). Even though, in the case of ARTigo, AI has not yet fully taken over the annotation task, new, more advanced problems could be addressed. For this new iteration of ARTigo, the platform, interfaces, and modes of engagement also needed to be adapted and developed further to remain interesting, engaging, and even ethical.

Furthermore, it could be argued that ARTigo, as an HC-based CS, *had to* undergo this development in order to remain an HC system, i.e., a system that "harness[es] human intelligence to solve computational problems that are beyond the scope of existing Artificial Intelligence (AI) algorithms" (Law and Von Ahn 2011). In the case of ARTigo, this was a specific problem at the intersection of computer vision and art. Progress on this immediate problem, for example, due to improved computational capabilities through more powerful AI models, meant that computers could increasingly take over the tasks performed by humans in the HC system. These intraversions, as in the case of both Stall Catchers and Foldit, are often largely driven by internal developments and based on the data collected or generated by the system itself. By contrast, ARTigo's intraversion was not primarily based on data collected as part of the project. This difference between ARTigo and the other systems studied does not imply that these developments should not be viewed as intraversions. Rather, intraversions in the HC systems' human–technology relations can be significantly influenced by both internal and external factors. HC systems often directly support or drive the AI advances that cause the intraversions, but this is not a necessary condition. This difference reinforces the notion that advances in computational capabilities do not typically lead to the human side of the system being considered redundant. Instead, as Michelucci explained, "humans can move on to the next task" (fieldnote, Sept. 30, 2021). The assemblages themselves typically do not cease to exist when the current purpose of the system is achieved but are repurposed toward a new goal, which only comes within reach due to the advances made previously. By adapting to new problems and needs, i.e., by intraversions of its human–technology relations, these assemblages themselves become something new (Deleuze and Guattari 2013, 7).

Contingent, Imagined, and Emergent Intraversions

In this chapter, I have analyzed different human–technology relations and how they intravert in the HC-based CS projects Foldit, Stall Catchers, and ARTigo in everyday life's instantaneity, over time, and on different levels.

The analysis shows that the relations are not fixed but open and dynamic and how the idea of "hybrid intelligence" and, thus, the target of HC, keeps moving. The target remains partly an empty shell which gets filled with the instantiations of continuously adjusted human–technology relations as such systems move forward. Role assignments here are never fixed but in flux not only in their everyday enactment but also by the design of the system itself. Built to be changed again, the nature of the tasks to be performed and the purpose of the system are constantly changing along with the intraverting relations. Given the continuous formation and alteration of human–AI or –technology relations in HC systems, it is not sufficient or possible to define once and for all the intended and realized relations between humans and AI. Instead, they must be traced and analyzed *along* and *with* their becoming. These observations of intraverting participant–technology relations not necessarily apply only to the examples studied but also to other HC-based CS systems, such as Galaxy Zoo,[44] in which participants first provide data for training ML models and then AI models, instead of replacing human participants, are introduced into

---

[44] The online CS project Galaxy Zoo was launched in 2007 with the goal of inviting participants to classify galaxies from the Sloan Digital Sky Survey (Lintott 2019, 41) because "the human brain is much better at recognising patterns than a computer" (Galaxy Zoo, n.d.). The project has seen many changes and modifications since its start. Despite its huge success regarding participant engagement and its purpose of classifying galaxies, the project faced the problem that they had too much data to be analyzed by human participants. To solve this problem, researchers started combining the participants' "classifications with those of machines, inspired by the idea that the combination of both automatic and human classification may be more powerful than either alone" (Zooniverse, n.d.). Similar to the approach in Stall Catchers, they used participant annotations to create CNNs "that make probabilistic predictions of Galaxy Zoo classifications" (Walmsley et al. 2020, 1555). Building on the work that some among the numerous images of galaxies are more significant than others (Walmsley et al. 2020, 1555), the Galaxy Zoo team introduced the new "enhanced" project mode in which participants are presented with those images to classify that have been selected by the classifier. With the "active learning approach" of using the human clas-

the same projects. The concept of intraversions helps to analyze these shifts and oscillations of positions, responsibilities, and tasks. Drawing on the example of user–technology and researcher–technology relations, I demonstrated how the intraverting relations reconfigure the tasks, practices, and forms of engagement of different actors within the sociotechnical assemblages.

In the example of Foldit, algorithms whose actions could be observed by human participants were first implemented to try to solve the protein structure prediction problem automatically. Participants' requests and involvement led to Foldit, in which computational tools assisted humans in manually manipulating proteins, which then evolved into a platform in which participants again relied primarily on automated algorithms to manipulate the protein but with a newly gained level of involvement and control. These evolutions of user–technology relations set the stage for entirely new human–AI relations to evolve in the form of supporting the development of Alphafold and its subsequent integration into the platform to review and comment on the protein structures developed by human–computational tool relations.

In the example of Stall Catchers, an analysis problem that could not be solved with current AI technologies led to the development of an HC-based CS in which volunteer participants were invited to fill in and assist by analyzing short research videos under the supervision of algorithms. This then facilitated the development of ML models, which were subsequently integrated into Stall Catchers to assist in the analysis. Here, the AI bots based on the ML models became both coworkers and competitors to human participants and could serve as preprocessors for humans in the future, thereby, also influencing the human participants' practices from analyzing videos of varying difficulty to focusing on the hard ones. This, in turn, would also impact the human participants' experience of contributing to Stall Catchers. If ML models can one day completely take over the analysis task, the designers and researchers believe that human participants could then move on to other tasks that are not yet computationally solvable, such as curating data before it is analyzed by AI bots, which would again intravert the human–AI relation.

Similarly, the researcher–technology relations were continuously changing in the HC system formed by the researchers of the Schaffer–Nishimura Lab and the Stall Catchers platform. They intraverted from the manual labor of analyzing data on the part of the researchers, to the development of Stall Catchers, accompanied by the introduction of computational tools and AI models to reduce this workload in the laboratory and the introduction of new tasks to be performed by researchers, such as masking data, correcting DeepVess' errors, and successfully reincorporating Stall Catchers' output into their scientific work.

Finally, the analysis of the long-term evolution of ARTigo revealed the interactions between intraversions of human–technology relations in HC systems and advances in AI. Due to advances in computational and algorithmic capabilities, participants' tasks and ways of engaging with the platform, as well as the purpose of the games themselves changed over the years. While participants initially provided text-based annotations about the images' content, they were then asked by the software to provide information

---

sifications to feed into the model, the latter keeps refining and, the assumption is, less and less labeled data will be required (Walmsley et al. 2020, 1555).

about their feelings when viewing the artworks and to specify image regions for existing annotations through tangible modes of interaction. These reconfigurations were necessary to ensure participation and to keep the HC system at the edge of technological development.

As the examples illustrate, intraversions are processes that, even when stabilized for a certain period of time, eventually present openings for new tweaks and improvements; the circumstances in which they occur tend to actively invite, almost require, such change. However, they are not arbitrary but contingent, imagined, and emergent. They are *contingent* in that their becoming always depends on previous relations and given materialities. They are also *imagined* in that they can occur through external deliberate design modifications, such as decisions made by developers, and *emergent*, in that they depend as much on and evolve through the underlying human–technology relations, which create new possibilities through their dynamic and partly unstable nature. In the Foldit example, participants requested to be involved in more active ways than just observing the automated program in Rosetta@Home, leading to the creation of the Foldit game, in which participants became decision-makers regarding the next steps to fold proteins and the automated tools assisted them in their attempts. In this way, participants can also be understood as designers of the intraversions of the human–technology relations in which they are involved. In the example of Stall Catchers, participants expressed their perspective on the AI bots as both assistants and competitors and, thus, actively related and contributed to the experimental research on new hybrid human–AI combinations in HC-based CS projects.[45] Participants accommodated the new presence of AI bots and, thereby, reflected on how they wanted to interact with AI bots in Stall Catchers. Following Dorrestijn, they "perform[ed] a transformation to their hybrid self" (2012a, 117).

At the same time, intraversions are still influenced and evoked by coincidental or accidental events, material breakdowns, or different relations intervening with each other. Finally, intraversions also go hand in hand with resistance and divergent understandings of the meanings of infrastructure, roles, and overall aims of CS games. On a more abstract and subject-focused level, these practices of resistance or tactics employed by different actors within intraverting human–technology relations can be understood as "coping with [the] influences" (Dorrestijn 2017, 318) of technology. In this sense, new subjectivities emerge from these intraverting relations (Foucault 1983; 1988).

Together, the HC-based assemblages studied are, thus, shaped by different actors, their intentions, and entangled human–technology relations that do not always align frictionlessly. Alignment or reterritorialization processes play an important role in bringing together the different interests, needs, visions, and material possibilities of the actors involved. In the next chapter, I discuss the example of trust building as one such alignment process, which became necessary due to the destabilization of established trust-building practices by intraversions. I show how trust emerges and must be adapted

---

45 As an example, even though AI bots redeeming points was being discussed as a potential future feature, the team decided to refrain from allowing it during the bot study in October 2021 due to concerns expressed by participants on the leaderboard.

alongside intraverting relations in HC-based CS. Trust, as I understand it in this next chapter, unfolds within human–technology relations.

7 Building Trust in and With Human Computation[1]

"[W]e have to trust the whole citizen science stuff" (fieldnote Jul. 27, 2021), explained researcher Anna during the first laboratory meeting I attended for my ethnographic fieldwork in Ithaca in 2021. In this meeting, the laboratory discussed recent improvements in the data pipeline that connects the laboratory's Alzheimer's disease research with the online game Stall Catchers. Although this statement struck me as remarkable at the time, I had not considered "trust" an empirical category of interest or an analytical concept for my research on HC systems. However, throughout my first three-month stay in Ithaca to learn about the researchers, developers, and designers' perspectives on their joint endeavor, Stall Catchers, the notion of trust came up repeatedly in conversations with laboratory members. Moreover, trust also emerged as a critical aspect of legitimizing the HC-based CS approach to knowledge production pursued at the Human Computation Institute and its mission to "engineer sustainable participatory systems that have a profound impact on health, humanitarian, and educational outcomes" (Human Computation Institute, n.d.).

What does it mean for biomedical researchers to "trust the whole citizen science stuff"? Trust, as I will show in this chapter, plays a crucial role in the formation and maintenance of HC-based CS assemblages. By analyzing not only the articulations but also the practices of researchers in human–technology relations, it becomes clear that trusting the "citizen science stuff" is not only a social relation but also includes sociomaterial[2] practices that continuously reestablish trust. The reappearance of the notion of trust in the field brought me a new analytical perspective on how these systems are continuously becoming in the interplay of the different human and nonhuman actors involved. I consider trust as reterritorialization processes that bring together and align various elements and relations to (re)configure these sociotechnical systems. Trust as a

---

1   This chapter builds on ideas presented in the talk "Between means and ends: Data infrastructures in biomedical research" at the RAIMed conference 2022 (Vepřek 2022a), and a poster presentation at the Spring School of the *Transformations in European Societies* Ph.D. program in Murcia, Spain, on March 25, 2022.

2   In this chapter, I use the term socio*material* practice, which includes socio*technical* practices, because the point I would like to make is not specific to technologies but refers more broadly to how trust is built in relations of humans and materialities.

sociomaterial practice is at the core of HC development as it emerges and needs to be adapted alongside the intraverting relations in HC-based CS due to new changes and shifts unsettling established trust-building mechanisms.[3] Trust, as I understand it in this chapter, is itself created within human–technology relations.

To unfold this perspective, I revisit some of the examples discussed previously through the lens of trust as a sociomaterial practice, while also discussing some new instances of HC-based CS and its human–technology relations. The chapter is organized as follows. First, I briefly discuss trust as an analytical concept before delving into various trusting practices and relations in the example of Stall Catchers. I then analyze trust within HC-based collaborations from the researchers' point of view and their understanding of scientific confidence. HC itself plays an essential role in building trust in the CS approach. At the same time, trust in HC systems must be programmed algorithmically, as I discuss from the perspective of the Human Computation Institute's team. Next, I turn to the participants' perspective and their trust in the researchers and the Stall Catchers team. Before concluding this chapter, I briefly discuss the question of proprietary software and trust. The aim of this chapter is not to disregard trust as a social relation but to open up the concept of trust to sociomaterial practice to gain a deeper understanding of human–technology relations in HC-based CS. The understanding of trust suggested in this chapter, thus, does not seek to define trust once and for all—which is an impossible undertaking anyway, as human computer interaction researcher Richard Harper shows in the edited volume *Trust, Computing, and Society* (2014a)—but rather to provide a valuable concept for analyzing the field at hand. While trust emerged from the field research on Stall Catchers and this chapter is, therefore, largely based on this example, it also draws on Foldit in some places as a comparative study to further reflect on the larger context of trust in the field.

Trust as an Analytical Concept

Trust, as a sociological concept, has received considerable attention.[4] In 2000, sociologist Piotr Sztompka described a "new wave of sociological interest in trust" (2000, 14),[5] which he attributes not only to a variety of reasons, such as the growing complexity and

---

3   My interest is in how trust is being built in HC-based CS. This exploration necessarily includes considering instances of lack of trust or mistrust. Anthropologist Florian Mühlfried suggests that mistrust, compared to trust, remains an understudied phenomenon (2018, 7). Mühlfried defines mistrust, following sociologist Niklas Luhmann, as a way of reducing the complexity of the world. Mistrust seeks out "defensive arrangements" (Luhmann 2014, 1), that, according to Mühlfried, are "ways to spread risks and weaken dependencies" (2018, 11). However, I do not delve further into conceptualizations of mistrust or distrust (on the relation between mistrust and distrust, see, for example, Carey [2017, 8]; *cf.* Mühlfried [2018]), but refer to the terms as they are used empirically by my research partners.

4   It is not my aim here to provide a comprehensive overview of the existing literature on trust. To point to further work beyond that discussed, see, for example, Garfinkel (1963), Braithwaite and Levi (1998), Gambetta (1988b), Apelt (1999), Endreß (2012), Schilcher, Will-Zocholl, and Ziegler (2012), Weichselbraun, Galvin, and McKay (2023).

5   For a review of sociological theories on trust in the twentieth century, see Sztompka (2000).

interdependence of the world and human relations but also to the indeterminacy of social roles or the increasing opaqueness of institutions and technological systems. Scholars, such as anthropologist Alberto Corsín Jiménez, have also observed a "crisis of trust" (2011) over the last few centuries. This crisis manifests itself in a decrease of trust not only in the state but also in institutions, and, as can be observed in the climate crisis and during the COVID-19 pandemic, in science. Contrary to earlier psychological understandings of trust as a personal attitude, according to Sztompka, trust is now conceived as "the trait of interpersonal relations, the feature of the socio-individual field in which people operate, the cultural resource utilized by individuals in their actions" (2000, 14). The interactive and interpersonal nature of trust (Weingardt 2011, 9) is an important aspect in the literature on trust, which is commonly understood as (a) social relations(hips) (e.g., Luhmann 1988; 2014; Hardin 2006; Weingardt 2011).[6] According to Sztompka, trust is something inherently human—it cannot be bound to natural phenomena (2000, 20). What characterizes trust is its future orientation and its association with the uncertainty and uncontrollability of the future (Sztompka 2000, 20). According to sociologist Niklas Luhmann, trust serves to reduce the complexity of the world and is both a risky investment and a "solution for specific problems of risk" (1988, 95).[7] However, trust does not refer to *any* future action but to those that shape our present decisions (Gambetta 1988a, 218–219). Trust, therefore, fills the gap between risky and unpredictable futures and the need to make decisions and take action. Sociologist Heinz Bude, following Luhmann, has aptly summarized the essence of trust:

> Trust opens up perspectives and enables action in complex and complicated situations. Wittgenstein provided a philosophical explanation for this when he explained how, through trust, the reason of life overrules the unreason of doubt. This is, as Niklas Luhmann has poignantly demonstrated, the logic of a risky advance, which is only ever rational in retrospect. From the feeling of trust, one dares to take the leap into the dark, which overcomes the hiatus between justification and decision. (Bude 2010, 11)

This sense of trust, sociologist Anthony Giddens contends, is not limited to social relationships or individuals but can also be extended to abstract systems (1990).[8] Giddens employs the term "abstract systems" to refer to symbolic tokens or expert systems that play a significant role in structuring life in the modern era. The subtle difference between these forms of trust is illustrated by the example of money: "it is money as such which is trusted, not only, or even primarily, the persons with whom particular transactions are carried out" (Giddens 1990, 26). As this quote indicates, however, trust in persons is

---

6 Some authors divide trust as social relations further into trust in oneself, trust related to certain networks of the lifeworld of which one is part, and trust in systems and institutions (Bude 2010, 13).

7 According to Luhmann (2014), distrust is the functional equivalent to trust. Therefore, they both function in the same way.

8 Sociologist Martin Endreß defines trust as a multidimensional phenomenon by considering three different modes of trust: reflexive, habitual, and pre-reflexive functioning mode (2012).

still, to some extent, involved in trust in systems regarding their *"proper* working" (Giddens 1990, 34, emphasis i.o.). Gidden's distinction is valuable for the following analysis because trust in the field includes the social dimension but also goes beyond it. However, like the sociological theories discussed previously, Giddens did not pay particular attention to the materialities (or sociomaterialities) for building trust, which play an essential role in understanding trust in sociotechnical systems, such as HC-based CS projects. In fact, most theories of trust in sociological and social theory, despite varying definitions of trust and related concepts, such as confidence (e.g., Luhmann 1988; Giddens 1990; Seligman 2000),[9] share the understanding of trust as a cognitive, social, and only human phenomenon. Yet, trust, as it was expressed and performed in the field, cannot be captured by a cognitive understanding of trust alone because, as Corsín Jiménez aptly describes, trust is "also distributed in a variety of human and nonhuman forms; it is as much a cognitive category as it is a material one; indeed, it belongs to the realm of the intersubjective in as much as it belongs to the interobjective" (2011, 179). Although Corsín Jiménez explicitly points out that his work does not aim to contribute to the existing literature on trust, it forms a fruitful basis for the following reflections since, as I will show below, trust is built *along* and *with* technology in the field of HC-based CS projects (Ingold 2007).[10] In postphenomenological terms, it is mediated by technology. Researchers in the laboratory are continuously working on establishing trust with technologies precisely because "trust [in] the whole citizen science stuff" (fieldnote Jul. 27, 2021) cannot be based entirely on social relations. In HC-based CS projects, scientists, developers, and participants enter new collaborations with each other and other actors, thereby introducing new relations and practices previously unknown to the individual parties and partly requiring the reevaluation of established processes and practices previously used to build and maintain trust.

---

9   Luhmann and other scholars have distinguished trust from other concepts, most prominently from the concept of confidence. According to Luhmann, trust and confidence share that they are both "modes of asserting expectations" (1988, 99), whereby these expectations can turn into disappointments (1988, 97). However, while confidence is the case when one has no other choice than to be confident in something, trust is connected to one's active previous engagement. Trust implies risk in a particular situation in which one decides to act in one way or the other (Luhmann 1988, 97–98). However, relations of trust or confidence can transform into the other. The distinction between trust and confidence is also common in everyday life, as the proverbs such as "trust, but verify" or the German form *Vertrauen ist gut, Kontrolle ist besser* demonstrate (*cf.* Seligman 2000, 17). While religious studies scholar Adam Seligman agrees on the distinction between trust and confidence, he argues that "trust" in its current understanding "as a solution to a particular type of risk" (2000, 7–8) is a phenomenon specific to modernity and not universally transferable to social organization in general (2000, 6). Confidence, by contrast, is required for any social organization to work. On the other hand, Giddens defines trust as "a particular type of confidence" (1990, 32) and, thus, not as something different from confidence. Similarly, during my field research, I could not observe a clear distinction between confidence and trust as suggested by Luhmann and Seligman. Instead, and as I will show in this chapter, trust, confidence, and control merged smoothly into each other and were sometimes interwoven to an extent that it was not possible to clearly differentiate between them.

10 Pink et al. have also shown how trust can be built through "familiar technologies" (2018, 11), such as paper documentation.

I analyze human–technology relations in practice to grasp how trust emerges and is maintained in collaboration and knowledge production in my case study. In this sense, and building upon recent anthropological approaches to trust (e.g., Pink, Lanzeni, and Horst 2018; Pink 2021; 2022; 2023; Weichselbraun, Galvin, and McKay 2023), I explore trust as a sociomaterial practice that is distributed between human and nonhuman actors, such as software tools and data flows. While the understanding of trust as a purely cognitive phenomenon seems too narrow for the analysis of HC-based CS (and, I think, sociotechnological or technologically mediated lifeworlds in general), the framework of trust as sociomaterial practice includes trust as social relations and opens it to human–technology relations in practice.[11] In this way, it also allows one to capture the understandings of trust that appear in the field studied itself without requiring a strict definition of the boundaries of trust—such as defining in advance where trust turns into confidence or belief—and, hence, remaining open to the meanings and practices that unfold in HC-based CS. Focusing on practices, moreover, follows a "processual theory of trust" (Pink, Lanzeni, and Horst 2018, 11), as suggested by Sarah Pink, anthropologist Deborah Lanzeni, and sociocultural anthropologist Heather Horst, which "maps out how people cope with the inevitable uncertainty and contingency of the emergent circumstances of everyday life" (2018, 11–12). Trust as sociomaterial practice, thus, resembles Pink's approach to "everyday trust," which understands trust as an anticipatory concept that "involves 'a sensory experience of feeling or disposition towards something' rather than an explicit cognitive decision made in relation to a specific technology" (Pink, Lanzeni, and Horst 2018; cited in Pink 2022, 47). In what follows, however, I aim to focus not so much on the feeling (as a sensory experience) of trust as a result of specific configurations (Pink 2021, 193) but on the trust-generating practices themselves unfolding in relations between humans and technology that are always situated in everyday and historical contexts. By focusing on practice, it also connects to approaches to trust that see trust as a "*doing* rather than a fixed point" (Garfinkel 1967; cited in Harper 2014b, 324, emphasis i.o.; *cf.* Watson 2014). Trust, then, is "ephemeral, emergent, contingent and shifting" (Pink 2023, 29) and, as such, is not independent of mistrust but interdependent with it (Pink 2023, 38; Mühlfried 2018, 11).

In the following subchapters, I will jump back and forth between both trust as an empirical category and an analytical concept to capture the different layers of trust involved in HC-based CS projects and foster the dialogue between them. Here, a basic tension between the analytical and empirical term may always remain, but I hope to turn it into a productive one.

---

11 Trust as a sociomaterial practice, thus, does not form an alternative or contrast to trust as social relations but, instead, a broader understanding that encompasses both human–human, human–technology, technology–human–technology, etc. relations.

Trust as a Sociomaterial Practice

"We Have to Trust the Whole Citizen Science Stuff"

Science is an inherently unpredictable adventure, and it is part of doing science to "wrestle with the unknown" (Schaffer, Dec. 7, 2021). Research pushes the boundaries of knowledge; therefore, it must deal with scientific uncertainty. Schaffer explained in our interview that "scientific uncertainty is something that [...] we naturally deal with" (Dec. 7, 2021). While this applies to research in general, researchers at the laboratory studied described Alzheimer's disease as a particularly complicated research subject. Very little is known about the mechanisms of the disease, and multiple factors seem to influence its onset and progression. A laboratory member summarized the challenges they face in studying Alzheimer's disease: "[w]hat a disaster of a disease" (fieldnote Sept. 07, 2021). Following Lock (2013), the laboratory's approach to Alzheimer's disease can be described with the term "localization theory" (2013, 5). Lock identifies two approaches to Alzheimer's disease research. Regarding localization theory, Lock writes, "neuropathological changes in the brain are assumed to be causal of specific behavioral changes in persons" (2013, 5). The second approach is the "entanglement" theory of dementia, in which advocates "favor theories about the way in which mind, persons, life events, aging, and environments interact to precipitate neurological and behavioral transformations that are pathological" (Lock 2013, 5). While localization theory was dominant in the twentieth century, today, there is a growing awareness of entanglements of the environment, the mind, and the body (Lock 2013, 5), as is evident in the laboratory member's quote about the disastrous disease. Despite the researchers' awareness of different entanglements at the studied laboratory, most of the experiments related to Alzheimer's disease revolved around the question of how stalled blood vessels could be resolved and blood flow restored. In our conversation in late October 2021, researcher Jada reflected on this research focus as a very small detail of Alzheimer's disease: "[E]verything we do is a minor aspect to the very complicated disease process that [...] may or may not turn out to be—even in years from now—to be really important or not. We just don't know at this point. So, it looks like now [...] it's likely going to be important, but we don't know that" (Oct. 27, 2021). Uncertainty about the future success of their current scientific efforts was part of the researchers' daily experience.

Apart from the general scientific uncertainties, there are other uncertainties specific to the research process of Alzheimer's disease research and the laboratory's collaboration with Stall Catchers. Research on Alzheimer's disease in the laboratory was marked on a daily basis by the unpredictability of experiments and their outcomes due to contingencies and the unruliness of the nonhuman actors involved. This included, for example, the material used for the chronic cranial windows, which were installed during surgery to allow subsequent *in vivo* imaging of the mouse brains, or the activity and life cycles of mice, which did not always align with the experimental procedures. How good will the image quality of an individual imaging session be? How long will a mouse survive with a new treatment? Conducting scientific research in this area necessarily included recognizing and accepting the unpredictability of experiments and the potential challenges that can arise during a research project.

The development of Stall Catchers and its introduction into the routines of researchers at the laboratory added even more previously unknown uncertainties. This was due in part to the nature of Stall Catchers as "a cutting-edge experiment," Schaffer explained (Dec. 7, 2021), a not-yet-well-established approach to analyzing scientific research data for which there were no solutions and which introduced questions about data quality and impact on the research at the laboratory which could not be anticipated at the beginning. Furthermore, unpredictable factors and, in particular, the large amount of work laboratory members had to invest in Stall Catchers without receiving any immediate benefit from the platform made it difficult to convince them of the value of the project, "[a]nd it took a little bit of pushing from Chris [Schaffer] to get the lab and everybody sort of buy-in on—because it took quite concerted effort without seeing payback for a couple of months," explained Nishimura (Dec. 7, 2021).

Although adapting work routines and methods to new technologies and tools was described as difficult for many laboratory members in general, researcher Isabel said it was especially difficult to convince some laboratory members of new steps and tools related to the CS game. She explained that this was also partly due to the distance between the laboratory and its collaborators: "Stall Catchers is a harder thing to do, it's a longer turnaround, and there is not as much of a direct feedback [...] [T]here is this lab, and then Stall Catchers and the Human Computation Institute is over there somewhere" (Isabel, Oct. 14, 2021).

The long turnaround and the novelty of the Stall Catchers approach added to the "usual" uncertainty researchers were used to. Moreover, and this will be the focus of the following pages, they could not rely on established practices and their familiar routines to work against uncertainties. Doing science was defined not only by the inherent or "natural" uncertainty of scientific research but also by the "commitment to [...] getting it right" (Nishimura, Dec. 7, 2021). Nishimura explained that her motivation to do science

> is that I really want [...] the answers that we find to be correct. And so in the field of Alzheimer's disease, it's really important to me that even if we're answering a small question or a big question, doesn't matter, and it does need to be correct, and [...] something that I really worry about is getting something wrong that is eventually going to feed into a drug that doesn't work or an idea that goes the wrong way or something like that. So that is something that I really do worry about and care about. And I think a lot of people in the lab are also motivated by what I call as a commitment to; I don't know if you want to personify like truth or something [...] that you really commit to. But getting it right. (Nishimura, Dec. 7, 2021)

The drive to be 100 percent certain of a new finding, to eliminate all uncertainty, was not a contradiction to the uncertainty that defines scientific inquiry but stood next to it. Scientific work unfolds in this space between uncertainty and "getting it right," and this is where the notion of trust comes into play.

During my first research visit to Ithaca in 2021, trust came up in most of my conversations with laboratory members when they discussed current problems in Alzheimer's disease research involving Stall Catchers. The term was most often used to describe what was not working well and why they could not always use data analyzed by Stall Catchers.

When I mentioned this observation in my conversation with the PIs, they explained that, when talking about "trust," they were not referring to trust "in people" but trust as "scientific confidence." It was about trust "in the process and in the system" (Schaffer, Dec. 7, 2021). This process- and system-oriented understanding of trust appears to be typical of HC systems as a whole, where the crowd is at a conceptual level, an anonymous, depersonalized group of individuals (see Chapter 4) and is also reflected in the development of these systems, as will be further explained later in this chapter. Trust has to be built differently when there is no person on the other side to address directly.

When laboratory members talked about ongoing problems with research building on Stall Catchers, they consistently emphasized that these were not Stall Catchers' fault. Researcher Emily stressed that "it's not their [Stall Catchers'] fault, it's, I think it's all on our side" (Sept. 8, 2021). The problems were rooted in the data pipeline, in the preprocessing of the research data to be sent to Stall Catchers. More specifically, as I discussed from the perspective of infrastructuring, members explained that the problem was that they could not trust the "absolute number" (Schaffer, Dec. 7, 2021) of stalls that resulted from the crowd annotations—not because they did not trust the crowd's accuracy in finding stalls, but because the preprocessing did not output the exact number of capillaries. Capillaries got lost in the data pipeline:

> [T]he holdup really is not so much [...] whether a vessel is stalled or not. Stall Catchers is really good at that. We're kind of getting stuck because we need to know the total vessel count and that's where ... it's ... getting a little more frustrating because Stall Catchers can, if we send a vessel that's not a vessel, Stall Catchers will mark it as not a vessel [...] most of the time. But that's also not fun for them if there's a whole bunch of garbage that we're sending [...]. So that's kind of where we're at right now. [...] I've been trying to get the data nicer for their sake and our sake. Using computers, not us, because initially, it was on us to do it manually. (Leander, Sept. 22, 2021)

This problem only became clear after the introduction and initial development of the CS project. Previously, one of the most important questions was whether a vessel was stalled. The question of how many vessels there are in the first place was introduced with Stall Catchers and along the intraverting researcher–technology relations in the pipeline. It had not arisen earlier because the manual research data analysis had been performed on the "raw" image data, where the vessel count could be relied upon. However, with the current state of the data pipeline and the preprocessing of the data, "the data we are sending, you had a whole ton of non-vessels and a whole bunch of broken-up vessels and things like that. So, we couldn't really trust that aspect of it" (Isabel, Oct. 14, 2021). Previous trust-building relations were disrupted with the introduction of Stall Catchers and the laboratory's new collaboration with the Human Computation Institute and Stall Catchers participants, as well as the new infrastructure; trust or scientific confidence had to be rebuilt.

In the following pages, I will show and discuss how trust had been established and maintained prior to the introduction of Stall Catchers and how it has evolved since the introduction of the CS platform in various sociomaterial practices.

In order to better understand the differences between the manual analysis of research data conducted in the laboratory and the analysis conducted by the crowd in Stall Catchers, the following excerpt from a field note from one of my laboratory visits in late September 2021 provides insights into the manual annotation of stalls.

> "I cannot promise that it will be exciting," explains James while plugging in a hard disk with the image stacks. He opens the folder containing another 18 folders with data of a different image session of a specific mouse, each including one image stack. James has been working on the analysis of this specific dataset for about two and a half months now. He explains that they have an extra spreadsheet with the depths of each image stack, so he knows how deep the imaging process went and how many frames he has to analyze. James picks a yellow post-it and writes down 10/20/30/40/50/60/ ... /470/480, each number representing a projection of ten slices. He will look at one projection, e.g., ten images/slices, at a time and cross out those numbers that he's already analyzed to ensure that he does not miss a slice.
>
> Before analyzing the images, James must first adjust the image program settings to fit the to-be-analyzed image stack. For example, he has to set the number of color channels [with two-photon microscopy, different color channels can be imaged simultaneously], and the depth of the image stack. James, furthermore, adjusts the brightness of the image to a contrast that is more pleasant for his eyes and sets the last channel to gray, which is easier for him to see. James keeps the little menu for the brightness setting open during the analysis session. "And from here, it's the same over and over again," he explains.
>
> As James progresses through the image stack, he frequently adjusts the brightness level to compensate for the increased graininess of the pictures at deeper levels. This helps to improve visibility and to discern finer details in the images. During the analysis, there are windows for each color channel with images from one image stack. James begins with the analysis of the first ten slices. He opens an additional window with the projection of the slices and looks for something "suspicious," a suspicious area or spot that could be a stall. If James finds something, "I keep an eye on it and go fast" through the image stack. Therefore, he switches to the combined channel, and by navigating with the right and left arrow buttons, he goes through the image stack, looking for stalls at this specific position. If he is not sure, he switches to one of the other channels depicting only specific structures or molecules. A stall, here, refers to a blockage in a capillary that doesn't start flowing within five slices. Whenever James finds one, he marks it in the channel and saves the image with the stall in an extra stalls folder.
>
> He explains that "you do everything you can to make it go faster." So, instead of analyzing one projection at a time, James opens two projections and works on them in parallel. Additionally, he uses hotkeys to speed up the clicking procedures. Not all experimentalists analyze projections of ten; others prefer to analyze projections of 20, and they all have their own pace and practices. After around 30 minutes, James completed today's first image stack. (fieldnote, Sept. 27, 2021)

Manually annotating images, as described in the fieldnote, can be summarized as dividing an image stack into projections of a few slices each, which are then compared with the scrollable image stack displaying certain channels in separate windows. Here, "scientific confidence" emerges in the manual annotation of stalls by relying not only on the experimenter's own eyes and practice to identify stalls but also through the "raw" image stacks being analyzed, which have not been manipulated by any processing algorithms—even though the data has already gone through different steps of translation (Callon 1984; Latour 1999) from excited photons captured by the microscope's objective to scrollable stacks of TIFF-format files depicting two-dimensional image slices. Interestingly, trust only became a concern after image generation. Researchers did not describe any gaps in the microscopic imaging process and the process of generating raw imaging data. To further reduce bias and error, the image stacks were analyzed by at least two laboratory members and were anonymized to "blind" the experimenters. They were not to know whether they were analyzing the data from an Alzheimer's mouse with a particular treatment or a control mouse.

Now, with the introduction of collaboration with the HC-based CS project and, thus, with the introduction of new data infrastructures that made the crowd analysis possible in the first place, these established trust-building practices no longer worked. As will be described below, this first created a lack of trust, requiring it to be built differently through new human–technology relations. Two different moments, which overlap in practice, can be distinguished analytically here: trust in the now outsourced analysis had to be (re)established and trust in the new automated preprocessing and manipulation of data—in short, the data pipeline—had to be built.

At the time of my fieldwork at the laboratory, some experimentalists still preferred to manually annotate stalls rather than "use" (James, Sept. 27, 2021) Stall Catchers. The use of Stall Catchers required researchers to rely on an anonymous crowd of people with no formal training to analyze the data. There remained a perceived gap between the researchers' own analysis and the crowd. One reason for this can be linked to a question of routine and the pleasure of working and engaging with the research data. Researcher Anna described the difficulty of handing over tasks to someone else in general and, as she described for other laboratory members, especially to an anonymous crowd of people doing the analysis:

> Where it's like, we are used to doing it this way, just manually, and when I do it manually, I trust it. But human computation, well, the wisdom of crowd type thing is like okay, well, you trust it, but is it right?! And statistically, it [manual analysis] is less likely to be right than getting a whole bunch of people that maybe don't know all the details of things, but you take an average of all those people, creating something that is probably more accurate than what you would do [...] the slow, painful way. [B]ut I think there is still a little bit of a mis... [...] I wouldn't say a lack of trust but just more of a strong belief in their own abilities than handing things off to other people and then taking those results and [...] taking those as a given. Definitely [...] with programs that they are used to [...] they can plug things in, and they get an answer from that, and they'll trust it if they've used it a few times. But the human computation stuff, they don't seem to think of it as another program. [...] [A]nd there is like this bit of [...] an inherent mistrust of "well it's just a bunch of people, how are they gonna do better than I could?!" And a lot of it is [...], I think, in the end, statistically, they will do better [...] than [researchers] could, but it's asking a large group of people to do a small repetitive task, and you don't have to invest as much time into that, but I think there is still this idea they get their results back and [...]

I need to double check this, I need to double-check and I can't just take it. (Oct. 14, 2021)

For many laboratory members, handing off the analysis step that had been part of the researchers' working routines to an anonymous crowd was difficult. At the beginning of the Stall Catchers project, the laboratory, together with the Human Computation Institute, had invested a lot of effort into validating the crowdsourced image analysis to the point where they were confident that the crowdsourced analysis was at least as statistically accurate as the researcher's results. Researcher Jada described this process as challenging because it involved a lot of extra and duplicative work:

> [I]n the beginning, it took at least [...] three to four years to really kind of get something out that works. And this is a long time and ... along this way, you did a lot of doubling of your work because I was counting [...] as before just to make sure that this sort of matches up. And then you [...], it's maybe kind of your baby, you're switching away to [...] trust it. It's like, sometimes you have a person working with you; it's kind of the same. So, you do sanity checks, but then you give it away. But those times just have been sometimes months of work to [...] confirm it on both hands. [...] [But] I think it was still critical for the data and for me personally, mostly. (Oct. 27, 2021)

Eventually, Jada explained, researchers had to give away the task and start trusting the outsourced analysis. Still, researchers had trouble relying on the crowd's answers. There was a gap between the statistics and the distance between the platform and the crowd, as perceived by laboratory members. Jada further described the difficulty of no longer interacting with the data, of not being able to build the relation with the images they would usually build if they had annotated the images themselves. Looking at the images themselves is sometimes "just [...] for [their] sanity" (Jada, Oct. 27, 2021). Jada explained, "I just would like to see these in the full stack" (Oct. 27, 2021). The reason they could not simply trust the result was in part because the preprocessing programs might still have difficulty identifying potential stalls in certain regions or edges (Jada, Oct. 27, 2021). However, Jada noted, "on a whole scale, you could say that doesn't matter, which is sort of true, but I still think it matters sometimes" (Oct. 27, 2021). Even if the difference was not statistically significant, it was important for the researcher's reason to analyze the images manually.

It was also a matter of routine. "[I]t's also the way I got trained myself to do it" (Jada, Oct. 27, 2021). Here, trust was built in and with routines and, as Pink and colleagues observed in their research on digital data anxiety and practices, "trust is invested in the routine, or a sense of trust is gained through the familiarity of the routine" (Pink, Lanzeni, and Horst 2018, 7).

Student researchers or those in their early scientific career "are all focused on future career and [...] building up to something," explained researcher Charles (Oct. 14, 2021), who was further along in his scientific career. Therefore, there was another reason for them not to simply trust Stall Catchers and new computational tools: "So [...] the ambition tends to make them a little bit distrustful of anything new, and I think that's both Stall Catchers and [new computational tools]. It takes a lot of work to get them to feel comfortable with it and to adopt it" (Charles, Oct. 14, 2021). Since their future scientific careers depended on their performance in the early years of scientific research, it was crucial for them to maintain direct control over the research processes. This became more difficult when some key processes were delegated to other (unknown) people or to computational tools that were not established research tools.

Recalling the "Closing the Loop" section of the last chapter, I would like to discuss how this lack of trust in the results of Stall Catchers and the preference for manual annotation and engagement with the images unfolded in the researchers' everyday practices with Stall Catchers results. Once Stall Catchers had completed the dataset analysis, the crowdsourced annotation results of individual vessel videos as flowing or stalled could be downloaded by the researchers as a spreadsheet. Researchers could go back to the videos and "recheck" (Emily, Sept. 08, 2021) the crowd results with the links to the corresponding Stall Catchers video files in the spreadsheet. Some researchers even returned to the "raw" image files to verify the data after reviewing the Stall Catchers vessel video.

Such a validation step was not intended to be necessary in the original design of the Stall Catchers' collaboration. However, Michelucci explained in a conversation in October 2021 that the initial idea had been that once they had the crowd answers, they would no longer need expert answers. The crowd's answer was supposed to replace the expert answer, allowing researchers to take the results and move on to the next step in their research. In practice, however, researchers still went back to check all the stalls in the images analyzed by the crowd.[12] Consequently, a second protocol that they used during the time of my research was implemented to at least facilitate the verification of Stall Catchers results. This was the procedure discussed in the previous chapter, where the results of Stall Catchers were ordered from high crowd confidence of a vessel being stalled to low crowd confidence, so that researchers only had to review the first 200 videos rather than all of them (fieldnote, Oct. 18, 2021). As the researchers reviewed the results, they added their own vessel video annotations to the spreadsheet. These new annotations—not the participants' aggregated annotations—were then considered the final "ground truth" data labels. The crowd answers, or more specifically, the crowd's confidence levels, were, thus, considered as benchmarks and guiding indicators but not valid answers the researchers would continue to work with directly. Nonetheless, the participants' annotations greatly reduced the researchers' workload, who only had to review about 200 vessels compared to the full manual annotation of about 50,000 vessels in a dataset (fieldnote, Nov. 4, 2022).

This example shows how new trust-building procedures and practices were implemented to allow researchers to manually review the results and connect the abstracted vessel videos, the *dry* data, back to the *wet* data, i.e., a mouse model with or without a specific treatment to create confidence in the data. Trust in the results generated by the crowd did not simply exist but was being built by interacting with the data—in its different states from imaged mice to "raw" image stacks to Stall Catchers videos—and the resulting numbers and data references presented in spreadsheets. At the same time,

---

[12] Similarly, in the example of Foldit, researchers reviewed all solutions submitted by participants by both manually looking at them and with computational analysis. Foldit researcher José explained that they did this "to try to see if I can find any errors just from looking at their models, but if they look good then we can test those in the lab" (Jan. 22, 2020).

trust in the CS collaboration was established in researcher–technology relations through infrastructuring, by working on the individual preprocessing steps, becoming familiar with the functionalities of new tools, and building a trustworthy data pipeline that would output the correct number of vessels (Chapter 6). Trust-building practices related to Stall Catchers were well underway but not yet established in 2021.

In the previous chapter, I showed how, during my second research visit in 2022, it was hoped that these would be established and, thus, the Stall Catchers loop closed once the new post-processing tool, which was supposed to further facilitate the researcher's experience with the pipeline and Stall Catchers, was in use (fieldnote Oct. 25, 2022). The hope was that once this new tool became part of the everyday research practice, all laboratory members would finally trust the whole project; building trust takes time (Endreß 2012, 91). The example of the data pipeline shows how trust in the collaboration with Stall Catchers "moves in and out of different idioms" (Corsín Jiménez 2011, 183), such as data quality, vessel count, and result-checking, and cannot be reduced to trust or lack of trust as a mere social relation.[13]

Another example of how previous trust-building practices had to be reconfigured with the introduction of Stall Catchers is the "blinding" procedure, which had to be translated from its original manual form into the virtual game environment. As described above, researchers would typically analyze image stacks manually without knowing the treatment or mouse model on which the data were based. However, despite efforts to de-identify the images, it was still possible for some researchers to identify the underlying mouse model:

> [T]ypically, I can recognize an Alzheimer [mouse] relatively easy. [...] [G]enerally, that doesn't mean I know the treatment. I try to not care at this stage about it [...] or [another researcher] is doing it and [this other researcher] definitely doesn't know what mouse is what. But for me [...] because I have done many other analyses with those mice before[ ], I typically know what it is. (Jada, Oct. 27, 2021)

The introduction of Stall Catchers, therefore, presented an opportunity to even improve anonymized analysis. At the same time, however, translating the anonymization process into the virtual game platform posed several challenges. For one, the crowdsourced data analysis approach was not an established research method with defined procedures. Therefore, the laboratory had to work with the Human Computation Institute to develop a new process to guarantee that participants would not know what research data they were looking at. It was essential that the original data, i.e., a mouse with or without Alzheimer's disease, be alienated. Nishimura explained:

---

13 Interestingly, when I returned to the laboratory in Fall 2022 and discussed my observations regarding trust with the biomedical researchers, some mentioned that they still preferred to sometimes annotate research data manually instead of sending it through Stall Catchers. When reflecting further about this preference, one of the laboratory members explained that there was no longer any reason for this, since the problems they used to have with ensuring the data quality with the Stall Catchers data pipeline had been resolved (see Chapter 6). However, the gap had not yet been closed, the new process of preparing data and sending it through the crowdsourced analysis had not yet become an established routine (fieldnote, Oct. 21, 2022).

> [I]t took a few iterations to come up with [...] this blinding, [...] that's the gold standard for data analysis, and sometimes we don't even quite do it in the lab as well as I would like. [...] So, it kind of takes a bit of [...] a leap of faith that you really have to do this right, but I do think it's important. (Dec. 7, 2021)

To get it right, in terms of scientific data quality standards, it was necessary to "sacrifice" user or play experience, as I discussed in Chapter 5. While they first thought about including a progress bar for individual mice, they later refrained from implementing it to ensure that "the whole dataset is blinded" (Nishimura, Dec. 7, 2021).

As the analysis of the intraverting researcher–technology relations showed, however, incorporating crowdsourced analysis of their data also required the development of a data pipeline, which was incomplete with the introduction of Stall Catchers and an ongoing endeavor. While I have already analyzed the process with a focus on the changing human–technology relations in the data pipeline, I here would like to return to this example with a focus on trust. In 2021, according to Schaffer, the laboratory had come a long way regarding the Stall Catchers project and now "already [had] a good confidence in Stall Catchers" (fieldnote Aug. 17, 2021). Compared to the early days of Stall Catchers, they now had "good buy-in" (Schaffer, Dec. 7, 2021). During our interview in December 2021, he noted that in the beginning,

> we had a period where people, I think, had not as good a buy-in on Stall Catchers because they saw it as a lot of upfront kind of work that they had to do. And then ... there was lingering uncertainties about the data quality. But I think with people being actively involved in fixing those problems, I think that's how you get by it. [...] And so now people can work with it and understand the limitations, understand the capabilities and not feel [...] so uncertain to or untethered, I guess, in their use of that capability. (Schaffer, Dec. 7, 2021)

Active involvement in the process, the data pipeline, and the individual tools created confidence and trust. Therefore, for example, laboratory members were required to complete training on laser alignment, which included learning how to set up the lasers for the microscopes. Researchers usually did not have to set up the laser parcourse from scratch for imaging, as it could be reused from previous imaging sessions and did not need to be changed for each individual one. Nevertheless, they should know what is going on behind the technologies supporting their work to be able to adjust them if necessary and, as Schaffer explained, to not "overtrust" the infrastructures (fieldnote Oct. 20, 2021).

In summary, prior to the introduction of Stall Catchers, building and maintaining trust in the "right" analysis of the research data and its results had already been characterized by sociomaterial practices, such as going through the image stacks, focusing one's eyes on the vessels in the raw image data, and, thereby, interacting with the data in a way that one could rely on one's own abilities and practice. Building trust was more straightforward here than with Stall Catchers, also because the data was considered "raw," or, at least, not manipulated by computational algorithms (although it had already been translated from mouse brains into digital images). Knowing that they were looking at the "original" data, there was no room for doubt about what they were seeing. The data representation displayed the "real" data, and no bug in any algorithms could have introduced distorted images or missed some vessels. Additionally, extra measures, such as the independent analysis of the same image stacks by multiple researchers and the anonymization of the research data, ensured "scientific confidence" (Schaffer, Dec. 7, 2021). With the introduction of Stall Catchers, not only did new collaborations with the Human Computation Institute and an anonymous crowd of people disrupt established practices and require new ones, but the introduction of new data pipeline steps for Stall Catchers, in particular, introduced new potential sources of error and, therefore, uncertainty suddenly multiplied.

Let us now return to the opening quote of this chapter after the analysis of how trust was built and maintained in scientific knowledge production with Stall Catchers at the biomedical engineering laboratory, namely, that they had "to trust the whole citizen science stuff" (fieldnote Jul. 27, 2021). It becomes clear that what was described in this rather plausible statement is not a mere social relation that is chosen instead of other approaches but unfolds in sociomaterial practices along intraverting human–technology relations. These practices, however, are to be located on the laboratory's side of the Stall Catchers project. Regarding Stall Catchers as the game, the software, and the HC system itself, the biomedical researchers were "confident" (fieldnote Aug. 17, 2021) and relied on the Human Computation Institute. Here at the institute, I observed other trust-building practices, which I discuss in the following section.

Building Trust With Human Computation and Algorithmic Evaluation

When I asked Paul, who had been part of the Stall Catchers team since its early days, how HC-based CS games like Stall Catchers (could) change science structurally, he explained that these projects "contribute—I hope they're contributing—to changing [...] the mentality, the way people think about science but not just normal people but the scientists themselves" (Oct. 14, 2020). Questions of trust in research with CS in general and how CS could change established structures and hierarchies within science have been discussed in scientific and public discourse (e.g., Haarmann 2013; Kosmala et al. 2016; Bedessem, Gawrońska-Nowak, and Lis 2021), and there are various proposals for frameworks to build trust in CS. For example, computer scientists Abdulmonem Alabri and Jane Hunter discuss a technological framework that uses trust models and filtering services to improve the reliability of and trust in CS (Alabri and Hunter 2010; *cf.* Hunter, Alabri, and van Ingen 2013). But despite the popularity of CS today and the support by major funding programs, such as the European Union's Horizon Europe 2021–2027 (European Commission 2021), recurring concerns remain, including those related to the evaluation of results. The authors of the Science Academies G7 Summit report on digital CS express concern about the risks associated with CS, "especially around the evaluation of results stemming from CBPR [Community-Based Participatory Research] and BTWR [Beyond The Walls Research]. These results are often disseminated through diverse channels outside the traditional peer-review system" (Gaffield et al. 2019, 1). Human Computation Institute team member Paul, who also observed such concerns, traced them back to an issue of mistrust on the part of some professional scientists. After his initial positive outlook on the potential of CS (quoted above), he clarified that today, the potential of CS has not been fully explored, in part because of this mistrust: "I think it's [...] an intimidating thing for most scientists to open up to the public [...] and involve the public [...] and they don't even understand what the public knows and can do and stuff like that. They, *they* mistrust the public" (Paul, Oct. 14, 2020). Even if CS projects were to follow standard scientific validation techniques, the problem of professional scientists not being familiar with the approach and methods of CS and the involvement of nonprofessionally trained scientists in research projects, in general, would remain. This is where projects like Stall Catchers, especially as they have spread in recent years, could play an important role in building trust, Paul argued (Oct. 14, 2020). With Stall Catchers, they "have actual evidence to support [...] this method [...] and to [...] demonstrate data quality and all the other stuff that scientists [...] worry about," continued Paul (Oct. 14, 2020). Developer Kate agreed with Paul in our November 2021 interview, clarifying that "we're proving that [...] you can do things differently" (Nov. 19, 2020). Although more and more institutions, libraries, political bodies, and universities were beginning to adopt a more open attitude toward CS, Paul did not expect a rapid change in thinking and academic scientific practice because academic institutions "change over [the course of] hundreds of years [...] but not several years" (Oct. 14, 2020). The COVID-19 pandemic has also played a role in accelerating the popularization and visibility of CS since early 2020, driven particularly by successful projects such as Foldit. "[T]heir visibility and their success, I think, is [...] helping scientists to understand what it is to open up and why do it, so ... I'm kind of ... finding myself sounding too optimistic [laughs], but I do wanna believe it's making that sort of a difference" (Paul, Oct. 14, 2020). Here, the Stall Catchers team argued that HC itself was "key" to gaining the trust of professional scientists in CS:

> I think that [...] this method [of HC] is the key to ensuring these things are done right. Cause you can, I mean, anybody can create a citizen science project [...] and just get people out there and do something. But unless you understand how to get the maximum value of individual contributions, then you're mostly just wasting your time and such a project I think risks to just become another science outreach project. [...] I'm sure there might be other methods, but I'm a bit biased, so I think human computation is sort of key in these types of projects, and [...] we will not open science unless scientists can trust the way it's done, which is why a ... solid scientific foundation like human computation methods must be behind it, I think. (Paul, Oct. 14, 2020)

HC was understood to provide the computational and statistical methods to bridge the gap between scientific knowledge production, which depends heavily on measurability and calculability in the natural sciences, and CS, as the engagement of the broader public without predefined knowledge. It, thus, linked people to numbers. Here, building trust with HC as a computational method carried the notion of legitimizing CS as a scientific approach.

But how did HC actually contribute to building trust within these projects? If biomedical researchers had to invest in infrastructuring and new working practices to reestablish trust in the results of analysis, what mechanisms, relations, and practices played a role within the human-in-the-loop system itself?

In a conversation with developer Samuel, he explained that developing Stall Catchers required a careful approach: "Because since we're working with the crowd, we're like asking everyone to connect to our system and […] all this then can be hacked and manipulated. So, we have to be careful" (Sept. 2, 2021). As an Internet platform that was meant to be accessible from all over the world and to everyone, Stall Catchers, like any other freely accessible website, was vulnerable to both targeted and random attacks. However, as Samuel explained, caution was especially important in the case of Stall Catchers for another reason: it worked with a crowd of people the team did not know. Stall Catchers developers also described this as a specificity of developing HC systems compared to other software projects they had worked on. According to Samuel, in other software projects, such as the development of online rating or booking platforms, it is often easier to identify the target audience and anticipate their needs on the platform. Moreover, since cheating is an inherent part of gameplay (see Chapter 5), the Stall Catchers source code included protection mechanisms to prevent cheating. Cheating, however, could not only break the game's functionalities and cause unfair play but could also lead to poor data quality.[14] This is probably the most significant aspect of the role of trust in HC-based CS projects and how trust is built and maintained at the source code level because poor data quality could harm the overall research. Since Stall Catchers contributes to Alzheimer's disease research, the team's top priority was to ensure that the quality of the analysis results met the requirements of scientific research as defined by the laboratory. To achieve these goals, the Human Computation Institute developed a customized "wisdom-of-the-crowds" method to calculate "final" Stall Catchers answers for a given data point. This method, as described previously, combined the responses of several Stall Catchers participants for that data point and applied a weighting mechanism to these responses according to the participants' "sensitivity" scores. This means that building trust in the collective answers and meeting the required scientific data quality requires customized algorithmic mechanisms. Before including the answers of new participants in the collective crowd answers, their skill level was assessed in the pregame tutorial. Once the tutorial was completed, participants could annotate "real" research videos for which there were no expert answers, and their annotations were incorporated into the calculated crowd answers according to their skill level. This skill level was not fixed, however, but constantly recalculated based on the participant's answer to "calibration movies," which were regularly sprinkled between research videos and to which the correct answer was known. These calibration videos were intended to be and usually were indistinguishable from the research videos. However, experienced participants and so-called "supercatchers" were sometimes able to tell the difference between calibration and research videos. Moreover, the frequency with which these "calibration movies" appeared was not random or the same for all participants but dependent on the individual's skill

---

14 Cheating in HC-based CS games can be categorized as either harmful or valuable for achieving the scientific purpose, as illustrated by the example of cheating discussed in Chapter 5, where a shortcut in the game discovered by a participant allowed the acceleration of data analysis. However, not knowing the intentions of individual participants and how cheating might influence data quality, cheating-prevention mechanisms had to be put in place. This distinction was also discussed on the Foldit forum in 2014 (v_mulligan 2014).

level. This shows how different participants and their contributions in Stall Catchers were not treated equally by the system but, on the contrary, how the system was tuned to individual participants. Different levels of trust were introduced algorithmically by continuously measuring the skill level and weighting answers accordingly.

Figure 11: Excerpt of the Stall Catchers source code with the saveNextMovie function

```
/*
 * Save next movie (for criminal clients who want skip hard known movies by
 * turning off browser
 * @param int $movieId
 * @param int/false $nextMovieIdToPreload // movie to preload in view
 * @param int $userId
 * @return bool
 */
public function saveNextMovie($movieId, $nextMovieIdToPreload, $userId)
{
    if (empty($userId)) {
        return false;
    }
    if ($nextMovieIdToPreload === false) {
        $updateArray = ['next_movie_id' => (int)$movieId];
    } else {
        $updateArray = ['next_movie_id'         => (int)$movieId,
                        'preload_next_movie_id' => (int)$nextMovieIdToPreload
        ];
    }
    $this->model->where('user_id', $userId)
                ->update($updateArray);
    return true;
}
```

Source: ©Human Computation Institute n.d.

In addition to the encoded trust-building practices to ensure scientific data quality, there were other algorithmic mechanisms in the game's source code to prevent cheating and, hence, harm to data quality, which I will discuss here with the example of the "saveNextMovie" function (see Figure 11). The purpose of this function is described in the comment above the function as "Save next movie (for criminal clients who want skip hard known videos by turning off browser." The function prevents users from switching to another vessel video without answering the current one. The algorithm, thus, ensured that all participants and their corresponding skill levels were evaluated on the same video data. The code here included the design assumption that some participants—specifically "criminal clients"—would actively try to circumvent the rules. Instead of assuming that participants would answer all videos as presented, the source code included this function to ensure that skipping videos was impossible. Furthermore, this fragment shows that it was assumed that "criminal" participants would strive for points and winning the game rather than focusing on contributing to research by skipping those videos that might be more difficult to answer correctly. Hence, decision power over the selection of the next video clearly rested with the algorithm, not with the participant.

The question of how to establish trust in HC systems is not answered once and for all but evolves alongside the continued development of participant–technology relations in HC systems. Intraversions in participant–AI relations, as with the introduction of AI bots in Stall Catchers, require new practices for building trust, as explored in the experimental study on human–AI partnerships in 2020 (Inkpen et al. 2023). This experiment, in which an AI agent was introduced into the Stall Catchers task to assist human participants with suggestions about whether a vessel video was flowing or stalled, explored the performance of hybrid human–AI teams depending on the skill level of the human participant and the sensitivity and bias of the AI agent. Michelucci explained the goal of this experiment in one of our conversations:

> [W]e [...] explored an actual collaboration [...] So I have this agent that's assigned to me as my partner. That AI agent has some level of sensitivity and some kind of bias with respect to answering about stalled and flowing. And the AI is giving me suggestions. And now, I have my own ideas about what's flowing and stalled and based on [...] seeing what the AI says, making my own decisions, seeing the outcome, I start to develop a trust maybe that the AI in certain cases knows better than I do, but in some cases, I know better than the AI. And if I could develop that predictive model about when the AI is going to be right or wrong and integrate that with my own responses and maybe collectively as a dyad, we can perform better than individually. And [...] our early result is that when the machine bias is opposite of the human bias, then we actually see better performance with the AI and the human working together than separately. So that was exciting. (Jan. 21, 2021)

While this experiment took place in a sandbox (a special environment designed to test an isolated feature or experimenting without all game functionalities and with no direct impact on the real game or Alzheimer's disease research), AI bots were later introduced into the actual Stall Catchers game, as I discussed in the previous chapter. Even before the introduction of AI bots, the idea of human participants and AI bots playing Stall Catchers side by side had already raised many questions and concerns. A few months before the AI bot GAIA was introduced to Stall Catchers, Paul described his expectations for the introduction of the bots:

> [There will be] some people who will be interested and [who] more or less understand what's happening and even be interested in competing with AI or whatever, and then there will be people who will be confused and feel replaced and not sure, mistrust the AI, and maybe the AI is now messing up the data, and their work will also be wasted. (Oct. 14, 2020)

Human–AI relations in the experimental bot studies unfolded in productive team spirit and competitive relations. How these relations will evolve in the future, if and when AI bots become a permanent part of Stall Catchers, remains to be seen. But this example shows that, much like human–technology relations, trust is not built once and for all but changes along intraversions. Finally, although the understanding of trust as a sociomaterial practice describes a shift away from the conceptualization of trust as a purely cognitive phenomenon, it does not exclude trust as a social relationship as defined by

Luhmann (1988; 2014) and others (e.g., Hardin 2006; Weingardt 2011). As a final example of trust from the perspective of the Human Computation Institute, I would like to show the importance of building trustworthy social relations for the development of new HC-based CS projects for researchers and developers. During my fieldwork in Ithaca, I accompanied Michelucci on a two-day trip to collaborative researchers in Washington D.C.[15] at the end of September 2021. The main goal of this trip was to "build relationships" (fieldnote Sept. 30, 2021) with researchers as partners for a new HC project. We left Ithaca in the late morning of September 29, arriving just in time for dinner with the researchers. To my surprise, the envisioned project was not discussed that evening. The next day, after breakfast, Michelucci and I reflected on the first impressions of the visit and discussed the importance of building social relations. Michelucci mentioned that he often had to first explain the Human Computation Institute's projects, such as Stall Catchers, and the organization of the Human Computation Institute because "the HCI itself is an unconventional institution" (fieldnote Sept. 30, 2021). Therefore, meeting in person and spending time together to get to know each other and build trustful relationships before developing a new HC-based CS project was important. While they had started working together and developing ideas for new projects more than a year ago, it had not really taken off before the September 2021 visit. The face-to-face meeting made it easier to discuss mutual expectations and clarify the approach, which could build upon the Human Computation Institute's previous experience in building Stall Catchers but for which there were no common procedures. A trusting relationship could be established through a personal meeting.

Trust the System, Trust Yourself

If we now turn to the participant's point of view, trust and mistrust unfold yet again differently. Trust came up in several conversations with participants when I asked them how important it was to them to learn about the scientific developments behind Stall Catchers. Here, trust in the Human Computation Institute and the researchers conducting the Alzheimer's disease research played an important role. At the same time, trust, or more accurately, mistrust in one's own abilities and in getting the answers right, was either explicitly or implicitly raised by some participants as one of their concerns. In this section, I focus on these two dimensions of trust.

Using the Human Computation Institute's blog, the Stall Catchers team provided insights into what was going on behind the UI at the institute, announced new features and special events, and provided updates on the research behind the game. Whenever a new dataset was uploaded to Stall Catchers for analysis, a blog post was published explaining the purpose of that particular dataset. Institute member Egle (seplute) explained in a July 22, 2017 post that "[t]he new dataset is focused on the effects of a **high fat diet on stalls in the brain in Alzheimer's disease**" (2017, emphasis i.o.). This brief statement was followed by an explanation of the research question or aim and its impact on Alzheimer's disease research: "We are seeking to understand the cellular mechanisms linking cardiovascular risk factors to Alzheimer's. Analyzing this dataset will be a big push towards un-

---

15   The destination is anonymized.

derstanding this long-debated link!" (Egle [seplute] 2017). For participants who wanted to learn more about the dataset, a more detailed scientific explanation was provided, citing research collaborator Schaffer. The Stall Catchers team aimed to make the research behind Stall Catchers more comprehensible and accessible with these blog posts, which were also shared and linked in the in-game chat.

Interestingly, the answers varied widely when I asked participants about the importance of learning about these developments and understanding the science behind Stall Catchers. It was important for many participants to learn about the scientific developments because they "like to know how playing the game is actually helping progress the research (Ebby, May 08, 2020). Or because "[t]hat's a reason […] for what you do. […] That's certainly something you can learn from" (Kamon, May 15, 2020). Akin even doubted that they "would engage in it at all if it weren't for that" (May 11, 2020). Reading about the progress created a sense of accomplishment and ownership: "Well, it's important from the standpoint that if I see the science of this making progress forward, I can feel a little piece of ownership that, hey, I think I've helped with making that progress forward" (Daan, May 26, 2020).

Other participants, however, expressed their interest in learning about the datasets they were analyzing and the scientific developments as a form of feedback that was not particularly important to them: "for me [it] is really not that important because I am confident that it will get sent there somewhere," Ellen explained (May 19, 2020). Caitlin said,

> it makes it more interesting if you know how the data is being used and what the scientists are learning from it. So although it wasn't my primary motivator early on, I just trusted that what we were doing was of use or we wouldn't be doing it, but it's good to get the feedback like with this NOX-inhibitor business, […] that's very positive feedback to be getting that […] we might be making some progress here and in some small way what we're doing may help advance progress towards a cure perhaps even for Alzheimer's. So, yeah, that's very good to have that kind of feedback. (Caitlin, May 5, 2020)

It was not particularly important to these participants to be kept informed of all scientific steps, or it was not the primary motivator for learning about Alzheimer's disease research because they were "confident" or "trusted" that their contributions would be meaningful and used for scientific purposes. The terms "trust" and "confidence" were used here to describe a similar relation. Although "trust" and "confidence" are commonly distinguished in sociological theories (among others, Luhmann 1988; 2014; Hardin 2006), the purpose of this work is not to apply a fixed definition of these terms to the empirical material but to better understand how trust unfolds. With this in mind, it is possible to see how trust navigates through various expressions in everyday life.

The distribution of knowledge in Stall Catchers and the role allocations of different actors can explain the trust relation. Longtime participant and frequent player John explained that

> you have to be able to trust the people […]. Hopefully, the people at the top would be willing to say "Hey, we've kind of reached the end of where we wanna go here;

maybe we gonna have to take some time moving to another area of research, or maybe it's time to pull this back"... or [...] if the people at the top are honest and if you believe they are—cause it's hard for maybe the average person to verify—[one would have to] read a bunch of articles [...], you put your head down...and it's not! I don't mean to make it sound like a struggle, but you dive in, and if you want to do it, then you do it. And if you do one film, that's good! If you do one million films, that's good! But we're all trying to go towards the end of hopefully—whether we know it or not—figuring this thing out. Even [...] if the research shows something contrary to what the scientists had thought, it was going to show and it's a dead end, that's good too because we know where not to look now and hopefully we spread that around saying you know "we did this and it is peer reviewed, and there is no point really going down this place anymore." (John, May 7, 2020)

While participants could read about the science behind Stall Catchers, they did not need to know the details about Alzheimer's disease research to contribute to the project. They had one specific task on the Stall Catchers platform: to analyze the data presented to them. While the Human Computation Institute took care of the algorithmic evaluation of participants' input and ensured that the required scientific data quality was achieved, the researchers in the Schaffer–Nishimura Lab directed the scientific approach, including decisions on which the research questions to investigate, and how to proceed.[16] Here, trust in the researchers filled the gap in scientific knowledge about Alzheimer's disease research and allowed participants to perceive their contribution as meaningful.

Despite trust in the scientists and the Human Computation Institute team to develop and maintain the platform and game in such a way that the participants' contributions were perceived as purposeful, I observed another dimension of trust in the participants' own abilities to analyze data on the Stall Catchers platform. Some participants expressed forms of mistrust in themselves or, more precisely, in their own ability to make meaningful contributions to Stall Catchers and Alzheimer's disease research. Some participants explicitly described their fears of answering research videos incorrectly in our conversations. Elle, for example, who began contributing to Stall Catchers in 2015, felt that she did not have the time to participate in Stall Catchers as much as she would like because her "level of emotional investment and commitment isn't always matched by [her] actual actions" (May 13, 2020). At one point in our conversation, she expressed fear of being presented with a research video that had not yet been annotated by many other participants (the UI includes an information box below the video frame indicating how many participants have answered the video presented and how): "when you start moving from the ones that are being done by multiple people into ones that you're the first person or the second person to check and you don't actually know whether other people agree with you or not, it's a little bit more scary" (Elle, May 13, 2020.).

Noemi, a participant who had not contributed to Stall Catchers for a while before our interview but who had gone back to Stall Catchers the night before our meeting, explained as she reflected on her Stall Catchers experience: "I was doing some last night, and I was like oooh now I have to get back into what I'm looking for ... cause I'm making

---

16   As described by Michelucci, in a few rare instances, though, the researchers took up suggestions from participants for specific studies (fieldnote Nov. 9, 2023).

mistakes, and I'm feeling bad, I'm making mistakes, that's wrong! But usually when I have long periods of time is when I will go and play" (Noemi, May 14, 2020).

Stall Catchers participants contributing to my research agreed that becoming "good" at Stall Catchers required practice. Noemi further explained that she would have to get used to playing Stall Catchers again and attributed her mistakes to her long absence from the project. It was very important for her to contribute in a meaningful way, and, therefore, she was afraid of getting videos wrong: "[I]t is challenging! It is challenging and sometimes [...] with this game I'm like 'oh no I don't wanna get it wrong'" (Noemi, May 14, 2020).

The fear of contributing "bad data" can be connected to the participants' understanding of how the HC system works and their practice, i.e., how often and how much they contributed. Participant Olav explained how participation helped build trust in one's annotations: "the longer I think you do it, [...] you [...] can then trust yourself after you've made a decision [...] that's the right decision because you've done it enough to know [...] what [...] clear, flowing [...] things should look like" (May 21, 2020). By regularly engaging with vessel videos, participants learned to distinguish between flowing and stalled vessels and developed an understanding of the system's functionalities. According to the Stall Catchers team, the implemented security measures and the "wisdom-of-the-crowds" methods made it virtually impossible for incorrect answers to harm the overall analysis. Most of the participants who were aware of this did not fear inputting incorrect data, although some also described missing stalls as a bad experience (see Chapter 5). In an interview with the Human Computation Institute published on the institute's blog, "supercatcher" Carol aka Mema described her "misses" as one of the things she disliked most about Stall Catchers because "I feel like I let myself and others down" (Carol aka Mema 2019).

Practicing and engaging with the system regularly, as well as reading up on the features in the FAQs or blog posts explaining the design of Stall Catchers and, thus, understanding how Stall Catchers as an HC system and the "wisdom-of-the-crowds" method work can, therefore, be interpreted as trust-building practices (*cf.* Pink, Lanzeni, and Horst 2018). Participants do not have to worry about their skill level in catching stalls, as participant Maya summarized: "[I]t doesn't matter if you're right or wrong because it's all statistics and [...] it's okay to be the wrong one, and it's okay to be the right one. And then depending on if you're right or wrong however a number, many of times, it starts to change what they give you" (May 13, 2020). By contrast, lacking this experience and the understanding of the game mechanics and crowd answer computation led to not trusting one's own analytical skills to contribute meaningfully to Stall Catchers.

The Question of Trust and Proprietary Software

Before turning to concluding thoughts on trust in and with HC-based CS, I would like to address a final aspect related to the participants' ability to understand the system and, hence, to verify that their contributions are meaningful. This aspect does not only apply to the example of Stall Catchers but was also discussed in the example of Foldit.

Both projects are based on proprietary software and the teams justify this in part because it is a game and people might try to modify the code in ways that are beneficial to the game but harmful to the science behind the projects.[17] Foldit team member bkoep explained in a forum discussion that "[p]art of the concern with open-sourcing the Foldit code is the potential for abuse—there are a lot of trivial ways that Foldit could be made more 'fun' as a game, but that would also undermine the scientific validity of Foldit players' work" (Bkoep 2018). In Foldit, proprietary software was further justified by the fact that Foldit was built on the code base of the protein-modeling software Rosetta. The Rosetta code was distributed under a Rosetta license, which was not open-source (but free for noncommercial use) and required registration with RosettaCommons (Bkoep 2018). Following Giddens, the "prime condition of requirements for trust is [...] lack of full information" (1990, 33). Participants in HC-based CS games that rely on proprietary code, such as Foldit and Stall Catchers, cannot verify scientific correctness even if they wanted to and had the technical knowledge to do so. Stall Catchers participants, for example, cannot verify that the statistical approach to combining individual answers is indeed correct. Instead, they must rely on the team of the Human Computation Institute and believe the system's explanations, which are simplified to make them accessible but, therefore, also incomplete. This is not to say that there is a reason to distrust the developers and researchers but to demonstrate the existence of this gap.

While the question of open-sourcing Stall Catchers had not been raised publicly on the Stall Catchers forum at the time of my research, in the case of Foldit, requests to open-source Foldit's source code had been recurring since the project's early days. Now and then, a new participant would (unknowingly) reopen the year-long discussion, prompting some other participants to leave comments such as "here we go again" (B_2, 2011), expressing their fatigue with the discussion. Nonetheless, these requests kept coming back and, most of the time, did not proceed satisfactorily for the requester. It is not surprising that such requests are more common at Foldit, as it attracts many programmers and participants interested in computer science. Since the code base of both projects is not open-source, it becomes impossible for participants or others interested in them to build trust as sociomaterial practice.

Distributed Trust

HC-based CS assemblages are multiplicities resulting from continuous reterritorialization processes in which different interests and human–technology relations are aligned, together forming the assemblages. Continuity is necessary due to deterritorializing processes that simultaneously act upon and carry away the assemblage. Examples of the latter are divergent interests, different logics at play (as discussed in Chapter 5), or material breakdowns. Furthermore, as Deleuze and Guattari write, an assemblage "necessarily changes in nature as it expands its connections" (2013, 7). Due to the need for

---

17   Interestingly, the ARTigo source code, by contrast, is open-source and published under the GNU General Public License on Github (Institute of Art History (Ludwig Maximilian University of Munich), n.d.).

HC-based CS systems to remain at the edge of AI, human–technology relations keep intraverting, and with them, the systems or assemblages themselves are changing. These intraversions and changes in the overall system require trust, an example of a reterritorialization process, to be continuously rebuilt. Trust, as I have aimed to show in this chapter, thereby, moves along various sociomaterial relations and practices. In the examples discussed, trust cannot be grasped by analyzing human social relations alone, even though they play an important role, as the example of the trust-building trip I accompanied demonstrated. Trust is built via human–technology relations. Thus, even if the explicit goal of the team members was to trust, for example, the participants contributing to Stall Catchers, this was achieved through the mediation of technologies and sociomaterial practices. It is, therefore, important not to overlook trust as a sociomaterial practice, which was the focus of this chapter.

The introduction of Stall Catchers in the biomedical engineering laboratory required the development of a data pipeline to prepare research data for subsequent analysis by Stall Catchers participants, which partly disrupted established research practices and work routines and even led to mistrust in the results. Trust-building practices, such as anonymized research data analysis, had to be translated into new practices. Trust emerged here through work on infrastructures and engagement with materialities.

Additionally, trust played a role in legitimizing CS as a not-yet-established approach to scientific knowledge production. Here, the HC system mediated trust in crowd answers via computational algorithms by preventing cheating and ensuring the required scientific data quality. At the same time, trustworthy relations with collaborators need to be built first, e.g., through in-person meetings and time spent together, to develop new HC-based CS projects.

Finally, a lower level of familiarity and understanding of Stall Catchers on the part of the participant tends to increase the importance of trust in the researchers and the team of the Human Computation Institute in order for the participants to perceive their contribution as meaningful. Similarly, regular participation and engagement with the system build trust in the participant's ability to catch stalls. Participants who were less familiar with the "wisdom-of-the-crowds" method correspondingly expressed mistrust in their competence to contribute to research.

Building and maintaining trust is essential for Stall Catchers to come into being in the interplay of the different human and nonhuman actors involved. As I have shown, the relations between, for example, infrastructure, software tools, and researchers, or participants and algorithms are just as important as those between human collaborators. Trust, therefore, unfolds in (sociomaterial) practices. Following social anthropologist Martin Holbraad's Truth in Motion: the Recursive Anthropology of Cuban Divination (2012), in which he argues for conceptualizing truth in "motile terms, as an event of collision—a meeting—between previously unrelated strands of meaning" (2012, xxiii), I suggest thinking about trust in motile terms. Through engaging with meanings of trust in the field, trust reveals itself as not just a static phenomenon but one that is constantly becoming through these sociomaterial relations and practices. It, thereby, depends on the very sociotechnical situation and "transforms itself in various stages or steps" (Harper 2014b, 307) in HC-based CS assemblages. "[T]rust is transitive; certainly a composite" (Harper 2014b, 307). In this way, trust unfolds as something distributed across HC-based CS projects, emerging within human–technology relations and along their intraversions, as much as it contributes to forming sociotechnical assemblages and holding them together. Finally, trust needs to be continuously maintained and rebuilt through the ongoing evolution of these HC assemblages.

8 Conclusions

> Hybridization [...] is not to be rejected, neither is it the greatest danger, but it does deserve the greatest care. (Dorrestijn 2012b, 240)

In 2021, for the first time in the history of Stall Catchers, an AI bot named GAIA was included alongside human participants in the HC-based CS game to analyze Alzheimer's disease research data. This first encounter between humans and AI bots in Stall Catchers presented at the beginning of this book initially appeared as a snapshot, independent of Stall Catchers' historical becoming and its future directions. Yet, the encounter merely represented one stage in the evolution of the Stall Catchers' assemblage and its continuously intraverting participant–technology relations. This human–AI bot encounter was first shaped by the Human Computation Institute's imaginaries of HC, the design of Stall Catchers as it was built in part upon the Stardust@home project, and the laboratory's practices for manually analyzing Alzheimer's disease research data. Importantly, the encounter was also formed by preexisting participant–technology relations in Stall Catchers, including the data annotation performed by human participants, while other algorithms, in turn, review and adjust for individual participants' skill levels. Human participants' annotations were then used to train ML algorithms, forming the basis for the AI bots' inclusion as artificial participants on the platform. At the same time, the HC system's intrinsic play/science entanglements created the encounter as a fun, yet serious situation, introducing competition among human and nonhuman participants, albeit mitigated through the common pursuit of advancing Alzheimer's disease research. However, the encounter was not solely defined by historical developments, path dependencies, and previous relations. Instead, with AI bots on the leaderboard, human participants actively related to the new configurations using tactics such as redeeming points and forming human alliances *against* the bots. Simultaneously, participants also shaped and created the possibility for new human–technology relations by seizing specific algorithmic affordances not actually intended by the system's design. Through their motivations and experiences that brought them to Stall Catchers, participants ascribed new meanings to the project while rejecting specific inscriptions, such as that of referring to Stall Catchers as a game. With all of this in mind, that initial encounter presents only one key moment in Stall Catchers' long chain of intraversions, one which specifically marks the emergence of a new dynamic in participant–technology relations. This new dynamic resulted in direct productive cooperation between humans and AI in pursuit of the CS project's overarching goal of bringing about a cure for Alzheimer's disease.

Although Stall Catchers and the Human Computation Institute served as the primary focus of my research, specific aspects regarding how human–technology relations in HC-based CS intravert only become clear when considering the additional case studies of Foldit and ARTigo.

More precisely, the analysis of Foldit showed particularly well how human–technology relations in HC-based CS continue to intravert over long time periods—over almost 20 years in this case. Beginning with the distributed computing project Rosetta@Home, Foldit underwent significant cycles of intraversions from its inception. Born from an idea regarding how to improve an automated system through the introduction of human creativity, the project was subsequently reconfigured to allow humans to create and deploy customized, automated tools. Foldit navigated through these major changes without ever becoming obsolete, even, and perhaps most notably, in the face of stark advances in AI capabilities, such as AlphaFold's solution to the problem of protein structure prediction. Instead of rendering it obsolete, AlphaFold's capabilities were soon subsumed into Foldit in the form of an additional tool, with Foldit's own purpose shifting from advancing protein structure prediction to, instead, furthering progress on protein design, as well as other unsolved problems. Notably, the most impactful intraversions in the case of Foldit often emerged out of the engagement of the participants themselves, rather than being introduced only by the system's developers. This provided a contrast between Foldit and Stall Catchers (where more, but not all, of the innovations were introduced by the Human Computation Institute team).

The analysis of ARTigo specifically highlighted how the overall purpose of HC-based CS assemblages transformed with its intraverting human–technology relations and by what means these changes unfold. By concentrating on the larger scale of the overall system's life cycle, I demonstrated how the platform required reimplementation and how participant–technology relations needed reconfiguration to remain both engaging for participants and at the edge of technological development. ARTigo persisted not only through the intraversions in its relations and alongside advancements in computer vision toward new goals. To an extent, it could be said that ARTigo *had to* undergo changes due to secondary factors, such as continued active participant engagement and motivation along with gradual personnel changes to the team developing and maintaining the system.

Various processes of territorialization and deterritorialization shape the complex interplay of different human and nonhuman actors in HC assemblages. As an example, I discussed trust as an ongoing territorialization process. Instead of understanding trust as a cognitive category, it must be considered a sociomaterial practice, since trust is negotiated in the intra-actions of humans and technologies. Given the changing nature of HC-based systems, new questions of trust emerge while disrupting existing trust-establishing practices. Ultimately, possibilities for new intraversions emerge at the horizon of human–technology relations in HC. Towards the conclusion of this study, for instance, the next intraversion of humans–AI relations in Stall Catchers remained open. However, it seems clear that even if AI eventually achieves the accuracy required to fully fill in for human participants in the analysis of research data, human participation in Stall Catchers is unlikely to become obsolete. New ideas regarding how to include human participants in other steps in Alzheimer's disease research already exist at the Human Computation Institute and in the biomedical laboratory.

As Gray and Suri (2019) argue, newly identified problems will continue to arise with the advancement of AI and the automation of increasing numbers of tasks: "Thus, there is an ever-moving frontier between what machines can and can't solve. We call this the paradox of automation's last mile: as machines progress, the opportunity to automate something else appears on the horizon" (Gray and Suri 2019, 175). The paradox lies in the continuous repetition of this process—taking over tasks performed by humans and simultaneously creating new ones humans can solve. "In other worlds, as machines solve more and more problems, we continue to identify needs for augmenting rather than replacing human effort" (Gray and Suri 2019, 175–176). It is clear for the Human Computation Institute that as ML models become more accurate in analyzing Stall Catchers data, the future involvement of humans must necessarily look differently than that of AI bots. If the intraversions of Foldit prove instructive here, this may result in human engagements moving to a higher level of abstraction and creativity, for example, by creatively combining different AI capabilities or solving a new, higher-order problem. As the frontier of AI shifts along with technological advancements, the system might even shift its purpose altogether as new projects become possible, addressing problems which cannot yet be solved in fully automated ways. In this way, HC systems can effectively remain at the edge of AI in pursuit of ever-evolving goals and capabilities.

This study aimed to investigate the formation of HC-based CS assemblages in the interplay between human and nonhuman actors. To do so, I analyzed how HC advocates and designers imagine them as new forms of HI, building a counter-imaginary to AGI and drawing their own understanding of how human–technology relations *ought* to evolve to create desirable futures. At the same time, however, these imaginaries are, nevertheless, rooted in shared AI paradigms. Imaginaries materialize in infrastructuring practices, which, in turn, shape HC designers' visions of the future. However, once other actors (including participants and materialities that bring with them the potential to fail) join and engage within HC, these systems are negotiated in practice through the interferences of play and science, the motivating role of the system's overarching purpose, and what human–technology relations in these systems *should* look like, just as they are shaped by different territorialization and deterritorialization processes. These HC-based CS assemblages, as I have shown, are temporally consistent but, nevertheless, fluctuating compositions of heterogeneous entangled human–technology relations. In turn, these relations are reconfigured within this complex interplay as they continuously transform through future-thinking, everyday adaptations, failures, and intentional redesigns. I introduced the concept of intraversions to analyze how these relations unfold and transform, and how subject/object positions, tasks, responsibilities, and, thus, power are distributed across these relations and along these developments. The concept of intraversions emerged from my co-laborative ethnographic observations and experiences in the field and is particularly suited for the investigation of HC systems developed as transitory, always open, and, in fact, intended, to change again. The transformations of human–technology relations in HC(-based CS) assemblages—that is, intraversions—can be described as forward circular movements that unfold along temporal developments, which are both instantaneous and gradual. As an analytical tool, intraversions allow one to account for both of these dimensions and combine them in the analysis. The concept helps to trace *how* the frontier of automation described by Gray and Suri (2019) moves by focusing on HC's continuously changing human–technology relations along their whole "life span" and beyond, as new relations and even entire sociotechnical systems emerge from existing ones, resembling assemblages that, as they extend, develop into new assemblages (Deleuze and Guattari 2013, 7).

Engaging at the Edge of Artificial Intelligence

I aimed in this work to not only contribute to the body of knowledge of digital anthropology and STS. It is also my hope that the results of this study and my engagement in the field, collaborative from the beginning, may serve as an example of how ethnographers can play active roles in informing and shaping the design, implementation, and maintenance of emerging technologies through constructively accompanying their everyday practices. To that end, in what follows, I highlight three examples of how this research can contribute to the development of HC systems, followed by more general arguments for why and how digital anthropologists and STS researchers should actively engage in the development of such technologies.

First, the analysis of HC imaginaries demonstrated that the *human* in the loop does not strictly refer to an embodied, and socially and culturally embedded human being, but rather to an information-processing unit understood primarily as part of a crowd, given motivation via the context of a game. My investigation of the becoming of assemblages demonstrated how cognition—just like agency, power, and responsibility—is distributed across human–technology relations and cannot be attributed to human actors alone. Furthermore, human actors cannot be reduced to their "intelligence" or "thinking," given that human experience relies on *being in the world*, which is always embodied and affective. Additionally, I discussed the importance of extending "the loop" when referring to the humans in the loop to include developers, researchers, and, depending on the specific example, other actors involved. If the goal of HC for human–AI hybridization is to form the future of AI, this analysis presents an attempt and further invitation to reflect on the understandings and imaginations of the human in the loop and the interplay of the humans and the technologies behind HC systems. As Dorrestijn describes, hybridization should be given the greatest care (2012b, 240).

Second, while the concept of intraversions primarily serves as analytical tool, it can, nevertheless, also function as a reflective tool aiding in the development of HC by providing an understanding of how humans and technology are configured and how these relations evolve over time. Such configurations and relations depend on existing relations, path dependencies, different actors' intentions, and various other forces such as serendipity acting upon, through, and with them. The concept of intraversions helps us to see how sociotechnical systems and their human–technology relations become what they are, suggesting that, while the specific potential paths along which they develop may vary significantly, they are, indeed, likely to change again in very particular ways in the future, possibly offering instructive insight for their design. It may, thus, also help the creators of these systems reflect upon and reconsider whether the path currently followed is desirable for the long-term evolution of the system.

Third, assemblage thinking can contribute to an understanding along the lines of what Bateson described in her foreword to the *Handbook of Human Computation* (2013). She argues that HC, with its crowd-based approach integrating a diversity of human and nonhuman actors, could contribute to a shift in "attitudes away from the fetish of individual autonomy and [teach] us, by implication, to recognize that we are connected parts of a larger whole, this is a goal to be pursued" (Bateson 2013, vii). Looking at HC systems as assemblages brings forth the various human and nonhuman elements and relations involved in the formation of these as well as the continuous forces and processes acting upon them, bringing them together and tearing them apart. In this context, intraversions, as a reflective tool, make the complex interplay and intra-actions between humans and technology in HC assemblages visible. Via this perspective, humans are not independent actors autonomously interacting with other humans and technology but a part of and situated within larger entanglements.

Cultural anthropological research, with its focus on everyday life and historical becoming, can connect past, present, and future. As Suchman (2007b) argues, plans do not unfold as linear sequences of action but, in practice, unfold rather unpredictably, given that they are always situated in sociocultural contexts. Ethnographic analysis can help us understand where AI systems come from, bringing together the different perspectives involved, and, thereby, pointing to potential futures. Cultural and digital anthropology can contribute by critically accompanying and contextualizing the developments of emerging technologies to steer them in directions preferable to all of the actors involved.

To do so, ethnographic analysis must be embedded in the development practice of emerging technologies (such as HC) itself. Collaborative approaches are, of course, well-established practice in ethnographic research. However, with a few (yet notable) exceptions, such as works by Suchman (2007b; 2021), Forsythe (2001f), and Pink (e.g., 2023),[1] participatory ethnographic research into the development of computational systems remains lacking, especially in the domain of AI research. Given how AI is becoming a greater presence in our everyday lives, and not only in the public discourse, I think it is crucial for digital anthropologists and STS researchers to not only accompany software development as advisors on cultural or ethical questions but to also actively engage in the very processes of creating socio-computational systems. Digital anthropologists and STS researchers must, thus, contribute to shaping the future. As Pink argues, "We must become players in the same futures-focused space as other stakeholders in the future

---

1   Additionally, two ongoing research projects should be mentioned here: Kinder-Kurlanda and colleague's ongoing work within NoBIAS: Artificial Intelligence without Bias (2020–2024), a Marie Skłodowska-Curie Innovative Training Network (ITN) (NoBIAS n.d.) and cultural anthropologist Sarah Thanner's collaborative research within the VIGITIA project (Physical-Digital Affordances Group University of Regensburg n.d.).

of emerging technologies, create new collaborations and bring different, diverse and everyday stories to the centre" (Pink 2023, 11).

The extent of domain-specific knowledge or training required for such work depends on the specific context and research field, but I argue that it does not always necessarily involve learning how to code, as often suggested. Instead, it can be much more important to understand a broader range of underlying conceptual levels more generally. In fields adjacent to AI, these may include concepts such as computation, the architecture and functioning of computers, the basic and emergent behavior of algorithms, or the cognitive ideas underlying the implementation of AI models. In other words, ethnographic research on computer science and AI cannot and should not treat the technical content of these fields as primarily about understanding and producing source code. Beck (2012) argues in his article on social neuroscience that anthropology must recognize that cognitive science eludes simple characterizations; instead, it combines various scientific fields, theoretical approaches, and methods. As a result, anthropology should join the various endeavors "that all somewhat fumble in the dark to better understand the distributed, self-organizing 'systematics' of cognitive phenomena" (Beck 2012, 114) and contribute to an ever-refining discourse:

> The unwanted alternative would be that an undifferentiated critique might deepen the split between the natural and social sciences and, even more important, might prevent possibly productive challenges that the new findings and modes of explanation in the neurosciences might have for established modes of thought and research practices cultivated by the social sciences and the humanities (Beck 2012, 114).

By adopting Beck's viewpoint and applying it to the field of AI, cultural anthropology ought not to confine AI to a singular discipline, theoretical concept, or method, and, consequently, should refrain from evaluating and criticizing it based on such limited perspectives. Instead, it should examine specific fields of research and applications and attempt to develop an understanding of both general and case-specific underlying concepts and assumptions. This will allow the discipline to engage in critical yet constructive ways (Suchman 2021, 70–71) in these fields to explore and contribute to shaping how we want to live and organize our everyday lives with AI.

Beyond the Edge

In this work, I analyzed HC as a phenomenon *at the edge* of AI, focusing on the specific and emerging branch of HC, more specifically considering its applications in the field of CS. I chose such a concrete focus on a specific area of AI because I believe grasping it as a whole is unfeasible and perhaps even counterproductive, given the diverse understandings, missions, and persistent lack of a cohesive definition for "intelligence," be it human or artificial. I think it is important for researchers in STS (and other areas) to broaden our focus beyond the imaginaries of and efforts to create "strong AI" or AGI but to also examine other existing and emerging approaches in the field, which increasingly inform and contribute to both developments and discourse in and around AI. The analysis of concrete fields of practice allows one to gain an in-depth understanding of the paradigms, visions, and intentions of the various actors, as well as the meanings and failings shaping the fields. Such an understanding is necessary to actively intervene, an undertaking which may be gaining in importance today given the vertigo-inducing (Dippel 2021) pace with which developments in AI are increasingly impacting our everyday lives. My research represents "a study of a culture in the making" (Turkle 2005b, 23). Much like in Turkle's investigation of computer culture in the late 1970s and 1980s, my research subject formed a moving (perhaps even accelerating) target. Today, the speed of developments continues at unprecedented levels, pushing the edge of AI further.

To conclude I turn *to the edge* of AI itself and situate my work in the broader context of AI research and the field's advancements. Taking a broader view, the edge of AI in recent years has been marked by significant advances driven largely by the emergence of highly powerful "foundation models" (Bommasani et al. 2022). This is perhaps most notable in OpenAI's advances in the domain of large language models with GPT-3 and, subsequently, GPT-4 (as well as ChatGPT, OpenAI's GPT-powered AI chatbot that popularized these developments in late 2022). These models significantly exhibit advanced capabilities in text comprehension, reasoning, problem-solving, and a host of other tasks, prompting some researchers to describe them as featuring first "Sparks of Artificial General Intelligence" (Bubeck et al. 2023), a characterization that invokes new imaginaries as well as reviving old ones. These advancements have reignited and accelerated debates about the automation and obsolescence of human labor, previously centered around menial tasks but now broadening their scope to include advanced domains of knowledge work, extending into medicine, finance, and even software engineering itself. While it is too soon to determine how these developments will unfold and what impact they might have, thus far, the deployment of these new AI systems has been marked primarily by what must be described as HI: "AI assistants" and "co-pilots" have been developed for myriad domains, including customer support, marketing, insurance, medicine, and software engineering, all with the promise of helping professionals to do *more*, rather than to help them do *nothing*.[2] In addition to such purpose-built AI assistants, another theme pointing in the direction of increasing human–AI hybridization can be found in the development of software frameworks that actively make affordances for building tools combining AI and human capabilities. One example of this is found in LangChain, an early but now highly popular Python framework for "building applications with LLMs through composability" (Chase 2022). One of the types of "agents" available in this framework is in fact *humans*—described as useful in helping other AI agents: "[h]uman[s] are AGI so they can certainly be used as a tool to help out [an] AI agent when it is confused" (Chase 2023), states the framework's documentation, introducing even newer and further configurations of human–AI relations in which humans and AI both take diverse subjective positions, sometimes simultaneously.

---

2   For specific examples, it is instructive to review the US-based technology startup accelerator Y Combinator's list of startup companies in its winter 2023 cohort (n.d.). A cursory glance suggests that the single biggest unifying theme is AI, typically represented as either domain-specific AI-powered software, functionally specific AI assistants or tools for the development of or with AI.

Today, it seems clear that some—perhaps many—specific currently existing jobs and tasks humans perform across a multitude of domains can indeed be automated using AI. While this book offered a deep dive into only a few specific domains, it is also clear that the primary theme has remained one of *enhancement*, not *replacement*, and my analysis suggests that humans will continue to play a role in these systems, turning to ever-newer tasks and goals to pursue. More precisely, the question of the "level of involvement" of humans or AI may be insignificant, if instead we think of these systems as truly hybrid human–AI assemblages. To that end, I offer intraversions as an analytical tool to trace and analyze how the relations between humans and technology unfold and transform alongside continued developments and progress in AI, and to engage constructively and with the greatest care with human–AI systems in the making.

Glossary

The following glossary entries are primarily intended to provide an overview of technical and functional aspects of concepts rather than analytical ones.

Algorithm	A structured set of instructions, typically in the form of steps, designed to solve a specific problem that can be implemented using programming languages to be executed as part of computer code.
AlphaFold	An **artificial intelligence** system for predicting protein folding, developed by the British-American company and Alphabet subsidiary DeepMind.
Artificial General Intelligence (AGI)	Broadly understood as **artificial intelligence** that performs a wide and generic range of tasks and behaviors at levels that are considered to be equal to or greater than that of humans. Its feasibility remains a subject of active debate.
Artificial Intelligence (AI)	Describes the broad field of research aimed at building "intelligent" machines. It consists of broad subsets or techniques such as **machine learning**, **deep learning**, natural language processing, reinforcement learning, computer vision, robotics, and expert systems.
Convolutional Neural Network (CNN)	A form of **deep neural network**, sometimes also referred to as **deep learning**, for processing data with a grid-like structure and non-local features. CNNs are frequently used in the field of computer vision, where they leverage chains of convolutional filters to automatically extract features from images at various levels of detail.
Deep Learning (DL)	Any form of **machine learning** that makes use of **deep neural networks**. *Cf.* **CNN.**
Deep Neural Network	A **neural network** with two or more layers.
Fluorescence microscopy	An imaging technique in which expressed fluorescent proteins or administered small molecule fluorophores are excited at a specific wavelength and emit a photon of a defined higher wavelength on recovery to their energy ground state.

Human Computation (HC)	In its current understanding, HC dates from the beginning of the twenty-first century and describes an interdisciplinary area of research that its advocates situate within the research field of **artificial intelligence**. HC addresses problems that cannot be solved by current **artificial intelligence** technologies or manual human approaches alone by combining the strengths of humans and machines to achieve superior capabilities. The term HC is sometimes used synonymously with **hybrid intelligence**.
Hybrid Intelligence (HI)	An emerging research field that focuses on the development of hybrid systems of humans and **artificial intelligence**, which are often considered to improve by learning from each other. The term is sometimes used synonymously with **human computation**.
Large Language Model (LLM)	A type of **deep neural network** trained on large amounts of text data to predict sequences of words, thereby building a model of the language. They are often able to generate output that resembles text produced by humans.
Machine Learning (ML)	Describes a subset or technique of **artificial intelligence** in which algorithmic prediction or decision models are built from large amounts of data without directly programmed instructions.
MATLAB	A programming language and platform for programming and numeric computation developed and owned by the private company MathWorks.
Multiphoton microscopy	A form of **fluorescence microscopy** in which multiple photons of lower energy are used to achieve excitation of the fluorophores. Cf. **two-photon microscopy**.
Neural Network (NN)	An NN combines different mathematical operations, whose composition is described by a directed acyclic graph, to form a complex parameterized function. Most often, this function simply consists of a linear chain of multiplications of so-called "weights matrices" with subsequent application of so-called "activation functions." The different components in the chain are also called "layers." A key component of a neural network is that its parameters (or weights) are "trainable," i.e., they are initialized to random values and then gradually optimized by applying the network to training data (inputs and expected outputs), iteratively updating the weights to improve the network's performance in mapping the inputs to the expected outputs, for example via gradient descent techniques.
Python	A widely used high-level programming language for web development, data analysis, **artificial intelligence**, and other applications.
Rosetta	A protein modeling and protein structure analysis software first developed by the Baker Lab of the Institute for Protein Design at the University of Washington. Today, it is developed and maintained by the members of the RosettaCommons collaborative, which includes academic institutes, government laboratories, and partner corporations.
RoseTTAfold	An advanced **deep learning**–based model that was developed by the Institute for Protein Design at the University of Washington.
Two-photon microscopy	A form of **fluorescence microscopy** in which two photons of lower energy, such as near-infrared light, are used to achieve excitation of the fluorophores. The use of low-energy near-infrared light, as well as the good tissue penetration of red light, allows for the imaging of thick specimens and even living tissue.
Voxel	A volumetric (3D) pixel depicting a digital value on a grid in 3D space.

References

Abend, Pablo, Sonia Fizek, Mathias Fuchs, and Karin Wenz, eds. 2020. *Laborious Play and Playful Work 1*. Digital Culture & Society 5, no. 2/2019. Bielefeld: transcript.

Abend, Pablo, Sonia Fizek, and Karin Wenz. 2020. "Introduction. The Boundaries of Play." In *Laborious Play and Playful Work 1*, edited by Pablo Abend, Sonia Fizek, Mathias Fuchs, and Karin Wenz, 5–12. Digital Culture & Society 5, no. 2/2019. Bielefeld: transcript. https://doi.org/10.14361/dcs-2019-0202.

Abt, Clark C. 1987. *Serious Games*. Lanham, MD: University Press of America.

Achterhuis, Hans, ed. 2001a. *American Philosophy of Technology: The Empirical Turn*. The Indiana Series in the Philosophy of Technology. Bloomington: Indiana University Press.

———. 2001b. "Introduction: American Philosophers of Technology." In *American Philosophy of Technology: The Empirical Turn*, edited by Hans Achterhuis, 1–9. Bloomington: Indiana University Press.

Adamowsky, Natascha. 2018. "Spiel/en." In *Philosophie des Computerspiels: Theorie – Praxis – Ästhetik*, edited by Daniel Martin Feige, Sebastian Ostritsch, and Markus Rautzenberg, 27–41. Stuttgart: J.B. Metzler. https://doi.org/10.1007/978-3-476-04569-0_3.

Ahrweiler, Petra. 1995. *Künstliche Intelligenz-Forschung in Deutschland. Die Etablierung Eines Hochtechnologie-Fachs*. Münster; New York: Waxmann.

Akata, Zeynep, Dan Balliet, Maarten De Rijke, Frank Dignum, Virginia Dignum, Guszti Eiben, Antske Fokkens, et al. 2020. "A Research Agenda for Hybrid Intelligence: Augmenting Human Intellect With Collaborative, Adaptive, Responsible, and Explainable Artificial Intelligence." Computer 53, no. 8 (Aug.): 18–28. https://doi.org/10.1109/MC.2020.2996587.

Akrich, Madeleine. 1995. "User Representations: Practices, Methods and Sociology." In *Managing Technology in Society. The Approach of Constructive Technology Assessment.*, edited by Rip Arie, Thomas J. Misa, and Johan Schot, 167–184. London; New York: Pinter Publishers.

Alabri, Abdulmonem, and Jane Hunter. 2010. "Enhancing the Quality and Trust of Citizen Science Data." In *2010 IEEE Sixth International Conference on e-Science*, 81–88. Brisbane, Australia: IEEE. https://doi.org/10.1109/eScience.2010.33.

Albawi, Saad, Tareq Abed Mohammed, and Saad Al-Zawi. 2017. "Understanding of a convolutional neural network." In *2017 International Conference on Engineering and Technology (ICET)*, 1–6. https://doi.org/10.1109/ICEngTechnol.2017.8308186.

Ali, Muhammad, Kaja Falkenhain, Brendah N Njiru, Muhammad Murtaza-Ali, Nancy E Ruiz-Uribe, Mohammad Haft-Javaherian, Stall Catchers, Nozomi Nishimura, Chris B. Schaffer, and Oliver Bracko. 2021. "Inhibition of peripheral VEGF signaling rapidly reduces leucocyte obstructions in brain capillaries and increases cortical blood flow in an Alzheimer's disease Mouse model." Preprint. https://doi.org/10.1101/2021.03.05.433976.

Alkhatib, Ali, Michael S. Bernstein, and Margaret Levi. 2017. "Examining Crowd Work and Gig Work Through The Historical Lens of Piecework." In *Proceedings of the 2017 CHI Conference on Human Factors in Computing Systems (CHI '17)*, 4599–4616. New York: ACM. https://doi.org/10.1145/3025453.3025974.

Allhutter, Doris. 2019. "Of 'Working Ontologists' and 'High-Quality Human Componants'. The Politics of Semantic Infrastructures." In *digitalSTS. A Field Guide for Science & Technology Studies*, edited by Janet Vertesi and David Ribes, 326–348. Princeton, Oxford: Princeton University Press.

Altenried, Moritz. 2020. "The platform as factory: Crowdwork and the hidden labour behind artificial intelligence." *Capital & Class* 44, no. 2 (Jun.): 145–158. https://doi.org/10.1177/0309816819899410.

———. 2022. *The Digital Factory: The Human Labor of Automation*. Chicago: University of Chicago Press.

Altenried, Moritz, Julia Dück, and Mira Wallis, eds. 2021. *Plattformkapitalismus und die Krise der sozialen Reproduktion*. Münster: Westfälisches Dampfboot.

Amann, Klaus, and Stefan Hirschauer. 1997. "Die Befremdung der eigenen Kultur. Ein Programm." In *Die Befremdung der eigenen Kultur. Zur ethnographischen Herausforderung soziologischer Empirie*, edited by Klaus Amann and Stefan Hirschauer, 7–52. Frankfurt am Main: Suhrkamp.

Amazon Mechanical Turk, Inc. n.d. "Amazon Mechanical Turk." Accessed Mar. 18, 2024. https://www.mturk.com/.

Amelang, Katrin. 2017. "Zur Sinnlichkeit von Algorithmen und ihrer Erforschbarkeit." In *Kulturen der Sinne. Zugänge Zur Sensualität der sozialen Welt*, edited by Karl Braun, Claus-Marco Dieterich, Thomas Hengartner, and Bernhard Tschofen, 358–367. Würzburg: Königshausen & Neumann.

Amelang, Katrin, and Susanne Bauer. 2019. "Following the algorithm: How epidemiological risk-scores do accountability." *Social Studies of Science* 49, no. 4 (Aug.): 476–502. https://doi.org/10.1177/0306312719862049.

Amoore, Louise. 2020. *Cloud Ethics: Algorithms and the Attributes of Ourselves and Others*. Durham: Duke University Press.

Amrute, Sareeta Bipin. 2016. *Encoding Race, Encoding Class: Indian IT Workers in Berlin*. Durham: Duke University Press.

Ananny, Mike. 2016. "Toward an Ethics of Algorithms: Convening, Observation, Probability, and Timeliness." *Science, Technology, & Human Values* 41, no. 1 (Jan.): 93–117.

Anderson, Ben, Matthew Kearnes, Colin McFarlane, and Dan Swanton. 2012. "On Assemblages and Geography." *Dialogues in Human Geography* 2, no 2 (Jul.): 171–189. https://doi.org/10.1177/2043820612449261.
Anthropic PBC. n.d. "Meet Claude." Accessed Mar. 18, 2024. https://www.anthropic.com/product.
Apelt, Maja. 1999. "Vertrauen in Organisationen und Netzwerken." In *Vertrauen in der zwischenbetrieblichen Kooperation*, by Maja Apelt, 7–42. Wiesbaden: Deutscher Universitätsverlag. https://doi.org/10.1007/978-3-322-99609-1_2.
Appadurai, Arjun. 1986. "Introduction: commodities and the politics of value." In *The Social Life of Things*, edited by Arjun Appadurai, 3–63. Cambridge: Cambridge University Press. https://doi.org/10.1017/CBO9780511819582.003.
———. (1990) 2002. "Disjuncture and Difference in the Global Cultural Economy." In *The Anthropology of Globalization: A Reader*, edited by Jonathan Xavier Inda and Renato Rosaldo, 46–64. Oxford: Blackwell Publishing.
Apple Inc. n.d. "Siri." Accessed Mar. 18, 2024. https://www.apple.com/siri/.
Association for Advancing Participatory Sciences. n.d. "Association for Advancing Participatory Sciences." Accessed Mar. 18, 2024. https://participatorysciences.org.
Association for the Advancement of Artificial Intelligence. n.d. "HCOMP AAAI Conference on Human Computation and Crowdsourcing." Accessed Mar. 18, 2024. https://aaai.org/aaai-conferences-and-symposia/.
Aytes, Ayhan. 2012. "Return of the Crowds. Mechanical Turk and Neoliberal States of Exception." In *DIGITAL LABOR. The Internet as Playground and Factory*, edited by Trebor Scholz, 79–97. Abingdon, Oxon; New York, NY: Routledge.
B_2. 2011. "Open source." Foldit forum. Sept. 12, 2011. https://fold.it/forum/suggestions/open-source.
Bachmann, Götz. 2018. "Dynamicland. Eine Ethnographie der Arbeit am Medium." *Zeitschrift für Volkskunde* 114, no. 1 (Jun.): 29–50.
Baek, Minkyung, Frank DiMaio, Ivan Anishchenko, Justas Dauparas, Sergey Ovchinnikov, Gyu Rie Lee, Jue Wang, et al. 2021. "Accurate prediction of protein structures and interactions using a three-track neural network." *Science* 373, no. 6557 (Aug.): 871–876. https://doi.org/10.1126/science.abj8754.
Baker, David. 2006. "Dr. David Baker from Rosetta@home. University of Washington. April 28, 2006 – Interview number 4. BOINC based project to determine protein structure." Apr. 28, 2006. https://web.archive.org/web/20090218192526/http://www.teampicard.com/profiles/Interview.php?id=4.
Baker Lab. 2021. "trRosetta yields structures for every protein family." Baker Lab. Mar. 3, 2021. https://www.bakerlab.org/index.php/2021/03/03/trrosetta-yields-structures-every-protein-family/.
Bakhtin, Michail M. 1981. "Discourse in the Novel." In *The Dialogic Imagination: Four Essays*, edited by Michael Holquist, 259–422. University of Texas Press Slavic Series, no. 1. Austin: University of Texas Press.
Ball, Andrew. 2018. "manuel delanda, *assemblage theory* (edinburgh university press, 2016)." *Parrhesia: A Journal of Critical Philosophy* 29: 241–247. http://parrhesiajournal.org/parrhesia29/parrhesia29.pdf.

Barad, Karen. 1996. "Meeting the Universe Halfway: Realism and Social Constructivism without Contradiction." In *Feminism, Science, and the Philosophy of Science*, edited by Lynn Hankinson Nelson and Jack Nelson, 161–194. Dordrecht; Boston; London: Kluwer Academic Publishers; Springer.

———. 2007a. "Agential Realism: How Material-Discursive Practices Matter." In *Meeting the Universe Halfway: Quantum Physics and the Entanglement of Matter and Meaning*, 132–185. Durham: Duke University Press. https://doi.org/10.1515/9780822388128-006.

———. 2007b. *Meeting the Universe Halfway: Quantum Physics and the Entanglement of Matter and Meaning*. Durham: Duke University Press. https://doi.org/10.1515/9780822388128-006.

———. 2015. *Verschränkungen*. Translated by Jennifer Sophia Theodor. Internationaler Merve-Diskurs 409. Berlin: Merve.

Bareis, Jascha, and Christian Katzenbach. 2022. "Talking AI into Being: The Narratives and Imaginaries of National AI Strategies and Their Performative Politics." *Science, Technology, & Human Values* 47, no. 5 (Jul.): 855–881. https://doi.org/10.1177/01622439211030007.

Bareither, Christoph (a.o.). 2013. "Alltag mit Facebook. Methodologische Überlegungen und ethnographische Beispiele." In *Update in Progress. Beiträge zu einer ethnologischen Medienforschung*, edited by Falk Blask, Joachim Kallinich, and Sanna Schondelmayer: 29–46. Berliner Blätter. Ethnographische und ethnologische Beiträge 64. Berlin: Panama.

———. 2019. "Doing Emotion through Digital Media: An Ethnographic Perspective on Media Practices and Emotional Affordances." *Ethnologia Europaea* 49, no. 1 (Apr.). https://doi.org/10.16995/ee.822.

———. 2020a. "Affordanz." In *Kulturtheoretisch Argumentieren*, edited by Timo Heimerdinger and Markus Tauschek, 32–55. UTB. Münster; New York: Waxmann.

———. 2020b. *Playful Virtual Violence: An Ethnography of Emotional Practices in Video Games*. Cambridge: Cambridge University Press. https://doi.org/10.1017/9781108873079.

———. 2022. "Kultur ist mehr ... Zum vielfältigen Kulturbegriff der EKW." In *Kultur ist. Beiträge der Empirischen Kulturwissenschaft in Tübingen*, edited by Ludwig-Uhland-Institut, 128: 11–45. Untersuchungen 128. Tübingen: EKW-Verlag.

———. 2023. "Museum-AI Assemblages: A Conceptual Framework for Ethnographic and Qualitative Research." In *AI in Museums*, edited by Sonja Thiel and Johannes C. Bernhardt: 99–114. Bielefeld: transcript. https://doi.org/10.1515/9783839467107-010.

Bateson, Mary Catherine. 2013. "Foreword: Making a Difference." In *Handbook of Human Computation*, edited by Pietro Michelucci, v–viii. New York: Springer.

Bauernschmidt, Stefan. 2014. "Kulturwissenschaftliche Inhaltsanalyse prozessgenerierter Daten." In *Methoden der Kulturanthropologie*, edited by Christine Bischoff, Karoline Oehme-Jüngling, and Walter Leimgruber, 415–430. UTB. Bern: Haupt.

Bechmann, Anja, and Geoffrey C Bowker. 2019. "Unsupervised by Any Other Name: Hidden Layers of Knowledge Production in Artificial Intelligence on Social Media." *Big Data & Society* 6, no. 1 (Jan.): 205395171881956. https://doi.org/10.1177/2053951718819569.

Beck, Stefan. 1997. *Umgang mit Technik. Kulturelle Praxen und kulturwissenschaftliche Forschungskonzepte*. Dissertation, University of Tübingen. Zeithorizonte 4. Berlin: Akademie Verlag zu Berlin.

———. 2012. "Interlacing the brain, contextualizing the body: Relational understandings in social neuroscience." In *The Atomized Body. The Cultural Life of Stem Cells, Genes and Neurons*, edited by Max Liljefors, Susanne Lundin, and Andréa Wiszmeg, 113–142. Lund: Nordic Academic Press.

———. 2019. "Von Praxistheorie 1.0 zu 3.0. Oder: wie analoge und digitale Praxen relationiert werden sollten." In *After Practice 2*, edited by the Laboratory: Anthropology of Environment | Human Relations: 9–27. Berliner Blätter. Ethnographische und ethnologische Beiträge 81. Panama: Berlin.

Beck, Stefan, Jörg Niewöhner, and Estrid Sørensen. 2012. "Einleitung. Science and Technology Studies – Wissenschafts- und Technikforschung aus sozial- und kulturanthropologischer Perspektive." In *Science and Technology Studies. Eine sozialanthropologische Einführung*, edited by Stefan Beck, Jörg Niewöhner, and Estrid Sørensen, 9–48. Bielefeld: transcript.

Becker, Christoph. 2023. *Insolvent: how to reorient computing for just sustainability*. Cambridge, Massachusetts; London, England: The MIT Press.

Beckert, Jens. 2018. *Imaginierte Zukunft: fiktionale Erwartungen und die Dynamik des Kapitalismus*. Translated by Stephan Gebauer. Berlin: Suhrkamp.

Bedessem, Baptiste, Bogna Gawrońska-Nowak, and Piotr Lis. 2021. "Can citizen science increase trust in research? A case study of delineating Polish metropolitan areas." *JCER. Journal of Contemporary European Research* 17, no. 2 (May): 305–321. https://doi.org/10.30950/jcer.v17i2.1185.

Bell, Genevieve. 2021. "Talking to AI: An Anthropological Encounter with Artificial Intelligence." In *The SAGE Handbook of Cultural Anthropology*, edited by Lene Pedersen and Lisa Cliggett, 442–458. London; New Delhi; Singapore; Washington DC; Melbourne: SAGE Publications Ltd. https://doi.org/10.4135/9781529756449.

Belliger, Andréa, and David J. Krieger. 2006. "Einführung in die Akteur-Netzwerk-Theorie." In *ANThology. Ein einführendes Handbuch zur Akteur-Netzwerk-Theorie*, edited by Andréa Belliger and David J. Krieger, 13–50. Bielefeld: transcript.

Bennett, Jane. 2004. "The Force of Things: Steps toward an Ecology of Matter." *Political Theory* 32, no. 3 (Jun.): 347–372. https://doi.org/10.1177/0090591703260853.

———. 2010. *Vibrant Matter: a political ecology of things*. Durham: Duke University Press. https://doi.org/10.1515/9780822391623.

Bennett Moses, Lyria, and Janet Chan. 2018. "Algorithmic prediction in policing: assumptions, evaluation, and accountability." *Policing and Society* 28, no. 7 (Nov.): 806–822. https://doi.org/10.1080/10439463.2016.1253695.

Bieler, Patrick, Milena D. Bister, and Christine Schmid. 2021. "Formate des Ko-laborierens. Geteilte epistemische Arbeit als katalytische Praxis." In *Kooperieren – Kollaborieren – Kuratieren. Positionsbestimmungen ethnografischer Praxis*, edited by Friederike Faust and Janine Hauer: 87–105. Berliner Blätter. Ethnographische und ethnologische Beiträge 83. Panama: Berlin. https://doi.org/10.18452/22407.

Bijker, Wiebe E., and John Law, eds. 1992. *Shaping Technology / Building Society: Studies in Sociotechnical Change*. Inside Technology. Cambridge, Massachusetts; London, England: The MIT Press. https://hdl.handle.net/2027/heb.01128.

Bijker, Wiebe E., and Trevor Pinch. 1984. "The Social Construction of Facts and Artifacts: or How the Sociology of Science and the Sociology of Technology Might Benefit Each Other." *Social Studies of Science* 14, no. 3 (Aug.): 399–441. https://doi.org/10.1177/030631 284014003004.

bkoep. 2018. "Unity 3D developer questions: Status of FoldIt? Access to source code?" Foldit Forum. Sept. 2, 2018. https://fold.it/forum/discussion/unity-3d-developer-q uestions-status-of-foldit-access-to-source-code#post_39733.

———. 2019. "Developer Chat." Chat on Foldit. Foldit. Accessed Mar. 18, 2024. https://w eb.archive.org/web/20211028044135/https://fold.it/portal/node/2008363.

———. 2021a. "The AlphaFold prediction tool in Foldit." Foldit forum. Jul. 31, 2021. https ://fold.it/forum/blog/the-alphafold-prediction-tool-in-foldit.

———. 2021b. "About Foldit and RoseTTAfold." Foldit. Aug. 2, 2021. https://web.archive. org/web/20220128190552/https://fold.it/portal/node/2011936.

Boellstorff, Tom, Bonnie A. Nardi, Celia Pearce, and T.L. Taylor. 2012. *Ethnography and Virtual Worlds: A Handbook of Method*. Princeton: Princeton University Press.

Bogner, Martin. n.d. "Modul 'Human Computation' (SS 2019)." Last modified Jul. 15, 2019, 9:43. https://www.pms.ifi.lmu.de/lehre/humancomputation/19ss/.

Bogner, Martin, François Bry, Niels Heller, Stephan Leutenmayr, Sebastian Mader, Alexander Pohl, Clemens Schefels, Yingding Wang, and Christoph Wieser. 2017. "Human Collaboration Reshaped: Applications and Perspectives." In *50 Jahre Universitäts-Informatik in München*, edited by Arndt Bode, Manfred Broy, Hans-Joachim Bungartz, and Florian Matthes, 47–73. Berlin; Heidelberg: Springer. https://doi.org/10.1007/97 8-3-662-54712-0_4.

Bommasani, Rishi, Drew A. Hudson, Ehsan Adeli, Russ Altman, Simran Arora, Sydney von Arx, Michael S. Bernstein, et al. 2022. "On the Opportunities and Risks of Foundation Models." arXiv. http://arxiv.org/abs/2108.07258.

Bond, Alan H., and Les Gasser. 1988. "An Analysis of Problems and Research in DAI." In *Readings in Distributed Artificial Intelligence*, edited by Alan H. Bond, and Les Gasser, 3–35. Burlington: Morgan Kaufmann. https://doi.org/10.1016/B978-0-934613-63-7.5 0006-1.

Bossen, Claus, and Randi Markussen. 2010. "Infrastructuring and Ordering Devices in Health Care: Medication Plans and Practices on a Hospital Ward." *Computer Supported Cooperative Work (CSCW)* 19, no. 6 (Nov.): 615–637. https://doi.org/10.1007/s10606-01 0-9131-x.

Bostrom, Nick. 2016. *Superintelligence: Paths, Dangers, Strategies*. Oxford, United Kingdom; New York, NY: Oxford University Press.

Bourdieu, Pierre. 1985. "The social space and the genesis of groups." *Theory and Society* 14, no. 6 (Nov.): 723–744. https://doi.org/10.1007/BF00174048.

Bowker, Geoffrey C. 1994. *Science on the Run: Information Management and Industrial Geophysics at Schlumberger, 1920–1940*. Inside Technology. Cambridge, Massachusetts; London, England: The MIT Press.

---. 2008. *Memory Practices in the Sciences*. Inside Technology. Cambridge, Massachusetts; London, England: The MIT Press.
Bowker, Geoffrey C., Karen Baker, Florence Millerand, and David Ribes. 2009. "Toward Information Infrastructure Studies: Ways of Knowing in a Networked Environment." In *International Handbook of Internet Research*, edited by Jeremy Hunsinger, Lisbeth Klastrup, and Matthew Allen, 97–117. Dordrecht: Springer. https://doi.org/10.1007/978-1-4020-9789-8_5.
Bowker, Geoffrey C., and Susan Leigh Star. 2008. *Sorting Things Out: Classification and Its Consequences*. Inside Technology. Cambridge, Massachusetts; London England: The MIT Press.
Bowker, Geoffrey C., Stefan Timmermans, Adele E. Clarke, and Ellen Balka, eds. 2015. *Boundary Objects and Beyond: Working with Leigh Star*. Infrastructures. Cambridge, Massachusetts; London England: The MIT Press.
Bowser, Anne, Michael Sloan, Pietro Michelucci, and Eleonore Pauwels. 2017. "Artificial Intelligence: A Policy-Oriented Introduction." *Wilson Briefs*. https://www.wilsoncenter.org/publication/artificial-intelligence-policy-oriented-introduction.
Boyd, Danah. 2009. "A Response to Christine Hine." In *Internet Inquiry: Conversations About Method*, edited by Annette Markham and Nancy Baym, 26–32. Thousand Oaks California: SAGE Publications, Inc. https://doi.org/10.4135/9781483329086.n3.
---. 2010. "Social Network Sites as Networked Publics: Affordances, Dynamics, and Implications." In *A Networked Self: Identity, Community, and Culture on Social Network Sites*, edited by Zizi Papacharissi, 39–58. Abingdon, Oxon; New York, NY: Routledge.
Bracko, Oliver, Lindsay K. Vinarcsik, Jean C. Cruz Hernández, Nancy E. Ruiz-Uribe, Mohammad Haft-Javaherian, Kaja Falkenhain, Egle M. Ramanauskaite, et al. 2019. "High fat diet worsens pathology and impairment in an Alzheimer's mouse model, but not by synergistically decreasing cerebral blood flow." Preprint. Neuroscience. https://doi.org/10.1101/2019.12.16.878397.
Braithwaite, Valerie, and Margaret Levi, eds. 1998. *Trust and Governance*. The Russel Sage Foundation Series on Trust 1. New York: Russell Sage Foundation.
Brayne, Sarah. 2017. "Big Data Surveillance: The Case of Policing." *American Sociological Review* 82, no. 5 (Aug.): 977–1008. https://doi.org/10.1177/0003122417725865.
Breidenstein, Georg, Stefan Hirschauer, Herbert Kalthoff, and Boris Nieswand. 2020. *Ethnografie: die Praxis der Feldforschung*. 3rd edition. UTB. München: UVK.
Brenner, Neil, David J. Madden, and David Wachsmuth. 2011. "Assemblage urbanism and the challenges of critical urban theory." *City* 15, no. 2 (Jun.): 225–240. https://doi.org/10.1080/13604813.2011.568717.
BrightFocus Foundation. n.d. "BrightFocus® Foundation. Cure in Mind. Cure in Sight." Accessed Mar. 18, 2023. https://www.brightfocus.org/.
Bruun Jensen, Casper. 2010. *Ontologies for Developing Things: Making Health Care Futures through Technology*. Rotterdam; Boston: Sense Publishers.
Bry, François, and Clemens Schefels. 2016. "An Analysis of the ARTigo Gaming Ecosystem With a Purpose." https://www.semanticscholar.org/paper/An-Analysis-of-the-ARTigo-Gaming-Ecosystem-With-a-Bry-Schefels/70a4a19d5fb3b218f8310e907f0a2cfc539ca5b7.

Bry, François, Clemens Schefels, and Christoph Wieser. 2018. "Human Computation." *It – Information Technology* 60, no. 1 (Feb.): 1–2. https://doi.org/10.1515/itit-2018-0007.

Bubeck, Sébastien, Varun Chandrasekaran, Ronen Eldan, Johannes Gehrke, Eric Horvitz, Ece Kamar, Peter Lee, et al. 2023. "Sparks of Artificial General Intelligence: Early experiments with GPT-4." arXiv. https://doi.org/10.48550/arXiv.2303.12712.

Buchanan, Ian. 2015. "Assemblage Theory and Its Discontents." *Deleuze Studies* 9, no. 3 (Aug.): 382–392. https://doi.org/10.3366/dls.2015.0193.

Bude, Heinz. 2010. "Quellen und Funktionen des Vertrauens." In *Vertrauen: Die Bedeutung von Vertrauensformen für das soziale Kapital unserer Gesellschaft*, edited by Heinz Bude, Karsten Fischer, and Sebastian Huhnholz, 10–15. Gedanken Zur Zukunft 19. Bad Homburg v.d. Höhe: Herbert-Quandt-Stiftung.

Burri, Regula Valérie. 2018. "Models of Public Engagement: Nanoscientists' Understandings of Science–Society Interactions." *NanoEthics* 12, no. 2 (May): 81–98. https://doi.org/10.1007/s11569-018-0316-y.

Caivano, Dean, and Sarah Naumes. 2021. *The Sublime of the Political: Narrative and Autoethnography as Theory*. Edition Politik 79. Bielefeld: transcript.

Callaway, Ewen. 2020. "'It will change everything': DeepMind's AI makes gigantic leap in solving protein structures." *Nature* 588, no. 7837 (Nov.): 203–204. https://doi.org/10.1038/d41586-020-03348-4.

Callon, Michel. 1984. "Some Elements of a Sociology of Translation: Domestication of the Scallops and the Fishermen of St Brieuc Bay." *The Sociological Review* 32, no. 1 (May): 196–233. https://doi.org/10.1111/j.1467-954X.1984.tb00113.x.

Cambridge University Press. n.d. "citizen." Cambridge Dictionary. Accessed Mar. 18, 2024. https://dictionary.cambridge.org/dictionary/english/citizen.

Campolo, Alexander, and Kate Crawford. 2020. "Enchanted Determinism: Power without Responsibility in Artificial Intelligence." *Engaging Science, Technology, and Society* 6 (Jan.): 1–19. https://doi.org/10.17351/ests2020.277.

Cantauw, Christiane, Michael Kamp, and Elisabeth Timm, eds. 2017. *Figurationen des Laien zwischen Forschung, Leidenschaft und politischer Mobilisierung: Museen, Archive und Erinnerungskultur in Fallstudien und Berichten*. Beiträge zur Volkskultur in Nordwestdeutschland, 127. Münster; New York: Waxmann.

Caprarom. 2021. "How did you experience catching stalls alongside GAIA?." Forum – Human Computation Institute. May 5, 2021, 3:50 AM. https://forum.hcinst.org/t/how-did-you-experience-catching-stalls-alongside-gaia/1089/2.

Carey, Matthew. 2017. *Mistrust: An ethnographic theory*. The Malinowski monographs. Chicago: HAU Books.

Carlsbergfondet. 2019. Skal vi frygte eller omfavne kunstig intelligens? Vimeo video, 2:57. https://vimeo.com/325155109.

Carlson, Rebecca, and Jonathan Corliss. 2007. "Rubble Jumping: From Paul Virilio's Techno-Dromology to Video Games and Distributed Agency." *Culture, Theory and Critique* 48, no. 2 (Nov.): 161–174. https://doi.org/10.1080/14735780701723207.

Carlson, Rebecca, Ruth Dorothea Eggel, Lina Franken, Sarah Thanner, and Libuše Hannah Veprek. 2021. "Approaching code as process: Prototyping ethnographic methodologies." *Code*. Kuckuck – Notizen Zur Alltagskultur, 1: 13–17. https://www.kuckucknotizen.at/kuckuck/index.php/1-21-code/216-1-21-code-leseprobe.

Carlson, Rebecca, and Libuše Hannah Vepřek. 2022. "At the 'hinge' of future fictions and everyday failings: Ethnographic interventions in AI systems." Conference presentation presented at the RAI2022 "Anthropology, AI and the Future of Human Society," Online, Jun. 6.

Carnegie Mellon University. n.d. "CAPTCHA: Telling Humans and Computers Apart Automatically." Accessed Mar. 18, 2024. http://www.captcha.net/.

Carol Aka Mema. 2019. "'If I Can Do This, Anyone Can, and We Need All the Help We Can Get!' – Interview with Supercatcher Carol Aka Mema!" Interview by Egle (Seplute). Human Computation Institute Blog. May 11, 2019. https://blog.hcinst.org/carol-aka-mema/.

Cassar, Robert. 2013. "Gramsci and Games." *Games and Culture* 8, no. 5 (Jul.): 330–53. https://doi.org/10.1177/1555412013493499.

Cassirer, Ernst. 1985. "Form und Technik." In *Symbol, Technik, Sprache. Aufsätze aus den Jahren 1927–1933*, edited by Ernst Wolfgang Orth and John Michael Krois: 39–91. Philosophische Bibliothek 372. Hamburg: Meiner.

Cave, Stephen, and Kanta Dihal. 2019. "Hopes and fears for intelligent machines in fiction and reality." *nature machine intelligence* 1, no. 2 (Feb.): 74–78. https://doi.org/10.1038/s42256-019-0020-9.

Cave, Stephen, Kanta Dihal, and Sarah Dillon. 2020. "Introduction: Imagining AI." In *AI Narratives: A History of Imaginative Thinking about Intelligent Machines*, edited by Stephen Cave, Kanta Dihal, and Sarah Dillon, 1–22. Oxford: Oxford University Press. https://doi.org/10.1093/oso/9780198846666.003.0001.

Center for Game Science (University of Washington), Institute for Protein Design (University of Washington), Cooper Lab (Northeastern University), Khatib Lab (University of Massachusetts, Dartmouth), Siegel Lab (University of California, Davis), Meiler Lab (Vanderbilt University), and Horowitz Lab (University of Denver). n.d.a "Foldit." Accessed Mar. 18, 2024. https://fold.it/.

———. n.d.b "About Foldit." Foldit. Accessed Mar. 18, 2024. https://fold.it/about_foldit.

———. n.d.c "Science." Accessed Mar. 18, 2024. https://fold.it/.

Certeau, Michel de. (1980) 2013. *The Practice of Everyday Life. 1.* 2nd edition. Berkeley, California: University of California Press.

Charmaz, Kathy. 2000. "Grounded Theory: Objectivist and Constructivist Methods." In *The Handbook of Qualitative Research*, edited by Norman K. Denzin and Yvonna S. Lincoln, 509–535. Thousand Oaks California: SAGE Publications.

———. 2009. "Recollecting good and bad days." In *ethnographies revisited. constructing theory in the field*, edited by Antony J. Puddephatt, William Shaffir, and Steven W. Kleinknecht, 48–62. Abingdon, Oxon; New York, NY: Routledge.

———. 2012. "The Power and Potential of Grounded Theory." *Medical Sociology Online* 6, no. 3 (May): 2–15. https://epicpeople.org/wp-content/uploads/2019/12/Charmaz-2012.pdf.

———. 2014. *Constructing Grounded Theory*. 2nd edition. Introducing Qualitative Methods. London; Thousand Oaks California: SAGE Publications.

Chase, Harrison. 2022. "LangChain README.md." Github. Accessed Mar. 18, 2024. https://github.com/hwchase17/langchain.

———. 2023. "Human as a tool." LangChain 0.0.188. Accessed Mar. 18, 2024. https://python.langchain.com/docs/integrations/tools/human_tools.

christiane. 2021. "'How Did You Experience Catching Stalls alongside GAIA?'" Forum – Human Computation Institute. May 11, 2021, 7:50 PM. https://forum.hcinst.org/t/how-did-you-experience-catching-stalls-alongside-gaia/1089/4.

Chun, Wendy Hui Kyong. 2008. "On 'Sourcery,' or Code as Fetish." *Configurations* 16, no. 3: 299–324. https://doi.org/10.1353/con.0.0064.

Citizen Science Games (2019), "ARTigo." CITIZEN SCIENCE GAMES. WHEN SCIENCE CONNECTS WITH GAMES. Accessed Mar. 18, 2024. https://citizensciencegames.com/games/artigo/.

Clickworker GmbH. n.d. "Clickworker." Accessed Mar. 18, 2024. https://www.clickworker.com/.

Code Ethnography Collective. n.d. "Code Ethnography Collective." Accessed Mar 18, 2024. https://codeethnographycollective-ceco.github.io/.

Coleman, E. Gabriella. 2013. *Coding Freedom. The Ethics and Aesthetics of Hacking*. Princeton, Oxford: Princeton University Press.

Consalvo, Mia. 2007. *Cheating: Gaining Advantage in Videogames*. Cambridge, Massachusetts; London England: The MIT Press.

Cooper, Caren B., Lisa M. Rasmussen, and Elizabeth D. Jones. 2021. "Perspective: The Power (Dynamics) of Open Data in Citizen Science." *Frontiers in Climate* 3 (Jul.): 637037. https://doi.org/10.3389/fclim.2021.637037.

Cooper, Seth, Firas Khatib, Ilya Makedon, Hao Lu, Janos Barbero, David Baker, James Fogarty, Zoran Popović, and Foldit players. 2011. "Analysis of social gameplay macros in the Foldit cookbook." In *Proceedings of the 6th International Conference on Foundations of Digital Games – FDG '11*, 9–14. Bordeaux, France: ACM Press. https://doi.org/10.1145/2159365.2159367.

Cooper, Seth, Firas Khatib, Adrien Treuille, Janos Barbero, Jeehyung Lee, Michael Beenen, Andrew Leaver-Fay, David Baker, Zoran Popović, and Foldit players. 2010. "Predicting protein structures with a multiplayer online game." *Nature* 466, no. 7307 (Aug.): 756–60. https://doi.org/10.1038/nature09304.

Cooper, Seth, Adrien Treuille, Janos Barbero, Andrew Leaver-Fay, Kathleen Tuite, Firas Khatib, Alex Cho Snyder, Michael Beenen, David Salesin, David Baker, et al. 2010. "The challenge of designing scientific discovery games." In *Proceedings of the Fifth International Conference on the Foundations of Digital Games*, 40–47. Monterey California: ACM. https://doi.org/10.1145/1822348.1822354.

Corsín Jiménez, Alberto. 2011. "Trust in anthropology." *Anthropological Theory* 11, no. 2 (Jun.): 177–196. https://doi.org/10.1177/1463499611407392.

Costa, Elisabetta. 2018. "Affordances-in-practice: An ethnographic critique of social media logic and context collapse." *New Media & Society* 20, no. 10 (Feb.): 3641–3656. https://doi.org/10.1177/1461444818756290.

Crouser, R. Jordan, Benjamin Hescott, and Remco Chang. 2014. "Toward Complexity Measures for Systems Involving Human Computation." *Human Computation* 1, no. 1 (Sept.). https://doi.org/10.15346/hc.v1i1.4.

Curtis, Vickie. 2015. "Motivation to Participate in an Online Citizen Science Game: A Study of Foldit." *Science Communication* 37, no. 6 (Oct.): 723–746. https://doi.org/10.1177/1075547015609322.

Darwin, Charles. 1859. *On the Origin of Species by Means of Natural Selection, or the Preservation of Favoured Races in the Struggle for Life*. London: J. Murray.

Das, Veena. 2020. *Textures of the Ordinary: Doing Anthropology After Wittgenstein*. New York: Fordham University Press.

DeLanda, Manuel. 2006. *A New Philosophy of Society: Assemblage Theory and Social Complexity*. London; New York: Continuum.

———. 2016. *Assemblage Theory*. Speculative Realism. Edinburgh: Edinburgh University Press. https://doi.org/10.1515/9781474413640.

Deleuze, Gilles, and Félix Guattari. 2000. *Anti-Oedipus: Capitalism and Schizophrenia*. 10th ed. Minneapolis: University of Minnesota Press.

———. 2013. *A Thousand Plateaus: Capitalism and Schizophrenia*. Bloomsbury Revelations Series. London: Bloomsbury.

Deleuze, Gilles, and Claire Parnet. 2007. *Dialogues II*. Rev. edition. European Perspectives. New York: Columbia University Press.

Dellermann, Dominik, Adrian Calma, Nikolaus Lipusch, Thorsten Weber, Sascha Weigel, and Philipp Ebel. 2019. "The Future of Human-AI Collaboration: A Taxonomy of Design Knowledge for Hybrid Intelligence Systems." In *Proceedings of the 52nd Hawaii International Conference on System Sciences*: 274–283. http://hdl.handle.net/10125/59468.

Dellermann, Dominik, Philipp Ebel, Matthias Söllner, and Jan Marco Leimeister. 2019. "Hybrid Intelligence." *Business & Information Systems Engineering* 61, no. 5 (Mar.): 637–643. https://doi.org/10.1007/s12599-019-00595-2.

DeMello, Margo. 2021. *Animals and Society. An introduction to human-animal studies*. 2nd edition. New York City: Columbia University Press.

De Munck, Bert. 2022. "Assembling Path Dependency and History: An Actor-Network Approach." *The Journal of Interdisciplinary History* 52, no. 4 (Mar.): 565–588. https://doi.org/10.1162/jinh_a_01767.

Denk, Winifried, James H. Strickler, and Watt W. Webb. 1990. "Two-Photon Laser Scanning Fluorescence Microscopy." *Science* 248, no. 4951 (Apr.): 73–76. https://doi.org/10.1126/science.2321027.

Detering, Sebastian, Rilla Khaled, Lennart E. Nacke, and Dan Dixon. 2011. "Gamification: Toward a Definition." In *CHI2011 Workshop: Gamification: Using Game Design Elements in Non-Gaming Contexts*: 1–4. http://gamification-research.org/wp-content/uploads/2011/04/02-Deterding-Khaled-Nacke-Dixon.pdf.

Deutsche Forschungsgemeinschaft. n.d. "iART: Ein Interaktives Analyse- und Retrieval-Tool zur Unterstützung von bildorientierten Forschungsprozessen." GEPRIS. Geförderte Projekte der DFG. Accessed Mar. 18, 2024. https://gepris.dfg.de/gepris/projekt/415796915?context=projekt&task=showDetail&id=41579691-5&.

Dev Josh. 2021a. "Foldit Newsletter, Jan. 15, 2021." Posted by agcohn821. Foldit forum. Jan. 15, 2021. https://fold.it/forum/discussion/foldit-newsletters/page-3#post_40361.

———. 2021b. "Newsletter August 27: Getting Edgy with Sheets." Posted by Agcohn821. Foldit forum. Sept. 6, 2021. https://fold.it/forum/discussion/foldit-newsletters/page-7#post_40394.

DeWalt, Kathleen Musante, and Billie R. DeWalt. 2011. *Participant Observation: A Guide for Fieldworkers*. 2nd edition. Lanham, Md; New York, Toronto, Plymouth, UK: AltaMira Press.

Díaz, Carlos, Marisa Ponti, Pinja Haikka, Rajiv Basaiawmoit, and Jacob Sherson. 2020. "More than data gatherers: exploring player experience in a citizen science game." *Quality and User Experience* 5: 1–19. https://doi.org/10.1007/s41233-019-0030-8.

Dietzsch, Ina. 2022. "Interview." Interview by and Carsten Wilhelm. *Interfaces Numériques* 11, no. 1 (Apr.). https://www.unilim.fr/interfaces-numeriques/4753.

Dippel, Anne. 2017. "Das Big Data Game." *NTM Zeitschrift für Geschichte der Wissenschaften, Technik und Medizin* 25, no. 4 (Nov.): 485–517. https://doi.org/10.1007/s00048-017-018 1-8.

———. 2018. "Arbeit." In *Philosophie des Computerspiels. Theorie – Praxis – Ästhetik*, edited by Daniel Martin Feige, Sebastian Ostritsch, and Markus Rautzenberg, 123–148. Stuttgart: J.B. Metzler.

———. 2019a. "Metaphors We Live By. Three Commentaries on Artificial Intelligence and the Human Condition." In *The Democratization of Artificial Intelligence. Net Politics in the Era of Learning Algorithms*, edited by Andreas Sudmann, 33–42. Bielefeld: transcript.

———. 2019b. "Ludopian Visions. On the Speculative Potential of Games in Times of Algorithmic Work and Play." In *Playing Utopia. Futures in Digital Games*, edited by Benjamin Beil, Gundolf S. Freyermuth, and Hanns Christian Schmidt, 235–252. Bielefeld: transcript. https://doi.org/10.1515/9783839450505-008.

———. 2020. "Spiel." In *Kulturtheoretisch Argumentieren. Ein Arbeitsbuch*, edited by Timo Heimerdinger, and Markus Tauschek, 467–489. Münster; New York: Waxmann.

———. 2021. "Schwindel in der Digitale. Re/Visionen einer Kulturanalyse des Alltags." *Code. Kuckuck – Notizen Zur Alltagskultur* 1: 6–10.

Dippel, Anne, and Sonia Fizek. 2017a. "Ludifizierung von Kultur. Zur Bedeutung des Spiels in alltäglichen Praxen der digitalen Ära." In *Digitalisierung. Theorien und Konzepte für sie empirische Kulturforschung*, edited by Gertraud Koch, 363–383. Köln: Herbert von Halem Verlag.

———. 2017b. "Playbouring Cyborgs. Renegotiating the human-machine ensembles." Conference presentation presented at the American Association of Geographers Annual Conference. Workshop "Human, Digital, Labour." Boston, Apr. 8. http://app.co re-apps.com/aagam2017/abstract/dcb181785f7fd70d2b1bd16d5ae6f8c5.

———. 2019. "Laborious playgrounds: Citizen science games as new modes of work/play in the digital age." In *The Playful Citizen: Civic Engagement in a Mediatized Culture*, edited by René Glas, Sybille Lammes, Michiel de Lange, Joost Raessens, and Imar O. de Vries, 255–271. Civic Engagement in a Mediatized Culture. Amsterdam: Amsterdam University Press.

Dörner, Ralf, Stefan Göbel, Wolfgang Effelsberg, and Josef Wiemeyer, eds. 2016. *Serious Games: Foundations, Concepts and Practice*. Cham: Springer. https://doi.org/10.1007/97 8-3-319-40612-1.

Dorrestijn, Steven. 2012a. *The design of our own lives: Technical mediation and subjectivation after Foucault*. PhD Thesis, University of Twente. Enschede: University of Twente. htt ps://doi.org/10.3990/1.9789036534420.

———. 2012b. "Technical Mediation and Subjectivation: Tracing and Extending Foucault's Philosophy of Technology." *Philosophy & Technology* 25, no. 2 (Dec.): 221–241. https://doi.org/10.1007/s13347-011-0057-0.

———. 2017. "The Care of Our Hybrid Selves: Ethics in Times of Technical Mediation." *Foundations of Science* 22, no. 2 (Oct.): 311–321. https://doi.org/10.1007/s10699-015-9440-0.

Dourish, Paul. 2016. "Algorithms and their others: Algorithmic culture in context." *Big Data & Society* 3, no. 2 (Aug.): 1–11. https://doi.org/10.1177/2053951716665128.

DrivenData Labs. n.d. "DRIVENDATA LABS." Accessed Mar. 18, 2024. https://drivendata.co/.

Dürr, Eveline, Moritz Ege, Johannes Moser, Christoph K. Neumann, and Gordon M. Winder. 2020. "Urban ethics: Towards a research agenda on cities, ethics and normativity." *City, Culture and Society* 20 (Mar.): 100313. https://doi.org/10.1016/j.ccs.2019.100313.

Egbert, Simon. 2017. "Siegeszug der Algorithmen? Predictive Policing im deutschsprachigen Raum." *Aus Politik und Zeitgeschichte* 67 (32–33): 17–23. https://www.bpb.de/shop/zeitschriften/apuz/253603/siegeszug-der-algorithmen-predictive-policing-im-deutschsprachigen-raum/.

Egbert, Simon, and Susanne Krasmann. 2019. "Predictive policing: not yet, but soon preemptive?" *Policing and Society* 30, no. 8 (May): 1–15. https://doi.org/10.1080/10439463.2019.1611821.

Ege, Moritz, and Johannes Moser. 2021a. "Introduction: Urban ethics – conflicts over the good and proper life in cities." In *Urban Ethics – Conflicts over the Good and Proper Life in Cities*, edited by Moritz Ege, and Johannes Moser, 3–27. Abingdon, Oxon; New York, NY: Routledge. https://tandfbis.s3-us-west-2.amazonaws.com/rt-files/docs/Open+Access+Chapters/9780429322310_oachapter1.pdf.

———, eds. 2021b. *Urban Ethics: Conflicts over the Good and Proper Life in Cities*. Routledge Studies in Urbanism and the City. Abingdon, Oxon; New York, NY: Routledge.

Egle (Seplute). 2017. "New dataset in Stall Catchers – high fat diet." Human Computation Institute Blog. Jul. 22, 2017. https://blog.hcinst.org/highfat-dataset/.

———. 2018. "The Science Behind EyesOnALZ." Human Computation Institute Blog. Mar. 30, 2018. https://blog.hcinst.org/science-behind-eyesonalz/.

———. 2019. "Congrats to *Dr.* Mohammad Haft-Javaherian for successfully defending his PhD thesis!." Human Computation Institute Blog. May 8, 2019. https://blog.hcinst.org/humans-of-stall-catchers-mohammad/.

———. 2020a. "Final results of the #CabinFever challege!" Human Computation Institute Blog. May 1, 2020. https://blog.hcinst.org/final-results-of-the-cabinfever/.

———. 2020b. "Stalls, machines and humans: an update ." Human Computation Institute Blog. Dec. 3, 2020. https://blog.hcinst.org/drivendata-competition-results/.

———. 2021a. "What's a Catchathon? (the pandemic edition)." Human Computation Institute Blog. Mar. 21, 2021. https://blog.hcinst.org/whats-a-catchathon-mar2021/.

———. 2021b. "The Stall Catchers Catchathon is TODAY!" Human Computation Institute Blog. Apr. 28, 2021. https://blog.hcinst.org/catchathon2021-is-today/.

———. 2021c. "Catchathon 2021: Full report!" Human Computation Institute Blog. May 4, 2021. https://blog.hcinst.org/catchathon-2021-full-report/.

Eiben, Christopher B, Justin B Siegel, Jacob B Bale, Seth Cooper, Firas Khatib, Betty W Shen, Foldit Players, Barry L Stoddard, Zoran Popovic, and David Baker. 2012. "Increased Diels-Alderase Activity through backbone remodeling guided by Foldit players." *nature biotechnology* 30, no. 2 (Jan.): 190–192. https://doi.org/10.1038/nbt.2109.

Eitzel, M. V., Jessica L. Cappadonna, Chris Santos-Lang, Ruth Ellen Duerr, Arika Virapongse, Sarah Elizabeth West, Christopher Conrad Maximillian Kyba, et al. 2017. "Citizen Science Terminology Matters: Exploring Key Terms." *Citizen Science: Theory and Practice* 2, no. 1 (Jun.): 1. https://doi.org/10.5334/cstp.96.

Elias, Norbert. (1970) 2012. *What is Sociology?* Translated by Grace Morrissey, Stephen Mennell, Edmund Jephcott. The Collected Works of Norbert Elias 5. Dublin: University College Dublin Press.

Elish, Madeleine Clare, and danah boyd. 2018. "Situating methods in the magic of Big Data and AI." *Communication Monographs* 85, no. 1: 57–80. https://doi.org/10.1080/03637751.2017.1375130.

Elmann, Merete. 2022. "Center for Hybrid Intelligence." SCHOOL OF BUSINESS AND SOCIAL SCIENCES. AARHUS UNIVERSITY. Mar. 7, 2024. https://mgmt.au.dk/center-for-hybrid-intelligence/.

EMBL-EBI. n.d. "AlphaFold. Protein Structure Database." Accessed Mar. 18, 2024. https://alphafold.ebi.ac.uk/.

Emerging Technology from the arXiv. 2015. "The Emerging Science of Human Computation." MIT Technology Review, Jun. 4, 2015. https://www.technologyreview.com/2015/06/04/248690/the-emerging-science-of-human-computation/.

Endreß, Martin. 2012. "Vertrauen und Misstrauen – Soziologische Überlegungen." In *Vertrauen und Kooperation in der Arbeitswelt*, edited by Christian Schilcher, Mascha Will-Zocholl, and Marc Ziegler, 81–102. Wiesbaden: Springer VS.

Engemann, Christoph, and Andreas Sudmann, eds. 2018. *Machine Learning: Medien, Infrastrukturen und Technologien der Künstlichen Intelligenz*. Digitale Gesellschaft, 14. Bielefeld: transcript.

European Commission. 2021. "HORIZON EUROPE. THE EU RESEARCH & INNOVATION PROGRAMME 2021–27." Accessed Mar. 18, 2024. https://research-and-innovation.ec.europa.eu/system/files/2022-06/ec_rtd_he-investing-to-shape-our-future_0.pdf.

Färber, Alexa. 2014. "Potenziale freisetzen: Akteur-Netzwerk-Theorie und Assemblageforschung in der Interdisziplinären kritischen Stadtforschung." *sub\urban. zeitschrift für kritische stadtforschung* 2, no. 1 (May): 95–103.

Farías, Ignacio, and Thomas Bender. 2010. *Urban Assemblages: How Actor-Network Theory Changes Urban Studies*. Questioning Cities. Abingdon, Oxon; New York, NY: Routledge.

Fassin, Didier, ed. 2012. "Introduction: Toward a Critical Moral Anthropology." In *A Companion to Moral Anthropology*, 1–17. Hoboken, New Jersey: Wiley-Blackwell.

———. 2014. "The ethical turn in anthropology: Promises and uncertainties." *HAU: Journal of Ethnographic Theory* 4, no. 1: 429–435. https://doi.org/10.14318/hau4.1.025.

———. 2015. "Troubled waters. At the confluence of ethics and politics." In *Four Lectures on Ethics. Anthropological Perspectives*, edited by Michael Lambek, Veena Das, Didier Fassin, and Keane Webb, 175–210. Chicago: HAU books.

Fassin, Didier, and Samuel Lézé, eds. 2014. *Moral Anthropology: A Critical Reader*. Abingdon, Oxon; New York, NY: Routledge.

Fassler, Manfred. 1996. *Mediale Interaktion: Speicher, Individualität, Öffentlichkeit*. Paderborn: Fink.

Faubion, James D. 2011. *An Anthropology of Ethics*. Cambridge: Cambridge University Press. https://doi.org/10.1017/CBO9780511792557.

Felstiner, Alek. 2011. "Working the Crowd: Employment and Labor Law in the Crowdsourcing Industry." *Berkeley Journal of Employment and Labor Law* 32, no. 1: 143–204.

Ferguson, Andrew Guthrie. 2017. *The Rise of Big Data Policing: Surveillance, Race, and the Future of Law Enforcement*. New York: New York University Press. https://doi.org/10.18574/nyu/9781479854608.001.0001.

Finke, Peter. 2014. *Citizen Science. Das unterschätzte Wissen der Laien*. München: oekom.

Fischer, Björn, Britt Östlund, and Alexander Peine. 2020. "Of robots and humans: Creating user representations in practice." *Social Studies of Science* 50, no. 2 (Feb.): 221–244. https://doi.org/10.1177/0306312720905116.

Fisher, Robert, Simon Perkins, Ashley Walker, and Erik Wolfart. 2003. "Skeletonization/Medial Axis Transform." https://homepages.inf.ed.ac.uk/rbf/HIPR2/skeleton.htm.

Fitsch, Hannah, and Kathrin Friedrich. 2018. "Digital Matters: Processes of Normalization in Medical Imaging." *Catalyst: Feminism, Theory, Technoscience* 4, no. 2 (Oct.): 1–31. https://doi.org/10.28968/cftt.v4i2.29911.

Fizek, Sonia. 2016. "All work and no play. Are games becoming the factories of the future?" *First Person Scholar*. Mar. 9, 2016. http://www.firstpersonscholar.com/all-work-and-no-play/.

Fjelland, Ragnar. 2020. "Why general artificial intelligence will not be realized." *Humanities and Social Sciences Communications* 7, no. 1 (Jun.): 10. https://doi.org/10.1057/s41599-020-0494-4.

Fleischhack, Julia. 2019. "Veränderte Bedingungen des *Sozialen*. Eine methodologische Betrachtung zur Digitalen Anthropologie / Digitalen Ethnographie." *Zeitschrift für Volkskunde* 115, no. 2 (Oct.): 196–215.

Foldit Contenders Group, Foldit Void Crushers Group, Firas Khatib, Frank DiMaio, Seth Cooper, Maciej Kazmierczyk, Miroslaw Gilski, Szymon Krzywda, Helena Zabranska, Iva Pichova, et al. 2011. "Crystal structure of a monomeric retroviral protease solved by protein folding game players." *nature structural & molecular biology* 18, no. 10 (Sept.): 1175–1177. https://doi.org/10.1038/nsmb.2119.

Foldit Wiki. 2017a. "Mutate." Fandom. Last modified Feb. 1, 2017, 03:36. https://foldit.fandom.com/wiki/Mutate.

———. 2017b. "Recipes." Fandom. Last modified Mar. 7, 2017, 08:55. https://foldit.fandom.com/wiki/Recipes.

———. 2017c. "Shake." Fandom. Last modified Mar. 17, 2017, 22:09. https://foldit.fandom.com/wiki/Shake.

———. 2017d. "Wiggle." Fandom. Last modified Mar. 19, 2017, 03:29. https://foldit.fandom.com/wiki/Wiggle.

———. 2017e. "De-Novo Puzzle." Fandom. Last modified May 27, 2017, 21:00. https://foldit.fandom.com/wiki/De-novo_puzzle.

———. 2018a. "Backbone." Fandom. Last modified Jan. 5, 2018, 06:34. https://foldit.fandom.com/wiki/Backbone.

———. 2018b. "Score." Fandom. Last modified Jan. 18, 2018, 23:14. https://foldit.fandom.com/wiki/Score.

———. 2019. "Revisiting Puzzle." Fandom. Last modified Jan. 12, 2019, 21:52. https://foldit.fandom.com/wiki/Revisiting_puzzle.

———. 2020a. "Sidechain." Fandom. Last modified Apr. 10, 2020, 22:08. https://foldit.fandom.com/wiki/Sidechain.

———. 2022. "Objective." Fandom. Last modified May 26, 2022, 03:29. https://foldit.fandom.com/wiki/Objective.

Forsythe, Diana. (1988) 2001a. "Artificial Intelligence Invents Itself: Collective Identity and Boundary Maintenance in an Emergent Scientific Discipline." In *Studying Those Who Study Us: An Anthropologist in the World of Artificial Intelligence*, by Diana Forsythe, edited by David J. Hess, 75–92. Writing Science. Stanford, California: Stanford University Press.

———. (1992) 2001b. "Blaming the User in Medical Informatics: The Cultural Nature of Scientific Practice." In *Studying Those Who Study Us: An Anthropologist in the World of Artificial Intelligence*, by Diana Forsythe, edited by David J. Hess, 1–15. Writing Science. Stanford, Calif: Stanford University Press.

———. (1993) 2001c. "Engineering Knowledge: The Construction of Work in Artificial Intelligence." In *Studying Those Who Study Us. An Anthropologist in the World of Artificial Intelligence*, by Diana Forsythe, edited by David J. Hess, 35–58. Stanford, California: Stanford University Press.

———. (1999) 2001d. "Ethics and Politics of Studying Up in Technoscience." In *Studying Those Who Study Us. An Anthropology in the World of Artificial Intelligence*, by Diana Forsythe, edited by David J. Hess, 119–131. Stanford, California: Stanford University Press.

———. (1996) 2001e. "New Bottles, Old Wine: Hidden Cultural Assumptions in a Computerized Explanation System for Migraine Sufferers." In *Studying Those Who Study Us: An Anthropologist in the World of Artificial Intelligence*, by Diana Forsythe, edited by David J. Hess, 93–118. Writing Science. Stanford, Calif: Stanford University Press.

———. 2001f. *Studying Those Who Study Us: An Anthropologist in the World of Artificial Intelligence*. Edited by David J. Hess. Writing Science. Stanford, Calif: Stanford University Press.

———. 2001g. "Studying Those Who Study Us: Medical Informatics Appropriates Ethnography." In *Studying Those Who Study Us: An Anthropologist in the World of Artificial Intelligence*, edited by David J. Hess, 132–145. Writing Science. Stanford, Calif: Stanford University Press.

———. (1993) 2001h. "The Construction of Work in Artificial Intelligence." In *Studying Those Who Study Us: An Anthropologist in the World of Artificial Intelligence*, edited by David J. Hess, 16–34. Stanford, Calif: Stanford University Press.

Foucault, Michel. 1972. *The Archaeology of Knowledge and the Discourse on Language*. Translated by Sheridan A. M. Smith. World of Man. New York: Pantheon Books.

———. 1983. "Afterword. The Subject and Power." In *Michel Foucault: Beyond Structuralism and Hermeneutics. With an Afterword by and an Interview with Michel Foucault*, edited by

Hubert L. Dreyfus and Paul Rabinow, 208–226. 2nd edition. Chicago: The University of Chicago Press.

———. 1988. *The Use of Pleasure*. Translated by Robert Hurley. The History of Sexuality 2. New York: Vintage Books.

———. 1995. *Discipline and Punish: The Birth of the Prison*. Translated by Alan Sheridan. 2nd Vintage Books edition. New York: Vintage Books.

———. 1998. *The Will to Knowledge*. The History of Sexuality 1. London: Penguin books.

Fraisl, Dilek, Gerid Hager, Baptiste Bedessem, Margaret Gold, Pen-Yuan Hsing, Finn Danielsen, Colleen B. Hitchcock, Joseph M. Hulbert, Jaume Piera, Helen Spiers, et al. 2022. "Citizen science in environmental and ecological sciences." *Nature reviews methods primers* 2, no. 1 (Aug.): 64. https://doi.org/10.1038/s43586-022-00144-4.

Franken, Lina. 2023. *Digitale Methoden für qualitative Forschung: Computationelle Daten und Verfahren*. UTB. Münster; New York: Waxmann.

Frissen, Valerie, Sybille Lammes, Michiel de Lange, Jos de Mul, and Joost Raessens. 2015. "Homo ludens 2.0: Play, media, and identity." In *Playful Identities: The Ludification of Digital Media Cultures*, edited by Valerie Frissen, Sybille Lammes, Michiel de Lange, Jos de Mul, and Joost Raessens, 9–50. MediaMatters. Amsterdam: Amsterdam University Press.

Froschauer, Ulrike, and Manfred Lueger. 2020. *Das qualitative Interview: zur Praxis interpretativer Analyse sozialer Systeme*. 2nd edition. UTB. Wien: facultas.

Fuller, Matthew, ed. 2008. *Software Studies: A Lexicon*. Cambridge, Massachusetts; London England: The MIT Press. https://doi.org/10.7551/mitpress/9780262062749.001.0001.

Future of Life Institute. n.d. "future of life INSTITUTE." Accessed Mar. 18, 2024. https://futureoflife.org/.

Gad, Christopher, and Casper Bruun Jensen. 2010. "On the Consequences of Post-ANT." *Science, Technology, & Human Values* 35, no. 1: 55–80. https://doi.org/10.1177/0162243908329567.

Gaffield, Chad, Pierre Corvol, Jörg Hacker, Giorgio Parisi, Juichi Yamagiwa, and Venkatraman «Venki» Ramakrishnan. 2019. "Citizen science in the Internet era." *Summit of the G7 science academies*. https://www.academie-sciences.fr/pdf/rapport/Citizen_G7_2019_EN.pdf.

Galaxy Zoo. n.d. "About." Galaxy Zoo. A Zooniverse Project Blog. Accessed Mar. 18, 2024. https://blog.galaxyzoo.org/about-2/.

Gambetta, Diego. 1988a. "Can We Trust Trust?" In *Trust. Making and Breaking Cooperative Relations*, edited by Diego Gambetta, 213–237. New York, NY: Blackwell.

———, ed. 1988b. *Trust: Making and Breaking Cooperative Relations*. New York, NY: Blackwell.

Garfinkel, Harold. 1963. "A Conception of and Experiments with 'Trust' as a Condition of Stable Concerted Actions." In *Motivation and social interaction: cognitive determinants*, edited by O. J. Harvey, 187–238. New York: Ronald.

———. 1967. *Studies in Ethnomethodology*. Cambridge, UK: Polity Press.

Geertz, Clifford. 1973. *The Interpretation of Cultures: Selected Essays*. New York: Basic Books.

———. 1983. *Local Knowledge: Further Essays in Interpretive Anthropology*. New York: Basic Books.

Geoghegan, Hilary, Alison Dyke, Rachel Pateman, Sarah West, and Glyn Everett. 2016. "Understanding Motivations for Citizen Science. Final Report on Behalf of the UK Environmental Observation Framework." Swindon, Wiltshire. https://www.ukeof.org.uk/resources/citizen-science-resources/MotivationsforCSREPORTFINALMay2016.pdf.

Gesing, Friederike, Michi Knecht, Michael Flitner, and Katrin Amelang, eds. 2019. *NaturenKulturen: Denkräume und Werkzeuge für neue politische Ökologien*. Edition Kulturwissenschaft 146. Bielefeld: transcript. https://doi.org/10.14361/9783839440070.

Gibson, James J. 1977. "The Theory of Affordances." In *Perceiving, Acting, and Knowing: Toward an Ecological Psychology*, edited by Robert E Shaw and John Bransford, 67–82. Hillsdale, NJ: Erlbaum.

———. 1979. *The Ecological Approach to Visual Perception*. Boston: Houghton Mifflin.

Giddens, Anthony. 1990. *The Consequences of Modernity*. Stanford, California: Stanford University Press.

Gieryn, Thomas F. 1983. "Boundary-Work and the Demarcation of Science from Non-Science: Strains and Interests in Professional Ideologies of Scientists." *American Sociological Review* 48, no. 6 (Dec.): 781–95. https://doi.org/10.2307/2095325.

Gillespie, Tarleton. 2014. "The Relevance of Algorithms." In *Media Technologies: Essays on Communication, Materiality, and Society*, edited by Tarleton Gillespie, Pablo Bockowski, and Kirsten Foot, 167–195. Cambridge, Massachusetts; London England: The MIT Press. https://doi.org/10.7551/mitpress/9780262525374.003.0009.

Gitelman, Lisa, and Virginia Jackson. 2013. "Introduction." In *"Raw Data" Is an Oxymoron*, edited by Lisa Gitelman, 1–14. Infrastructures Series. Cambridge, Massachusetts; London England: The MIT Press. https://doi.org/10.7551/mitpress/9302.003.0002.

GitHub, Inc. n.d. "GitHub." Accessed Mar 18, 2024. https://github.com.

Glaser, Barney G., and Anselm L. Strauss. (1967) 1971. *The Discovery of Grounded Theory. Strategies for Qualitative Research*. 4th edition. Chicago; New York: Aldine.

Gonzalez, Laura Lynn. 2007. Rosetta@home. YouTube video, 7:00. https://www.youtube.com/watch?v=GzATbET3g54.

Goodfellow, Ian, Yoshua Bengio, and Aaron Courville. 2016. *Deep Learning*. Adaptive Computation and Machine Learning. Cambridge, Massachusetts; London England: The MIT Press.

Google. n.d. "Google reCAPTCHA." Accessed Mar. 18, 2024. https://www.google.com/recaptcha/about/.

Görsdorf, Alexander. 2007. *Die "Weisheit der Laien" als politische Ressource? Ethnographie eines Szenario-Workshops zur Bürgerbeteiligung am Diskurs um die Biomedizin*. Saarbrücken: VDM.

Graham, Stephen, ed. 2009. *Disrupted Cities: When Infrastructure Fails*. Abingdon, Oxon; New York, NY: Routledge.

Gray, Mary L., and Siddharth Suri. 2019. *Ghost Work: How to Stop Silicon Valley from Building a New Global Underclass*. Boston: Houghton Mifflin Harcourt.

Greth, Nicola. 2019. "Automatic Semantic Categorization of Image Annotations Generated by Games With a Purpose." Master's thesis, LMU Munich. Munich, Germany: LMU Munich. https://www.en.pms.ifi.lmu.de/publications/diplomarbeiten/Nicola.Greth/MA_Nicola.Greth.pdf.

Grier, David Alan. 2013. *When Computers Were Human:* Princeton: Princeton University Press. https://doi.org/10.1515/9781400849369.

Grint, Keith, and Steve Woolgar. 1997. *the Machine at Work: Technology, Work, and Organization.* Oxford: Polity Press.

Gutekunst, Miriam, and Alex Rau. 2017. "Das ethnographische Porträt. Ein Plädoyer für eine didaktische Auseinandersetzung mit dem Schreiben als Lernprozess". In *Facetten des Alter(n)s: ethnografische Porträts über Vulnerabilitäten und Kämpfe älterer Frauen*, edited by Alexandra Rau and Irene Götz: 119–128. Münchner ethnographische Schriften 25. München: Herbert Utz Verlag.

Haarmann, Tim. 2013. "Citizen Science: Zocken für die Forschung." *Die Zeit*, no. 38 (September): 44. https://www.zeit.de/2013/38/citizen-science-eyewire-seti.

Haft-Javaherian, Mohammad, Linjing Fang, Victorine Muse, Chris B. Schaffer, Nozomi Nishimura, and Mert R. Sabuncu. 2019. "Deep convolutional neural networks for segmenting 3D *in vivo* multiphoton images of vasculature in Alzheimer disease mouse models." *PLOS ONE* 14, no. 3 (Mar.): 1–21. https://doi.org/10.1371/journal.pone.0213539.

Hallinan, Blake, and James N. Gilmore. 2021. "Infrastructural politics amidst the coils of control." *Cultural Studies* 35, no. 4–5 (Mar.): 617–640. https://doi.org/10.1080/09502386.2021.1895259.

Hamm, Marion. 2011. "Zur ethnografischen Ko-Präsenz in digitalen Forschungsfeldern." *Feldforschung@Cyberspace.de*, edited by Victoria Hegner and Dorothee Hemme. Kulturen 5, no. 2: 27–33.

Hannerz, Ulf. 1992. *Cultural Complexity: Studies in the Social Organization of Meaning.* New York: Columbia University Press.

Hansen, Lara, and Gertraud Koch. 2022. "Assemblage – Constructing the Social for Empirical Cultural Research." *Assemblage – Constructing the Social for Empirical Cultural Research*: 3–15. Hamburger Journal Für Kulturanthropologie (HJK), no. 14 (Aug.). https://journals.sub.uni-hamburg.de/hjk/article/view/1955/1777.

Haraway, Donna. 1985. "Manifesto for Cyborgs: Science, Technology, and Socialist Feminism in the 1980s." *Socialist Review* 80: 65–108. https://monoskop.org/images/4/4c/Haraway_Donna_1985_A_Manifesto_for_Cyborgs_Science_Technology_and_Socialist_Feminism_in_the_1980s.pdf.

———. 1988. "Situated Knowledges: The Science Question in Feminism and the Privilege of Partial Perspective." *Feminist Studies* 14, no. 3: 575–599. https://doi.org/10.2307/3178066.

———. 1991. "A Cyborg Manifesto: Science, Technology, and Socialist-Feminism in the Late Twentieth Century." In *Simians, Cyborgs and Women: The Reinvention of Nature*, by Donna Haraway, 141–181. Abingdon, Oxon; New York, NY: Routledge.

———. 2003. *The Companion Species Manifesto: Dogs, People, and Significant Otherness.* Paradigm 8. Chicago: Prickly Paradigm Press.

Hardin, Russell. 2006. *Trust.* Cambridge, UK: Polity.

Harper, Richard H. R., ed. 2014a. *Trust, Computing, and Society.* Cambridge: Cambridge University Press. https://doi.org/10.1017/CBO9781139828567.

———. 2014b. "Reflections on Trust, Computing, and Society." In *Trust, Computing, and Society*, edited by Richard H. R. Harper, 299–338. Cambridge: Cambridge University Press. https://doi.org/10.1017/CBO9781139828567.018.

Hartman, Björn, and Eric Horvitz. 2013. "Preface." *Proceedings of the AAAI Conference on Human Computation and Crowdsourcing* 1, no. 1 (Nov.): xi-x. https://doi.org/10.1609/hcomp.v1i1.13066.

Hassabis, Demis 2020. "DeepMind co-founder: Gaming inspired AI breakthrough." Interview by Nick Robinson. BBC News. Dec. 2, 2020. https://www.bbc.com/news/technology-55157940.

Haydon, Ian. 2022. Foldit Evolved. YouTube video, 1:36. https://www.youtube.com/watch?v=CtI7qpsoFqM.

Hayles, N. Katherine. 2017. *Unthought: The Power of the Cognitive Nonconscious*. Chicago; London: The University of Chicago Press.

Hecker, Susanne, Mordechai Haklay, Anne Bowser, Zen Makuch, Johannes Vogel, and Aletta Bonn. 2018. *Citizen Science: Innovation in Open Science, Society and Policy*. London: UCL Press. http://www.jstor.org/stable/10.2307/j.ctv550cf2.

Heidegger, Martin. 1996. *Being and Time*. Translated by Joan Stambaugh. Albany, NY: State University of New York Press.

Heimerdinger, Timo, and Markus Tauschek. 2020. "Einführung. Kulturtheoretisch argumentieren." In *Kulturtheoretisch argumentieren*, edited by Timo Heimerdinger and Markus Tauschek, 7–31. UTB. Münster; New York: Waxmann.

Heller, Kevin Jon. 1996. "Power, Subjectification and Resistance in Foucault." *SubStance* 25, no.1: 78–110. https://doi.org/10.2307/3685230.

Hengartner, Thomas. (2001) 2007. "Volkskundliches Forschen im, mit dem und über das Internet." In *Methoden Der Volkskunde. Positionen, Quellen, Arbeitsweisen Der Europäischen Ethnologie*, edited by Silke Göttsch and Albrecht Lehmann, 189–218. 2nd edition. Berlin: Reimer.

———. 2012. "Technik – Kultur – Alltag. Technikforschung Als Alltagsforschung." *Schweizerisches Archiv Für Volkskunde* 106: 117–139. https://www.zora.uzh.ch/id/eprint/75932/9/SAVk2_2012_117_139.pdf.

Hepp, Andreas, Juliane Jarke, and Leif Kramp, eds. 2022. *New Perspectives in Critical Data Studies: The Ambivalences of Data Power*. Transforming Communications – Studies in Cross-Media Research. Cham: Palgrave Macmillan; Springer. https://doi.org/10.1007/978-3-030-96180-0.

Hesmondhalgh, David. 2021. "The infrastructural turn in media and internet research." In *The Routledge Companion to Media Industries*, edited by Paul McDonald, 132–142. Abingdon, Oxon; New York, NY: Routledge.

Hilgartner, Stephen. 2015. "Capturing the imaginary. Vanguards, visions and the synthetic biology revolution." In *Science and Democracy. Making knowledge and making power in the biosciences and beyond*, edited by Stephen Hilgartner, Clark A. Miller, and Rob Hagendijk, 33–55. Abingdon, Oxon; New York, NY: Routledge.

Hine, Christine. 2006. "Databases as Scientific Instruments and Their Role in the Ordering of Scientific Work." *Social Studies of Science* 36, no. 2 (Apr.): 269–298. https://doi.org/10.1177/0306312706054047.

Hinrichs, Peter, Martina Röthl, and Manfred Seifert, eds. 2021. *Theoretische Reflexionen: Perspektiven der Europäischen Ethnologie*. Berlin: Reimer.

Hirschauer, Stefan. 2004. "Praktiken und ihre Körper. Über materielle Partizipanden des Tuns." In *Doing Culture. Neue Positionen zum Verhältnis von Kultur und sozialer Praxis.*, edited by Karl H. Hörning and Julia Reuter, 73–91. Bielefeld: transcript. https://doi.org/10.14361/9783839402436-005.

Holbraad, Martin. 2012. *Truth in Motion: The Recursive Anthropology of Cuban Divination*. Chicago; London: The University of Chicago Press.

Holohan, Anne. 2013. *Community, Competition and Citizen Science: Voluntary Distributed Computing in a Globalized World*. Global Connections. Farnham, Surrey; Burlington, Vermont: Ashgate.

Holt, Nathalia. 2016. *Rise of the Rocket Girls: The Women Who Propelled Us, from Missiles to the Moon to Mars*. New York; Boston; London: Little, Brown and Company.

Horowitz, Scott, Brian Koepnick, Raoul Martin, Agnes Tymieniecki, Amanda A. Winburn, Seth Cooper, Jeff Flatten, et al. 2016. "Determining crystal structures through crowdsourcing and coursework." *nature communications* 7, no. 1 (Sept.): 1–9. https://doi.org/10.1038/ncomms12549.

Howe, Jeff. 2006. "The Rise of Crowdsourcing." *Wired Magazine* 14, no. 6 (Jun.): 1–4. https://www.wired.com/2006/06/crowds/.

Huang, Ming-Hui, and Roland Rust. 2019. "The Feeling Economy: Managing in the Next Generation of Artificial Intelligence (AI)." *California Management Review* 61, no. 4 (Jul.): 43–65. https://doi.org/10.1177/0008125619863436.

Huizinga, Johan. (1938) 2016. *Homo Ludens: A Study of the Play-Element in Culture*. Kettering, OH: Angelico Press.

Hultin, Lotta. 2019. "On becoming a sociomaterial researcher: Exploring epistemological practices grounded in a relational, performative ontology." *Information and Organization* 29, no. 2 (Jun.): 91–104. https://doi.org/10.1016/j.infoandorg.2019.04.004.

Human Computation Institute. 2017. Stall Catchers – a citizen science game fighting Alzheimer's. YouTube video, 0:31. https://www.youtube.com/watch?v=wog7cVTMuNM.

———. 2018. Where do Stall Catchers vessel movies come from? Interview with Prof. Chris Schaffer. Youtube video, 16:56. https://www.youtube.com/watch?v=_RyNaohCFsM&list=PLOXMOfnh9jPloOuujXB6EjMSIbyu J1Kr1&t=182s.

———. 2021. Stall Catchers Catchathon 2021 – Final Hour. YouTube video, 1:03:03. https://www.youtube.com/watch?v=ZEzHFXIhj4E.

———. n.d. "About. Stall Catchers." Accessed Mar. 19, 2024. https://stallcatchers.com/about#stall-catchers.

———. n.d. "COLLECTIVE SOLUTIONS TO SOCIETAL PROBLEMS." Accessed Mar. 19, 2024. https://humancomputation.org/.

———. n.d. "Dedications." Accessed Mar. 19, 2024. https://stallcatchers.com/dedications.

———. n.d. "The Humans of Stall Catchers." Accessed Jun. 1, 2023. https://blog.hcinst.org/tag/the-humans-of-stall-catchers/.

Humane AI Net; The Hybrid Intelligence Centre. n.d. "International Conference Series on Hybrid Human-Artificial Intelligence." Accessed Mar. 19, 2024. https://hhai-conference.org/.

Hunter, Jane, Abdulmonem Alabri, and Catharine van Ingen. 2013. "Assessing the quality and trustworthiness of citizen science data." *Concurrency and Computation: Practice and Experience* 25, no. 4 (Sept.): 454–466. https://doi.org/10.1002/cpe.2923.

Hutchby, Ian. 2001. "Technologies, Texts and Affordances." *Sociology* 35, no. 2 (May): 441–456. https://doi.org/10.1177/S0038038501000219.

Hutchins, Edwin. 1995a. *Cognition in the Wild*. Cambridge, Massachusetts; London, England: The MIT Press.

———. 1995b. "How a Cockpit Remembers Its Speeds." *Cognitive Science* 19, no. 3 (Jul. – Sept.): 265–288. https://doi.org/10.1016/0364-0213(95)90020-9.

Iacovides, Ioanna, Charlene Jennett, Cassandra Cornish-Trestrail, and Anna L. Cox. 2013. "Do games attract or sustain engagement in citizen science? a study of volunteer motivations." In *CHI'13 Extended Abstracts on Human Factors in Computing Systems*, 1101–1106. New York: ACM.

Ihde, Don. 1975. "The Experience of Technology: Human-Machine Relations." *Philosophy & Social Criticism* 2, no. 3 (Oct.): 267–279. https://doi.org/10.1177/019145377500200304.

———. 1990. *Technology and the Lifeworld: From Garden to Earth*. The Indiana Series in the Philosophy of Technology. Bloomington: Indiana University Press.

———. 2015. "Preface: Positioning Postphenomenology." In *Postphenomenological Investigations: Essays on Human–Technology Relations*, edited by Robert Rosenberger and Peter-Paul Verbeek, vii–xvi. Lanham: Lexington Books.

"iNaturalist." n.d. "iNaturalist." Accessed Mar. 19, 2024. https://www.inaturalist.org/.

Ingold, Tim. 2000. "From trust to domination: an alternative history of human-animal relations." In *The Perception of the Environment. Essays on Livelihood, Dwelling and Skill*, by Tim Ingold, 61–76. Abingdon, Oxon; New York, NY: Routledge.

———. 2007. *Lines: A Brief History*. Abingdon, Oxon; New York, NY: Routledge.

———. 2014. "That's enough about ethnography!" *HAU: Journal of Ethnographic Theory* 4, no. 1: 383–395. https://doi.org/10.14318/hau4.1.021.

———. 2020. "Ecocriticism and 'Thinking with Writing': An Interview with Tim Ingold." Interview by Antonia Spencer. 2020 *Ecocriticism: In Europe and Beyond*. Ecozon@: European Journal of Literature, Culture and Environment 11, no. 2 (Sept.): 208–215. https://doi.org/10.37536/ECOZONA.2020.11.2.3666.

Inkpen, Kori, Shreya Chappidi, Keri Mallari, Besmira Nushi, Divya Ramesh, Pietro Michelucci, Vani Mandava, Libuše Hannah Vepřek, and Gabrielle Quinn. 2023. "Advancing Human-AI Complementarity: The Impact of User Expertise and Algorithmic Tuning on Joint Decision Making." *ACM Transactions on Computer-Human Interaction*, (Mar.): 1–29. https://doi.org/10.1145/3534561.

Institute of Art History (Ludwig Maximilian University of Munich). n.d. "artigo." Github. Accessed Mar. 19, 2024. https://github.com/arthist-lmu/artigo.

Introna, Lucas D. 2016. "Algorithms, Governance, and Governmentality: On Governing Academic Writing." *Science, Technology, & Human Values* 41, no. 1 (Jun.): 17–49. https://doi.org/10.1177/0162243915587360.

Ipeirotis, Panagiotis G., Raman Chandrasekar, and Paul Bennett. 2009. "A report on the Human Computation Workshop (HComp 2009)." https://www.microsoft.com/en-us/research/wp-content/uploads/2009/01/HComp2009ReportFinal.pdf.

Irwin, Alan. 1995. *Citizen Science: A Study of People, Expertise, and Sustainable Development*. Environment and Society. Abingdon, Oxon; New York, NY: Routledge.

Jackson, Steven J. 2014. "Rethinking Repair." In *Media Technologies*, edited by Tarleton Gillespie, Pablo J. Boczkowski, and Kirsten A. Foot, 221–240. Cambridge, Massachusetts; London, England: The MIT Press. https://doi.org/10.7551/mitpress/9042.003.0015.

Jarrahi, Mohammad Hossein, Christoph Lutz, Karen Boyd, Carsten Oesterlund, and Matthew Willis. 2023. "Artificial intelligence in the work context." *Artificial Intelligence and Work*. Journal of the Association for Information Science and Technology 74, no. 3 (Feb.): 303–310. https://doi.org/10.1002/asi.24730.

Jasanoff, Sheila. 2015a. "Future Imperfect: Science, Technology, and the Imaginations of Modernity." In *Dreamscapes of Modernity: Sociotechnical Imaginaries and the Fabrication of Power*, edited by Sheila Jasanoff and Sang-Hyun Kim, 1–33. Chicago; London: The University of Chicago Press.

———. 2015b. "Imagined and Invented Worlds." In *Dreamscapes of Modernity: Sociotechnical Imaginaries and the Fabrication of Power*, edited by Sheila Jasanoff and Sang-Hyun Kim, 321–341. Chicago; London: The University of Chicago Press.

Jasanoff, Sheila, and Sang-Hyun Kim. 2009. "Containing the Atom: Sociotechnical Imaginaries and Nuclear Power in the United States and South Korea." *Minerva* 47, no. 2 (Jun.): 119–146. https://doi.org/10.1007/s11024-009-9124-4.

———, eds. 2015. *Dreamscapes of Modernity: Sociotechnical Imaginaries and the Fabrication of Power*. Chicago; London: The University of Chicago Press.

Jaton, Florian. 2021. *The Constitution of Algorithms: Ground-Truthing, Programming, Formulating*. Cambridge, Massachusetts; London, England: The MIT Press. https://doi.org/10.7551/mitpress/12517.001.0001.

Jeggle, Utz. 1995. "Volkskunde." In *Handbuch Qualitative Sozialforschung. Grundlagen, Konzepte, Methoden Und Anwendungen*, edited by Uwe Flick, Ernst von Kardorff, Heiner Keupp, Stephan Wolff, and Lutz Rosenstiel, 56–59. Weinheim: Beltz Verlagsgruppe.

Jewett, Tom, and Rob Kling. 1991. "The Dynamics of Computerization in a Social Science Research Team: A Case Study of Infrastructure, Strategies, and Skills." *Social Science Computer Review* 9, no. 2 (Jul.): 246–275. https://doi.org/10.1177/089443939100900205.

joshmiller. 2021. "2020 Snowflake Challenge Results." Foldit. Jan. 19, 2021. https://fold.it/forum/blog/2020-snowflake-challenge-results-blog.

Joyce, Kelly, Laurel Smith-Doerr, Sharla Alegria, Susan Bell, Taylor Cruz, Steve G. Hoffman, Safiya Umoja Noble, and Benjamin Shestakofsky. 2021. "Toward a Sociology of Artificial Intelligence: A Call for Research on Inequalities and Structural Change." *Socius: Sociological Research for a Dynamic World* 7 (Jan.): 1–11. https://doi.org/10.1177/2378023121999581.

Jumper, John, Richard Evans, Alexander Pritzel, Tim Green, Michael Figurnov, Olaf Ronneberger, Kathryn Tunyasuvunakool, et al. 2021. "Highly accurate protein structure prediction with AlphaFold." *nature* 596 (Aug.): 583–589. https://doi.org/10.1038/s4158 6-021-03819-2.

Jung, Matthias. 2015. "'Citizen Science' – eine Programmatik zur Rehabilitierung des Handelns wissenschaftlicher Laiinnen und Laien und ihre Implikationen für die Archäologie." *Forum Kritische Archäologie* 4: 42–54. https://doi.org/10.6105/JOURNAL.FKA.2015.4.6.

Kamar, Ece. 2016a. "Directions in Hybrid Intelligence: Complementing AI Systems with Human Intelligence." In *Proceedings of the Twenty-Fifth International Joint Conference on Artificial Intelligence (IJCAI-16)*: 4070–73.

———. 2016b. "Hybrid Workplaces of the Future." *XRDS: Crossroads, The ACM Magazine for Students* 23, no. 2 (Dec.): 22–25. https://doi.org/10.1145/3013488.

Khatib, Firas, Seth Cooper, Michael D. Tyka, Kefan Xu, Ilya Makedon, Zoran Popović, David Baker, and Foldit Players. 2011. "Algorithm discovery by protein folding game players." *Proceedings of the National Academy of Sciences* (PNAS) 108, no. 47 (Nov.): 18949–18953. https://doi.org/10.1073/pnas.1115898108.

Khoury, George A., Adam Liwo, Firas Khatib, Hongyi Zhou, Gaurav Chopra, Jaume Bacardit, Leandro O. Bortot, et al. 2014. "WeFold: A coopetition for protein structure prediction." *Proteins: Structure, Function, and Bioinformatics* 82, no. 9 (Feb.): 1850–1868. https://doi.org/10.1002/prot.24538.

Kieseberg, Peter, Johannes Schantl, Peter Frühwirt, Edgar Weippl, and Andreas Holzinger. 2015. "Witnesses for the Doctor in the Loop." In *Brain Informatics and Health. BIH 2015*, edited by Yike Guo, Karl Friston, Faisal Aldo, Sean Hill, and Hanchuan Peng, 369–78. Cham: Springer International Publishing. https://doi.org/10.1007/978-3-319-23344-4_36.

Kimura, Aya H., and Abby Kinchy. 2016. "Citizen Science: Probing the Virtues and Contexts of Participatory Research." *Engaging Science, Technology, and Society* 2 (Dec.): 331–361. https://doi.org/10.17351/ests2016.99.

Kitchin, Rob. 2016. "Thinking critically about and researching algorithms." *Information, Communication & Society* 20, no. 1 (Feb.): 14–29. https://doi.org/10.1080/1369118X.2016.1154087.

Kitchin, Rob, and Martin Dodge. 2011. *Code/Space: Software and Everyday Life*. Software Studies. Cambridge, Massachusetts; London, England: The MIT Press. https://doi.org/10.7551/mitpress/9780262042482.001.0001.

Klausner, Martina. 2015. *Choreografien psychiatrischer Praxis: eine ethnografische Studie zum Alltag in der Psychiatrie*. VerKörperungen 22. Bielefeld: transcript.

Klausner, Martina, Milena D. Bister, Jörg Niewöhner, and Stefan Beck. 2015. "Choreografien klinischer und städtischer Alltage: Ergebnisse einer ko-laborativen Ethnografie mit der Sozialpsychiatrie." *Zeitschrift für Volkskunde* 111, no. 2 (Nov.): 214–235.

Klausner, Martina, and Jörg Niewöhner. 2020. "Integrierte Forschung – ein ethnographisches Angebot zur Ko-Laboration." In *Das geteilte Ganze*, edited by Bruno Gransche and Arne Manzeschke, 153–169. Wiesbaden: Springer. https://doi.org/10.1007/978-3-658-26342-3_8.

Kleemann, Frank, Günter G. Voß, and Kerstin Rieder. 2008. "Crowdsourcing und der Arbeitende Konsument." In *Arbeits- und Industriesoziologische Studien* 1, no. 1: 29–44. https://doi.org/10.21241/ssoar.64725.

Knecht, Michi. 2012. "Ethnographische Praxis im Feld der Wissenschafts-, Medizin- und Technikanthropologie." In *Science and Technology Studies: Eine sozialanthropologische Einführung*, edited by Stefan Beck, Jörg Niewöhner, and Estrid Sørensen, 245–274. Bielefeld: transcript.

———. 2013. "Nach *Writing Culture*, mit *Actor-Network*: Ethnografie/Praxeografie in der Wissenschafts-, Medizin- und Technikforschung." In *Europäisch-ethnologisches Forschen: Neue Methoden und Konzepte*, edited by Sabine Hess, Johannes Moser, and Maria Schwertl, 79–106. Berlin: Reimer.

Knorr-Cetina, Karin. 1999. *Epistemic Cultures: How the Sciences Make Knowledge*. Cambridge, Massachusetts: Harvard University Press.

Koch, Gertraud. 2005. *Zur Kulturalität der Technikgenese: Praxen, Policies und Wissenskulturen der künstlichen Intelligenz*. Wissen – Kultur – Kommunikation 1. St. Ingbert: Röhrig.

———. 2015. "Empirische Kulturanalyse in digitalisierten Lebenswelten." *Zeitschrift für Volkskunde* 111, no. 2 (Nov.): 179–200.

———. 2017a. "Einleitung: Digitalisierung als Herausforderung der empirischen Kulturanalyse." In *Digitalisierung. Theorien und Konzepte für die empirische Kulturforschung*, edited by Gertraud Koch, 7–18. Köln: Herbert von Halem.

———. 2017b. "Ethnografie digitaler Infrastrukturen." In *Digitalisierung. Theorien Und Konzepte Für Die Empirische Kulturforschung*, edited by Gertraud Koch, 107–124. Köln: Herbert von Halem.

Koch, Gertraud, and Lina Franken. 2020. "Filtern als digitales Verfahren in der wissenssoziologischen Diskursanalyse: Potenziale und Herausforderungen der Automatisierung im Kontext der Grounded Theory." In *Soziale Medien*, edited by Samuel Breidenbach, Peter Klimczak, and Christer Petersen, 121–138. ars digitalis. Wiesbaden: Springer. https://doi.org/10.1007/978-3-658-30702-8_6.

Koepnick, Brian. 2020. "Mit Foldit gegen COVID-19: Brian Koepnick über das Zusammenspiel von Games und Wissenschaft." Interview by Manouchehr Shamsrizi. Deutscher Computerspielpreis. Oct. 19, 2020. https://deutscher-computerspielpreis.de/mit-foldit-gegen-covid-19-brian-koepnick-ueber-das-zusammenspiel-von-games-und-wissenschaft/.

Koepnick, Brian, Jeff Flatten, Tamir Husain, Alex Ford, Daniel-Adriano Silva, Matthew J. Bick, Aaron Bauer, et al. 2019. "De novo protein design by citizen scientists." *nature* 570 (Jun.): 390–394. https://doi.org/10.1038/s41586-019-1274-4.

Kohle, Hubertus. 2016. "The wisdom of crowds." *On_Culture: the Open Journal for the Study of Culture*, no. 1 (May). http://geb.uni-giessen.de/geb/volltexte/2016/12072/.

———. 2018. "Artigo – Eine Crowdsourcing-Anwendung zur Generierung von Beschreibungsdaten für Kunstwerke." In *BBE-Newsletter für Engagement und Partizipation in Deutschland*, edited by Bundesnetzwerk Bürgerschaftliches Engagement (BBE), no. 20: 1–3. https://www.b-b-e.de/fileadmin/Redaktion/05_Newsletter/01_BBE_Newsletter/2018/newsletter-20-kohle.pdf.

———. "KUNST IM AUGE DES DIGITALEN BETRACHTERS. INTERVIEW MIT HUBERTUS KOHLE." Interview by Kristin Oswald. Bürger Künste Wissenschaft. Jun. 10, 2019. https://bkw.hypotheses.org/1509.

Kosmala, Margaret, Andrea Wiggins, Alexandra Swanson, and Brooke Simmons. 2016. "Assessing data quality in citizen science." *Frontiers in Ecology and the Environment* 14, no. 10 (Dec.): 551–560. https://doi.org/10.1002/fee.1436.

Kücklich, Julian. 2005. "Precarious Playbour: Modders and the Digital Games Industry" In *The Fibreculture Journal* 5: 025. https://five.fibreculturejournal.org/fcj-025-precarious-playbour-modders-and-the-digital-games-industry/.

Kunzelmann, Daniel. 2015. "Die stille Politik der Algorithmen: Das Beispiel Facebook." *Politiken*. Kuckuck – Notizen Zur Alltagskultur 30, no. 2: 30–35.

Kurzweil, Ray. 2005. *The Singularity Is near: When Humans Transcend Biology*. New York: Viking.

Lambek, Michael. 2010. "Toward an Ethics of the Act." In *Ordinary Ethics: Anthropology, Language, and Action*, edited by Michael Lambek, 39–63. New York: Fordham University Press.

———. 2015. "Living as If It Mattered." In *Four Lectures on Ethics. Anthropological Perspectives*, edited by Michael Lambek, Veena Das, Didier Fassin, and Webb Keane, 5–51. Chicago: HAU books.

Lambek, Michael, Veena Das, Didier Fassin, and Webb Keane, eds. 2015. *Four Lectures on Ethics. Anthropological Perspectives*. Chicago: HAU books.

Landau, Michael. 2018. "Catcher Michael Landau: 'Playing the Game Makes Me Feel Less Powerless.'" Interview by Egle (Seplute). Human Computation Institute Blog. Oct. 31, 2018. https://blog.hcinst.org/catcher-michael-landau/.

Land-Zandstra, Anne, Gaia Agnello, and Yaşar Selman Gültekin. 2021. "Participants in Citizen Science." In *The Science of Citizen Science*, edited by Katrin Vohland, Anne Land-Zandstra, Luigi Ceccaroni, Rob Lemmens, Josep Perelló, Marisa Ponti, Roeland Samson, and Katherin Wagenknecht, 243–259. Cham: Springer. https://doi.org/10.1007/978-3-030-58278-4_13.

Land-Zandstra, Anne, Mara van Beusekom, Carl Koppeschaar, and Jos van den Broek. 2016. "Motivation and learning impact of Dutch flu-trackers." In *JCOM Journal of Science Communication* 15, no. 1 (Jan.): 1–26. https://doi.org/10.22323/2.15010204.

Lange, Anna-Christina, Marc Lenglet, and Robert Seyfert. 2019. "On Studying Algorithms Ethnographically: Making Sense of Objects of Ignorance." *Organization* 26, no. 4 (Oct.): 598–617. https://doi.org/10.1177/1350508418808230.

Lanzeni, Débora, Karen Waltorp, Sarah Pink, and Rachel Charlotte Smith, eds. 2023. *An Anthropology of Futures and Technologies*. Abingdon, Oxon; New York, NY: Routledge.

Larkin, Brian. 2008. *Signal and Noise: Media, Infrastructure, and Urban Culture in Nigeria*. Durham: Duke University Press. https://doi.org/10.1515/9780822389316.

———. 2013. "The Politics and Poetics of Infrastructure." *Annual Review of Anthropology* 42, no. 1 (Aug.): 327–343. https://doi.org/10.1146/annurev-anthro-092412-155522.

Larson, Lincoln R., Caren B. Cooper, Sara Futch, Devyani Singh, Nathan J. Shipley, Kathy Dale, Geoffrey S. LeBaron, and John Y. Takekawa. 2020. "The diverse motivations of citizen scientists: Does conservation emphasis grow as volunteer participation progresses?" *Biological Conservation* 242 (Feb.): 108428. https://doi.org/10.1016/j.biocon.2020.108428.

Larsson, Simon, and Martin Viktorelius. 2022. "Reducing the contingency of the world: magic, oracles, and machine-Learning technology." *AI & SOCIETY* 39 (Feb.): 1–11. https://doi.org/10.1007/s00146-022-01394-2.

Latimer, Joanna, and Mara Miele. 2013. "Naturecultures? Science, Affect and the Non-Human." *Theory, Culture & Society* 30, no. 7–8 (Oct.): 5–31. https://doi.org/10.1177/0263276413502088.

Latour, Bruno. 1987. *Science in Action: How to Follow Scientists and Engineers through Society*. Cambridge, Massachusetts: Harvard University Press.

———. 1992. "Where are the missing masses? The sociology of a few mundane artifacts." In *Shaping Technology / Building Society: Studies in Sociotechnical Change*, edited by Wiebe E. Bijker and John Law, 225–258. Cambridge, Massachusetts; London, England: The MIT Press.

———. (1988) 1993. *The Pasteurization of France*. Translated by Alan Sheridan. Cambridge, Massachusetts: Harvard University Press.

———. 1993. *We Have Never Been Modern*. Cambridge, Massachusetts: Harvard University Press.

———. 1996. "On actor-network theory: A few clarifications." *Soziale Welt* 47, no. 4: 369–381. https://www.jstor.org/stable/40878163.

———. 1999. *Pandora's Hope: Essays on the Reality of Science Studies*. Cambridge, Massachusetts: Harvard University Press.

———. 2005. *Reassembling the Social: An Introduction to Actor-Network-Theory*. Clarendon Lectures in Management Studies. Oxford; New York: Oxford University Press.

Latour, Bruno, and Steve Woolgar. (1979) 1986. *Laboratory Life: The Construction of Scientific Facts*. Princeton, N.J: Princeton University Press.

Laurel, Brenda. 2014. *Computers as Theatre*. 2nd edition. Upper Saddle River, NJ: Addison-Wesley.

Law, Edith. 2011. "Defining (Human) Computation." In *CHI 2011*. Vancouver, BC: ACM. https://www.humancomputation.com/crowdcamp/chi2011/papers/law.pdf.

Law, Edith, and Luis Von Ahn. 2011. *Human Computation*. Synthesis Lectures on Artificial Intelligence and Machine Learning 13. San Rafael, Calif: Morgan & Claypool.

Law, John. 1984. "On the Methods of Long-Distance Control: Vessels, Navigation and the Portuguese Route to India." *The Sociological Review* 32, no. 1 (May): 234–263. https://doi.org/10.1111/j.1467-954X.1984.tb00114.x.

———. 2004. *After Method: Mess in Social Science Research*. Abingdon, Oxon; New York, NY: Routledge.

Law, John, and John Hassard, eds. 1999. *Actor Network Theory and After*. The Sociological Review. Oxford, England; Malden, Massachusetts: Blackwell.

Law, John, and Wen-yuan Lin. 2020. "Care-Ful Research: Sensibilities from STS." http://heterogeneities.net/publications/LawLin2020CarefulResearchSensibilitiesFromSTS.pdf.

Lazar, Jonathan, Jinjuan Heidi Feng, and Harry Hochheiser. 2017. "Online and ubiquitous HCI research." In *Research Methods in Human Computer Interaction*, 411–453. 2nd edition. Cambridge, Massachusetts: Morgan Kaufmann. https://doi.org/10.1016/B978-0-12-805390-4.00014-5.

LeCun, Yann, Koray Kavukcuoglu, and Clement Farabet. 2010. "Convolutional networks and applications in vision." In *Proceedings of 2010 IEEE International Symposium on Circuits and Systems*, 253–256. Paris, France: IEEE. https://doi.org/10.1109/ISCAS.2010.5537907.

Lemke, Matthias. 2017. "Blended Reading." *Sozialwissenschaftliche Methodenberatung. Blog mit Beiträgen zu qualitativen sozialwissenschaftlichen Methoden*. 2017. https://sozmethode.hypotheses.org/139.

Lemke, Matthias, and Gregor Wiedemann, eds. 2016. *Text Mining in den Sozialwissenschaften: Grundlagen und Anwendungen zwischen qualitativer und quantitativer Diskursanalyse*. Wiesbaden: Springer. https://doi.org/10.1007/978-3-658-07224-7.

Lepczyk, Christopher A., Owen D. Boyle, and Timothy L. V. Vargo, eds. 2020. *Handbook of Citizen Science in Ecology and Conservation*. Oakland, California: University of California Press. https://doi.org/10.2307/j.ctvz0h8fz.

Liboiron, Max. 2019. "The Power (Relations) of Citizen Science." *CLEAR*. Mar. 19, 2019. https://civiclaboratory.nl/2019/03/19/the-power-relations-of-citizen-science/.

Licklider, Joseph Carl Robnett. 1960. "Man-Computer Symbiosis." *IRE Transactions on Human Factors in Electronics* HFE-1, no. 1 (Mar.): 4–11. https://doi.org/10.1109/THFE2.1960.4503259.

Light, Ben, Jean Burgess, and Stefanie Duguay. 2018. "The Walkthrough Method: An Approach to the Study of Apps." *New Media & Society* 20, no. 3 (Mar.): 881–900. https://doi.org/10.1177/1461444816675438.

Light, Jennifer S. 1999. "When Computers Were Women." *Technology and Culture* 40, no. 3 (Jul.): 455–483. https://www.jstor.org/stable/25147356.

Lindner, Rolf. 1981. "Die Angst des Forschers vor dem Feld. Überlegungen zur Teilnehmenden Beobachtung als Interaktionsprozeß." *Zeitschrift für Volkskunde* 77: 51–66. https://www.digi-hub.de/viewer/fulltext/DE-11-001938281/63/.

Link, Jürgen. 2014. "Crisis between 'Denormalization' and the 'New Normal': Reflections on the Theory of Normalism Today." In *Norms, normality and normalization: Papers from the Postgraduate Summer School in German Studies, Nottingham, July 2013*, edited by Matthias Uecker, Dirk Göttsche, Helen Budd, and Gesine Haberlah, 7–17. Nottingham: University of Nottingham.

Lintott, Chris. 2019. *The Crowd & the Cosmos: Adventures in the Zooniverse*. Oxford: Oxford University Press.

Lipstein, Greg. 2020. "MEET THE WINNERS OF THE CLOG LOSS CHALLENE FOR ALZHEIMER'S RESEARCH." DRIVENDATA LABS. Sept. 10, 2020. https://www.drivendata.co/blog/clog-loss-alzheimers-winners.

LociOiling. 2017. "Lua v2 Recipe in Foldit Recipe Editor." Foldit Wiki. Jan. 16, 2017. https://web.archive.org/web/20221103123551/https://foldit.fandom.com/wiki/Recipes.

Lock, Margaret M. 2013. *The Alzheimer Conundrum: Entanglements of Dementia and Aging*. Princeton, New Jersey: Princeton University Press.

Löfgren, Orvar. 1994. "Consuming Interests." In *Consumption and Identity*, edited by Jonathan Friedman, 47–70. Studies in anthropology and history 15. Chur, Switzerland: Harwood Academic Publishers.

Ludwig-Maximilians-Universität. n.d.a. "ARTIGO." Accessed Mar. 19, 2024. https://www.artigo.org/en.

———. n.d.b. "DFG Research Group Urban Ethics." Accessed Mar. 19, 2024. https://www.en.urbane-ethiken.uni-muenchen.de/index.html.

Luhmann, Niklas. 1988. "Familiarity, Confidence, Trust: Problems and Alternatives." In *Trust: Making and Breaking Cooperative Relations*, edited by Diego Gambetta, 94–107. New York, NY: Blackwell.

———. 2014. *Vertrauen. Ein Mechanismus der Reduktion sozialer Komplexität*. 5th edition. UTB. Konstanz; München: UVK.

Lund, Arwid. 2015. "A Contribution to a Critique of the Concept Playbour." In *Reconsidering Value and Labour in the Digital Age*, edited by Eran Fisher and Christian Fuchs, 63–79. London: Palgrave Macmillan. https://doi.org/10.1057/9781137478573_4.

Lynch, Michael. 1985. *Art and Artifact in Laboratory Science: A Study of Shop Work and Shop Talk in a Research Laboratory*. Studies in Ethnomethodology and Conversation Analysis. Abingdon, Oxon; New York, NY: Routledge.

Mackenzie, Adrian. 2005. "Algorithmic convolutions and hidden states in living systems and telecommunications: the case of the Viterbi algorithm." Presented at the Creative Evolution, Goldsmiths College, London, Feb. 13.

———. 2006. *Cutting Code: Software and Sociality*. Digital Formations 30. New York: Peter Lang.

———. 2017. *Machine Learners: Archaeology of a Data Practice*. Cambridge, Massachusetts; London, England: The MIT Press.

Malaby, Thomas M. 2007. "Beyond Play: A New Approach to Games." *Games and Culture* 2, no. 2 (Apr.): 95–113. https://doi.org/10.1177/1555412007299434.

———. 2009. "Anthropology and Play: The Contours of Playful Experience." *New Literary History* 40, no. 1: 205–218. https://www.jstor.org/stable/20533141.

———. 2012. "Digital Gaming, Game Design and Its Precursors." In *Digital Anthropology*, edited by Daniel Miller and Heather A. Horst, 288–305. London; New York: Berg.

Malinowski, Bronislaw. (1922) 2013. *Argonauts of the Western Pacific. An Account of Native Enterprise and Adventure in the Archipelagoes of Melanesian New Guinea*. Malinowski Collected Works 2. London: Routledge. https://doi.org/10.4324/9781315014463.

Malone, Thomas W., and Michael S. Bernstein. 2015. *Handbook of Collective Intelligence*. Cambridge, Massachusetts; London, England: The MIT press.

Malone, Thomas W., Robert Laubacher, and Chrysanthos N. Dellarocas. 2009. "Harnessing Crowds: Mapping the Genome of Collective Intelligence." *MIT Sloan School of Management Research Paper No. 4732-09*. http://dx.doi.org/10.2139/ssrn.1381502.

Marcus, George. 2009. "Multi-sited Ethnography: Notes and Queries." In *Multi-sited Ethnography Theory, Praxis and Locality in Contemporary Research*, edited by Mark-Anthony Falzon, 181–196. Aldershot; Burlington; Farnham: Ashgate.

Marino, Mark C. 2016. "Why We Must Read the Code: The Science Wars, Episode IV." In *Debates in the Digital Humanities 2016*, edited by Matthew K. Gold and Lauren F. Klein. Minneapolis; London: University of Minnesota Press. https://dhdebates.gc.cuny.edu/read/untitled/section/879bc64b-93ba-4d9a-9678-9a7239fc41e4#ch13.

———. 2018. "Reading Culture through Code." *Routledge Companion to Media Studies and Digital Humanities*, edited by Jenterey Sayers, 472–482. Abingdon, Oxon; New York, NY: Routledge.

———. 2020. *Critical Code Studies*. Software Studies. Cambridge, Massachusetts; London, England: The MIT Press.

Markham, Annette N. 2005. "The Methods, Politics, and Ethics of Representation in Online Ethnography." In *The Sage Handbook of Qualitative Research*, edited by Norman K. Denzin, 793–820. Thousand Oaks, California: SAGE. http://citeseerx.ist.psu.edu/viewdoc/summary?doi=10.1.1.508.444.

Marshall, Philip J., Chris J. Lintott, and Leigh N. Fletcher. 2015. "Ideas for Citizen Science in Astronomy." *Annual Review of Astronomy and Astrophysics* 53, no. 1 (Aug.): 247–278. https://doi.org/10.1146/annurev-astro-081913-035959.

Martín Abadi, Ashish Agarwal, Paul Barham, Eugene Brevdo, Zhifeng Chen, Craig Citro, Greg S. Corrado, et al. 2015. "TensorFlow: Large-Scale Machine Learning on Heterogeneous Systems." TensorFlow. https://www.tensorflow.org/.

Mathar, Tom. 2012. "Akteur-Netzwerk Theorie." In *Science and Technology Studies. Eine sozialanthropologische Einführung*, edited by Stefan Beck, Jörg Niewöhner, and Estrid Sørensen, 173–190. Bielefeld: transcript.

MBF Bioscience. n.d. "ScanImage®." Accessed Mar. 19, 2024. https://www.mbfbioscience.com/products/scanimage.

McDaniel, John, and Ken Pease, eds. 2021. *Predictive Policing and Artificial Intelligence*. Abingdon, Oxon; New York, NY: Routledge. https://doi.org/10.4324/9780429265365.

McFarlane, Colin. 2011a. "Assemblage and critical urbanism." *City: Analysis of Urban Change, Theory, Action* 15, no. 2 (Jun.): 204–224. https://doi.org/10.1080/13604813.2011.568715.

———. 2011b. "The City as Assemblage: Dwelling and Urban Space." *Environment and Planning D: Society and Space* 29, no. 4 (Jan.): 649–671. https://doi.org/10.1068/d4710.

McGonigal, Jane. 2012. *Reality Is Broken: Why Games Make Us Better and How They Can Change the World*. London: Vintage Books.

McKinsey & Company. 2022. "Why hybrid intelligence is the future of artificial intelligence at McKinsey." Apr. 29, 2022. https://www.mckinsey.com/about-us/new-at-mckinsey-blog/hybrid-intelligence-the-future-of-artificial-intelligence.

Memory Cafe Directory. n.d. "Memory Cafe Directory." Accessed Mar. 19, 2024. https://www.memorycafedirectory.com/.

Merleau-Ponty, Maurice. 1962. *Phenomenology of Perception: An Introduction*. Translated by Colin Smith. London; New York: Routledge.

Meta. 2023. "Introducing LLaMA: A foundational, 65-billion-parameter large language model." Feb. 24, 2023. https://ai.facebook.com/blog/large-language-model-llama-meta-ai/.

Michelucci, Pietro, ed. 2013a. *Handbook of Human Computation*. New York: Springer.

———. 2013b. "Introduction." In *Handbook of Human Computation*, edited by Pietro Michelucci, xxxvii-xxxxli. New York: Springer.

———. 2013c. "Organismic Computing." In *Handbook of Human Computation*, edited by Pietro Michelucci, 475–501. New York: Springer. https://doi.org/10.1007/978-1-4614-8806-4_36.

———. 2013d. "Synthesis and Taxonomy of Human Computation." In *Handbook of Human Computation*, edited by Pietro Michelucci, 83–86. New York: Springer. https://doi.org/10.1007/978-1-4614-8806-4_9.

———. 2016. "Human Computation and Convergence." In *Handbook of Science and Technology Convergence*, edited by William Sims Bainbridge and Mihail C. Roco, 455–474. Cham: Springer. https://doi.org/10.1007/978-3-319-07052-0_35.

———. 2017a. "What does stardust have to do with curing Alzheimer's disease?" Becoming Human. Exploring Artificial Intelligence & What it Means to be Human. Medium. Jun. 24, 2017. https://becominghuman.ai/what-does-stardust-have-to-do-with-curing-alzheimers-disease-61a84c6a470b.

———. 2017b. "Science of Stall Catchers: Our new Magic Number." Human Computation Institute Blogs. Dec. 7, 2017. https://blog.hcinst.org/our-new-magic-number/.

———. 2019a. "Crowd, Cloud and the Future of Work: Updates from human AI computation." Microsoft Research Faculty Summit, Jul. 19, 2019. https://www.microsoft.com/en-us/research/video/crowd-cloud-and-the-future-of-work-updates-from-human-ai-computation/.

———. 2019b. "How do we create a sustainable thinking economy?" Towards Data Science. Medium. Oct. 22, 2019. https://towardsdatascience.com/how-do-we-create-a-sustainable-thinking-economy-4d77839b031e.

———. 2019c. "Early peek at the #Dreamathon research results." Human Computation Institute Blog. Nov. 9, 2019. https://blog.hcinst.org/early-peek-at-the-dreamathon-research-results/.

———. 2020. "Citizen Science and ethical review." Forum – Human Computation Institute. Sep. 12, 2020, 9:34 AM. https://forum.hcinst.org/t/citizen-science-and-ethical-review/1010/10.

Michelucci, Pietro, and Egle [Seplute]. 2020. "The machines are coming! (but the humans are staying)." Human Computation Institute Blog. May 22, 2020. https://blog.hcinst.org/dd-ml-challenge/.

Michelucci, Pietro, and Janis L. Dickinson. 2016. "The power of crowds. Combining humans and machines can help tackle increasingly hard problems." *Science* 351, no. 6268 (Jan.): 32–33. https://doi.org/10.1126/science.aad6499.

Michelucci, Pietro, and Ujwal Gadiraju. n.d. "Human Computation." Accessed Mar. 19, 2024. https://hcjournal.org/index.php/jhc/about/editorialTeam.

Michelucci, Pietro, Lea Shanley, Janis Dickinson, and Haym Hirsh. 2015. "A U.S. Research Roadmap for Human Computation." Computer Community Consortium; Computing Research Association. https://cra.org/ccc/wp-content/uploads/sites/2/2015/05/Final-HC-Report.pdf.

Michelucci, Pietro, and Elena Simperl. 2014. "From the Editors." *Human Computation* 1, no. 1 (Oct.): 1–3. https://doi.org/10.15346/hc.v1i1.1.

Miller, Daniel, and Heather A. Horst. 2012. "The Digital and the Human: A Prospectus for Digital Anthropology." In *digital anthropology*, edited by Daniel Miller and Heather A. Horst, 3–35. London; New York: Berg.

Miller, Josh Aaron, Uttkarsh Narayan, Matthew Hantsbarger, Seth Cooper, and Magy Seif El-Nasr. 2019. "Expertise and engagement: re-designing citizen science games with players' Minds in mind." In *Proceedings of the 14th International Conference on the Founda-*

tions of Digital Games, 1–11. San Luis Obispo, California: ACM. https://doi.org/10.1145/3337722.3337735.

Miller, Joshua Aaron, Libuše Hannah Vepřek, Sebastian Deterding, and Seth Cooper. 2023. "Practical recommendations from a multi-perspective needs and challenges assessment of citizen science games." *PLOS ONE* 18, no. 5 (May): e0285367. https://doi.org/10.1371/journal.pone.0285367.

Miller, Toby. 2006. "Gaming for Beginners." *Games and Culture* 1, no. 1 (Jan.): 5–12. https://doi.org/10.1177/1555412005281403.

Mol, Annemarie. 2002a. "Cutting Surgeons, Walking Patients: Some Complexities Involved in Comparing." In *Complexities: Social Studies of Knowledge Practices*, by John Law, edited by Annemarie Mol, 218–258. Durham: Duke University Press. https://doi.org/10.1515/9780822383550-009.

———. 2002b. *The Body Multiple: Ontology in Medical Practice*. Science and Cultural Theory. Durham: Duke University Press.

Mol, Annemarie, and John Law. 2002. "Complexities: An Introduction." In *Complexities: Social Studies of Knowledge Practices*, by John Law, edited by Annemarie Mol, 1–22. Durham: Duke University Press. https://doi.org/10.1215/9780822383550-001.

Monarch, Robert. 2021. *Human-in-the-Loop Machine Learning: Active Learning and Annotation for Human-Centered AI*. Shelter Island, NY: Manning Publications Co.

Moretti, Franco. 2016. *Distant Reading*. Translated by Christine Pries. Konstanz: Konstanz University Press.

Mosqueira-Rey, Eduardo, Elena Hernández-Pereira, David Alonso-Ríos, José Bobes-Bascarán, and Ángel Fernández-Leal. 2022. "Human-in-the-loop machine learning: a state of the art." *Artificial Intelligence Review* 56, (Aug.): 3005–3054. https://doi.org/10.1007/s10462-022-10246-w.

Mousavi Baygi, Reza, Lucas D. Introna, and Lotta Hultin. 2021. "Everything Flows: Studying Continuous Socio-Technological Transformation in a Fluid and Dynamic Digital World." *MIS Quarterly* 45, no. 1b: 423–452. https://doi.org/10.25300/MISQ/2021/15887.

Mühlfried, Florian. 2018. "Introduction: Approximating Mistrust." In *Mistrust: Ethnographic Approximations*, edited by Florian Mühlfried, 7–22. Culture and Social Practice. Bielefeld: transcript.

Mühlhoff, Rainer. 2020. "Human-aided artificial intelligence: Or, how to run large computations in human brains? Toward a media sociology of machine learning." *New Media & Society* 22, no. 10: 1868–1884. https://doi.org/10.1177/1461444819885334.

Müller, Martin. 2015. "Assemblages and Actor-Networks: Rethinking Socio-material Power, Politics and Space." *Geography Compass* 9, no. 1 (Jan.): 27–41. https://doi.org/10.1111/gec3.12192.

———. n.d. "Scalable Reading." Accessed Mar. 19, 2024. https://sites.northwestern.edu/scalablereading/scalable-reading/.

Nam, Sang-Hui. 2019. "Qualitative Analyse von Chats und anderer usergenerierter Kommunikation." In *Handbuch Methoden der empirischen Sozialforschung*, edited by Nina Baur and Jörg Blasius, 1041–1051. Wiesbaden: Springer. https://doi.org/10.1007/978-3-658-21308-4_74.

Nando de Freitas (@NandoDF). 2022. "Someone's opinion article. My opinion: It's all about scale now! The Game is Over! It's about making these models bigger, safer, compute efficient, faster at sampling, smarter memory, more modalities, INNOVATIVE DATA, on/offline, ... 1/N." Twitter. May 14, 2022. https://twitter.com/NandoDF/status/1525397036325019649.

National Institute of Health. n.d. "ImageJ. Image Processing and Analysis in Java." Accessed Mar. 19, 2024. https://imagej.net/ij/index.html.

Newman, Greg. 2014. "Citizen CyberScience New Directions and Opportunities for Human Computation." *Human Computation* 1, no. 2 (Dec.): 103–109. https://doi.org/10.15346/hc.v1i2.2.

Neyland, Daniel. 2015. "On Organizing Algorithms." *Theory, Culture & Society* 32, no. 1 (May): 119–132. https://doi.org/10.1177/0263276414530477.

Niewöhner, Jörg. 2014. "Perspektiven der Infrastrukturforschung: care-full, relational, ko-laborativ." In *Schlüsselwerke der Science & Technology Studies*, edited by Diana Lengersdorf and Matthias Wieser, 341–352. Wiesbaden: Springer. https://doi.org/10.1007/978-3-531-19455-4_28.

———. 2015. "Infrastructures of Society, Anthropology of." In *International Encyclopedia of the Social & Behavioral Sciences*, 119–125. 2nd edition. Oxford: Elsevier. https://doi.org/10.1016/B978-0-08-097086-8.12201-9; Secondary publication on the edoc server of the Humboldt-Universität zu Berlin: https://edoc.hu-berlin.de/handle/18452/20133.

———. 2016. "Co-Laborative Anthropology: Crafting Reflexivities Experimentally." In *Etnologinen Tulkinta Ja Analyysi: Kohti Avoimempaa Tutkimusprosessia*, edited by Jukka Jouhki and Tytti Steel: 81–122. Helsinki: Ethnos; English translation used in this work available under: https://www.researchgate.net/profile/Joerg-Niewoehner/publication/304248438_Co-laborative_Anthropology_Crafting_Reflexivities_Experimentally_published_in_Finnish_as_Niewohner_J_2016_Co-laborative_anthropology_crafting_reflexivities_experimentally_Etnologinen_tulkinta_ja_analy/links/5877b2b308ae6eb871d15f05/Co-laborative-Anthropology-Crafting-Reflexivities-Experimentally-published-in-Finnish-as-Niewoehner-J-2016-Co-laborative-anthropology-crafting-reflexivities-experimentally-Etnologinen-tulkinta-ja-ana.pdf.

———, (Stefan Beck). 2017. "Phänomenographie: Sinn-volle Ethnographie jenseits des menschlichen Maßstabs." In *Kulturen der Sinne: Zugänge zur Sensualität der sozialen Welt*, edited by Karl Braun, Claus-Marco Dieterich, Thomas Hengartner, and Bernhard Tschofen, 78–95. Würzburg: Königshausen & Neumann.

———. 2019a. "Introduction | After Practice. Thinking through Matter(s) and Meaning Relationally." In: *After Practice 1*, edited by The Laboratory: Anthropology of Environment | Human Relations: 10–26. Berliner Blätter. Ethnographische und ethnologische Beiträge 81. Panama: Berlin.

———. 2019b. "Situierte Modellierung: Ethnografische Ko-Laboration in der Mensch-Umwelt-Forschung." In *Zusammen Arbeiten*, edited by Stefan Groth and Christian Ritter, 23–50. Bielefeld: transcript. https://doi.org/10.14361/9783839442951-002.

Niewöhner, Jörg, Patrick Bieler, Maren Heibges (née Klotz), and Martina Klauser. 2016. "Phenomenography: Relational Investigations into Modes of Being-in-the-World." *Cyprus Review* 28, no. 1: 67–84. https://cyprusreview.org/index.php/cr/article/view/35/6.

Nilsson, Nils J. 2010. *The Quest for Artificial Intelligence*. Cambridge: Cambridge University Press. https://doi.org/10.1017/CBO9780511819346.

NoBIAS. n.d. "NoBIAS. Artificial Intelligence without Bias." NoBIAS. Accessed Mar. 19, 2024. https://nobias-project.eu/.

Noble, Wiliam G. 1981. "Gibsonian Theory and the Pragmatist Perspective." *Journal for the Theory of Social Behaviour* 11, no. 1 (Mar.): 65–85. https://doi.org/10.1111/j.1468-5914.1981.tb00023.x.

O'Donnell, Casey. 2014. *Developer's Dilemma: The Secret World of Videogame Creators*. Inside Technology. Cambridge, Massachusetts; London, England: The MIT Press. https://doi.org/10.7551/mitpress/9035.001.0001.

Oechslen, Anna. 2020. "Grenzenlose Arbeit? Eine Exploration der Arbeitskulturen von Crowdwork." *Digitale Arbeitskulturen: Rahmungen, Effekte, Herausforderungen*, edited by Dennis Eckhardt, Sarah May, Martina Röthl, and Roman Tischberger: 83–94. Berliner Blätter. Ethnographische und ethnologische Beiträge 82. https://edoc.hu-berlin.de/bitstream/handle/18452/22819/Oechslen.pdf?sequence=1.

Ong, Aihwa, and Stephen J. Collier, eds. 2005. *Global Assemblages: Technology, Politics, and Ethics as Anthropological Problems*. Malden, Massachusetts: Blackwell Publishing.

OpenAI. n.d. "DALL-E2." Accessed Mar. 19, 2024. https://openai.com/dall-e-2/.

———. n.d. "Introducing ChatGPT." Accessed Mar. 19, 2024. https://openai.com/blog/chatgpt.

Oudshoorn, Nelly, and Trevor Pinch, eds. 2005. *How Users Matter: The Co-Construction of Users and Technology*. Inside Technology. Cambridge, Massachusetts; London, England: The MIT Press.

Palikaras, Konstantinos, and Nektarios Tavernarakis. 2015. "Multiphoton Fluorescence Light Microscopy." In *Encyclopedia of Life Sciences*, edited by John Wiley & Sons, Ltd, 1–8. Chichester: John Wiley & Sons, Ltd. https://doi.org/10.1002/9780470015902.a0002991.pub3.

Peeters, Marieke M. M., Jurriaan van Diggelen, Karel van den Bosch, Adelbert Bronkhorst, Mark A. Neerincx, Jan Maarten Schraagen, and Stephan Raaijmakers. 2021. "Hybrid collective intelligence in a human–AI society." *AI & SOCIETY* 36, no. 1 (Jun.): 217–238. https://doi.org/10.1007/s00146-020-01005-y.

Petryna, Adriana. 2022. *Horizon Work: At the Edges of Knowledge in an Age of Runaway Climate Change*. Princeton: Princeton University Press.

Pfaffenberger, Bryan. 1992. "Technological Dramas." *Science, Technology, & Human Values* 17, no. 3 (Jul.): 282–312. https://www.jstor.org/stable/690096.

Phillips, John. 2006. "Agencement/Assemblage." *Theory Culture & Society* 23, no. 2–3 (May): 108–109. 10.1177/026327640602300219.

Physical-Digital Affordances Group University of Regensburg. n.d. "Project: BMBF-Verbundprojekt VIGITIA – Vernetzte Intelligente Gegenstände durch, auf und um interaktive Tische im Alltag." Physical-Digital Affordances Group University of Regensburg. Accessed Mar. 19, 2024. https://hci.ur.de/projects/vigitia.

Pickering, Andrew. 2010. "Material Culture and the Dance of Agency." in *The Oxford Handbook of Material Culture Studies*, edited by Dand Hicks and Mary C. Beaudry, 191–208. Oxford: Oxford University Press.

Pink, Sarah. 2018. "Afterword. Refiguring Collaboration and Experimentation." In *EXPERIMENTAL COLLABORATIONS: Ethnography through Fieldwork Devices*, edited by Adolfo Estalella and Tomas Sanchez Criado, 201–212. Easa Series. New York: Berghahn Books.

———. 2021. "Sensuous futures: re-thinking the concept of trust in design anthropology." *Senses & Society* 16, no. 2 (Jan.): 193–202. https://doi.org/10.1080/17458927.2020.1858655.

———, ed. 2022. *Everyday Automation: Experiencing and Anticipating Emerging Technologies*. Abingdon, Oxon; New York, NY: Routledge.

———. 2023. *Emerging Technologies: Life at the Edge of the Future*. Abingdon, Oxon; New York, NY: Routledge.

Pink, Sarah, Heather A. Horst, John Postill, Larissa Hjorth, Tania Lewis, and Jo Tacchi, eds. 2016. *Digital Ethnography: Principles and Practice*. Los Angeles: SAGE.

Pink, Sarah, Debora Lanzeni, and Heather Horst. 2018. "Data anxieties: Finding trust in everyday digital mess." *Big Data & Society* 5, no. 1 (Jan.): 1–14. https://doi.org/10.1177/2053951718756685.

Ploder, Andrea, and Johanna Stadlbauer. 2013. "Autoethnographie und Volkskunde? Zur Relevanz wissenschaftlicher Selbsterzählungen für die volkskundlich-kulturanthropologische Forschungspraxis." *Österreichische Zeitschrift für Volkskunde* 116, no. 3–4: 373–404.

Plontke, Sandra. 2018. "If {battleState = BattleState.standby}: Bringing the Gamer Into Play in Computer Game Development." In *Cultures of Computer Game Concerns. The Child Across Families, Law, Science and Industry*, edited by Estrid Sørensen, 39–66. Bielefeld: transcript.

Podjed, Dan, and Rajko Muršič. 2021. "To be or not to be there. Remote ethnography during the crisis and beyond." *Etnolog* 31: 35–51. https://erepo.uef.fi/handle/123456789/27429.

Poel, Ibo van de. 2020. "Three philosophical perspectives on the relation between technology and society, and how they affect the current debate about artificial intelligence." *Human Affairs* 30, no. 4 (Oct.): 499–511. https://doi.org/10.1515/humaff-2020-0042.

Polop, Carlos. n.d. "Captcha Bypass." In *HackTricks*. Accessed Mar. 19, 2024. https://book.hacktricks.xyz/pentesting-web/captcha-bypass.

Ponti, Marisa, Laure Kloetzer, Grant Miller, Frank O. Ostermann, and Sven Schade. 2021. "Can't we all just get along? Citizen scientists interacting with algorithms." *Human Computation* 8, no. 2 (Jul.): 5–14. https://doi.org/10.15346/hc.v8i2.128.

Ponti, Marisa, and Alena Seredko. 2022. "Human-machine-learning integration and task allocation in citizen science." *Humanities & Social Sciences Communications* 9 (Feb.): 1–15. https://doi.org/10.1057/s41599-022-01049-z.

Ponti, Marisa, Igor Stankovic, Wolmet Barendregt, Bruno Kestemont, and Lyn Bain. 2018. "Chefs Know More than Just Recipes: Professional Vision in a Citizen Science Game." *Human Computation* 5, no. 1 (Jul.): 1–12. https://doi.org/10.15346/hc.v5i1.1.

Postill, John. 2017. "Remote Ethnography: Studying Culture from Afar." In *The Routledge Companion to Digital Ethnography*, edited by Larissa Hjorth, Heather Horst, Anne Galloway, and Genevieve Bell, 61–69. Abingdon, Oxon; New York, NY: Routledge.

Prestopnik, Nathan R., and Kevin Crowston. 2012. "Citizen science system assemblages: understanding the technologies that support crowdsourced science." In *Proceedings of the 2012 iConference*, 168–176. New York: ACM. https://doi.org/10.1145/2132176.2132198.

Przybylski, Liz. 2021. *Hybrid Ethnography: Online, Offline, and In Between*. Qualitative Research Methods 58. Los Angeles: SAGE.

Quamen, Harvey, and Jon Bath. 2016. "Databases." In *Doing Digital Humanities: Practice, Training, Research*, edited by Constance Crompton, Richard J. Lane, and Ray G. Siemens, 145–162. Abingdon, Oxon; New York, NY: Routledge.

Quinn, Alexander J., and Benjamin B. Bederson. 2011. "Human computation: a survey and taxonomy of a growing field." In *CHI '11: Proceedings of the SIGCHI Conference on Human Factors in Computing Systems*, 1403–1412. Vancouver, BC, Canada: ACM. https://doi.org/10.1145/1978942.1979148.

Rabinow, Paul, George E. Marcus, James D. Faubion, and Tobias Rees. 2008. *Designs for an Anthropology of the Contemporary*. Durham: Duke University Press.

Rackwitz, Roman. 2015. "Gamification. Spielen ist keine Erfindung der Unterhaltungsindustrie." In *New Media Cultures: Mediale Phänomene der Netzkultur*, edited by Christian Stiegler, Patrick Breitenbach, Thomas Zorbach, 217–236. Bielefeld: transcript. https://doi.org/10.14361/9783839429075-013.

Rafner, Janet, Miroslav Gajdacz, Gitte Kragh, Arthur Hjorth, Anna Gander, Blanka Palfi, Aleks Berditchevskaia, François Grey, Kobi Gal, Avi Segal, et al. 2021. "Revisiting Citizen Science Through the Lens of Hybrid Intelligence." https://doi.org/10.48550/ARXIV.2104.14961.

Rafner, Janet, Miroslav Gajdacz, Gitte Kragh, Arthur Hjorth, Anna Gander, Blanka Palfi, Aleksandra Berditchevskiaia, François Grey, Kobi Gal, Avi Segal, et al. 2022. "Mapping Citizen Science through the Lens of Human-Centered AI." *Human Computation* 9, no. 1 (Nov.): 66–95. https://doi.org/10.15346/hc.v9i1.133.

Rahwan, Iyad, Manuel Cebrian, Nick Obradovich, Josh Bongard, Jean-François Bonnefon, Cynthia Breazeal, Jacob W. Crandall, Nicholas A. Christakis, Iain D. Couzin, Matthew O. Jackson, et al. 2019. "Machine behaviour." *nature* 568 (Apr.): 477–486. https://doi.org/10.1038/s41586-019-1138-y.

Ramanauskaite, Egle M. 2016. "WeCureALZ – crowdsourcing a cure for Alzheimer's." TECHNOLOGY.ORG. SCIENCE & TECHNOLOGY NEWS. Feb. 2, 2016. https://www.technology.org/2016/02/02/wecurealz-crowdsourcing-a-cure-for-alzheimers/.

———. 2020. "Dream Catchers." Human Computation Institute. Feb. 24, 2020. https://humancomputation.org/dream-catchers/.

Ramesh, Aditya, Prafulla Dhariwal, Alex Nichol, Casey Chu, and Mark Chen. 2022. "Hierarchical Text-Conditional Image Generation with CLIP Latents." arXiv. https://doi.org/10.48550/arXiv.2204.06125.

Ramesh, Aditya, Mikhail Pavlov, Gabriel Goh, Scott Gray, Chelsea Voss, Alec Radford, Mark Chen, and Ilya Sutskever. 2021. "Zero-Shot Text-to-Image Generation." arXiv. http://arxiv.org/abs/2102.12092.

Rasmussen, Lisa M., and Caren Cooper. 2019. "Citizen Science Ethics." *Citizen Science: Theory and Practice* 4, no. 1 (Mar.): 5. https://doi.org/10.5334/cstp.235.

Reckwitz, Andreas. 2016. *Kreativität und soziale Praxis: Studien zur Sozial- und Gesellschaftstheorie*. Sozialtheorie. Bielefeld: transcript. https://doi.org/10.14361/9783839433454.

Reed, Edward S. 1991. "James Gibson's ecological approach to cognition." In *Against Cognitivism: Alternative Foundations for Cognitive Psychology*, edited by Arthur Still and Alan Costall, 171–197. Hemel Hempstead: Harvester Wheatsheaf.

Reed, Scott, Konrad Zolna, Emilio Parisotto, Sergio Gómez Colmenarejo, Alexander Novikov, Gabriel Barth-maron, Mai Giménez, Yury Sulsky, Jackie Kay, Jost Tobias Springenberg, et al. 2022. "A Generalist Agent." *Transactions on Machine Learning Research* (Nov.). https://openreview.net/forum?id=1ikK0kHjvj.

Resnik, David B., Kevin C. Elliott, and Aubrey K. Miller. 2015. "A framework for addressing ethical issues in citizen science." *Environmental Science & Policy* 54 (Dec.): 475–481. https://doi.org/10.1016/j.envsci.2015.05.008.

Rhee, Jennifer. 2018. *The Robotic Imaginary: The Human and the Price of Dehumanized Labor*. Minneapolis: University of Minnesota Press.

Ritterfeld, Ute, Michael Cody, and Peter Vorderer. 2009a. "Introduction." In *Serious Games: Mechanisms and Effects*, edited by Ute Ritterfeld, Michael Cody, and Peter Vorderer, 3–9. Abingdon, Oxon; New York, NY: Routledge. https://doi.org/10.4324/9780203891650.

———, eds. 2009b. *Serious Games: Mechanisms and Effects*. Abingdon, Oxon; New York, NY: Routledge. https://doi.org/10.4324/9780203891650.

Robinson, Andrew. 2019. "Why Citizen Scientists Should be Paid." Humans Are the Artificial Intelligence of Plants. Medium. Nov. 3, 2019. https://medium.com/questanotes/why-citizen-scientists-should-be-paid-78262f4e7331.

Rombach, Robin, Andreas Blattmann, Dominik Lorenz, Patrick Esser, and Björn Ommer. 2022. "High-Resolution Image Synthesis with Latent Diffusion Models." *Proceedings of the IEEE/CVF Conference on Computer Vision and Pattern Recognition (CVPR)*, 10684–10695. https://openaccess.thecvf.com/content/CVPR2022/papers/Rombach_High-Resolution_Image_Synthesis_With_Latent_Diffusion_Models_CVPR_2022_paper.pdf.

Rose, Geena de. 2022. "The 20th Annual Wiley Prize in Biomedical Sciences Awarded for Protein Structure Predictions." WILEY. Sept. 3, 2022. https://johnwiley2020news.q4web.com/press-releases/press-release-details/2022/The-20th-Annual-Wiley-Prize-in-Biomedical-Sciences-Awarded-for-Protein-Structure-Predictions/.

Rosenberger, Robert, and Peter-Paul Verbeek. 2015a. "A Field Guide to Postphenomenology." In *Postphenomenological Investigations: Essays on Human–Technology Relations*, edited by Robert Rosenberger and Peter-Paul Verbeek, 8–41. Lenham: Lexington Books.

———. 2015b. "Introduction." In *Postphenomenological Investigations: Essays on Human–Technology Relations*, edited by Robert Rosenberger and Peter-Paul Verbeek, 1–6. Lanham: Lexington Books.

———, eds. 2015c. *Postphenomenological Investigations: Essays on Human-Technology Relations*. Postphenomenology and the Philosophy of Technology. Lanham: Lexington Books.

Rueckert, Martin, and Martin Riedl. 2022. "Human-in-the-Loop: Wie Mensch und KI Aufgaben besser lösen." DIGITALE WELT. SCIENCE MEETS INDUSTRY. Jun. 13,

2022. https://digitalweltmagazin.de/fachbeitrag/human-in-the-loop-wie-mensch-und-ki-aufgaben-besser-loesen/.

Rust, Roland T., and Ming-Hui Huang. 2021. "The Thinking Economy." In *The Feeling Economy*, by Roland T. Rust and Ming-Hui Huang, 23–39. Cham: Springer. https://doi.org/10.1007/978-3-030-52977-2_3.

Salazar, Juan Francisco, Sarah Pink, Andrew Irving, and Johannes Sjöberg, eds. 2017. *Anthropologies and Futures: Researching Emerging and Uncertain Worlds*. London; New York; New Delhi; Sydney: Bloomsbury Academic.

Santander, Paz. 2022. "Libuše's poster won! A look at how human computation systems align in citizen science." Human Computation Institute Blog. Jun. 16, 2022. https://blog.hcinst.org/libuses-poster-won-a-look-at-how-human-computation-systems-align-in-citizen-science/.

Sartori, Laura, and Giulia Bocca. 2022. "Minding the gap(s): public perceptions of AI and socio-technical imaginaries." *AI & SOCIETY* 38 (Mar.). https://doi.org/10.1007/s00146-022-01422-1.

Savage, Neil. 2012. "Gaining wisdom from crowds." *Communications of the ACM* 55, no. 3 (Mar.): 13–15. https://doi.org/10.1145/2093548.2093553.

Schager, Ben, and Craig E. Brown. 2020. "Susceptibility to capillary plugging can predict brain region specific vessel loss with aging." *Journal of Cerebral Blood Flow & Metabolism* 40, no. 12 (Jan.): 2475–2490. https://doi.org/10.1177/0271678X19895245.

Schefels, Clemens. n.d. "RESEARCH ACTIVITIES." CLEMENS SCHEFELS. Accessed Mar. 19, 2024. https://schefels.de/#research.

Schemainda, Corina. 2014. "Qualitative Analysis of the ARTigo Gaming Ecosystem." Bachelor's thesis, LMU Munich. Munich, Germany: LMU Munich. https://www.en.pms.ifi.lmu.de/publications/projektarbeiten/Corina.Schemainda/PA_Corina.Schemainda.pdf.

Schilcher, Christian, Mascha Will-Zocholl, and Marc Ziegler, eds. 2012. *Vertrauen und Kooperation in der Arbeitswelt*. Wiesbaden: Springer VS.

Schiller, Maria. 2018. "The 'Research Traineeship': The Ups and Downs of Para-siting Ethnography." In *EXPERIMENTAL COLLABORATIONS: Ethnography through Fieldwork Devices*, edited by Adolfo Estalella and Tomas Sanchez Criado, 53–70. Easa Series. New York: Berghahn Books.

Schirmer, Dominique, Nadine Sander, and Andreas Wenninger, eds. 2015. *Die qualitative Analyse internetbasierter Daten: Methodische Herausforderungen und Potenziale von Online-Medien*. Soziologische Entdeckungen. Wiesbaden: Springer VS.

Schmidt-Lauber, Brigitta. 2001. "Das qualitative Interview oder: Die Kunst des Reden-Lassens." In *Methoden Der Volkskunde: Positionen, Quellen, Arbeitsweisen der Europäischen Ethnologie*, edited by Silke Göttsch and Albrecht Lehmann, 165–186. Berlin: Reimer.

Schneider, Stefanie, and Hubertus Kohle. 2017. "The Computer as Filter Machine: A Clustering Approach to Categorize Artworks Based on a Social Tagging Network." *Visualizing Networks: Approaches to Network Analysis in Art History*. Artl@s Bulletin 6, no. 3: 80–89. https://epub.ub.uni-muenchen.de/41319/1/The%20Computer%20as%20Filter%20Machine.pdf.

Schneider, Stefanie, Maximilian Kristen, and Ricarda Vollmer. 2023. "Re: ARTigo. Neuentwurf eines Social-Tagging-Frameworks aus funktionalen Programmbausteinen."

In *DHd 2023 Open Humanities Open Culture. 9. conference of the association of "Digital Humanities im deutschsprachigen Raum" (DHd 2023)*, Trier; Belval. https://zenodo.org/record/7715482.

Scholz, Trebor, ed. 2013. *Digital Labor: The Internet as Playground and Factory*. Abingdon, Oxon; New York, NY: Routledge.

Schönberger, Klaus. 2007. "Technik als Querschnittsdimension. Kulturwissenschaftliche Technikforschung am Beispiel von Weblog-Nutzung in Frankreich und Deutschland." *Zeitschrift für Volkskunde* 103, no. 2: 197–222. https://www.waxmann.com/artikelART101066.

Schubert, Matthias. 2007. *Datenbanken. Theorie, Entwurf und Programmierung relationaler Datenbanken*. Wiesbaden: Teubner Verlag.

Schwertl, Maria. 2013. "Vom Netzwerk zum Text." In *Europäisch-ethnologisches Forschen. Neue Methoden und Konzepte*, edited by Sabine Hess, Johannes Moser, and Maria Schwertl, 107–126. Berlin: Reimer.

Scientistt. 2020. "Oliver Bracko: Using Crowd-Sourced Science to Study Alzheimer's Disease. von The Scientistt Podcast." The Scientistt Podcast. Jul. 1, 2020, 31:12. https://anchor.fm/scientistt/episodes/Oliver-Bracko-Using-crowd-sourced-science-to-study-Alzheimers-disease-eg5p3k/a-a2jme6c.

Scistarter.org. n.d. "scistarter. Science we can do together." Accessed Mar. 19, 2024. https://scistarter.org/.

Seaver, Nick. 2017. "Algorithms as culture: Some tactics for the ethnography of algorithmic systems." *Big Data & Society* 4, no. 2 (Nov.): 1–12. https://doi.org/10.1177/2053951717738104.

Seligman, Adam B. 2000. *The Problem of Trust*. Princeton, New Jersey: Princeton University Press.

Service, Robert F. 2020. "'The game has changed.' AI triumphs at solving protein structures." *Science*. Nov. 30, 2020. https://doi.org/10.1126/science.abf9367.

Seung Lab, Princeton University. n.d. "Eyewire." Accessed Mar. 19, 2024. https://eyewire.org/explore.

Seyfert, Robert, and Jonathan Roberge. 2017. "Was sind Algorithmuskulturen?" In *Algorithmuskulturen: Über die rechnerische Konstruktion der Wirklichkeit*, edited by Robert Seyfert and Jonathan Roberge, 7–40. Kulturen der Gesellschaft 26. Bielefeld: transcript.

Shmerling, Robert H. 2015. "The myth of the Hippocratic Oath." Harvard Health. November 25, 2015. https://www.health.harvard.edu/blog/the-myth-of-the-hippocratic-oath-201511258447.

Simon, Herbert Alexander. 1996. *The Sciences of the Artificial*. 3rd edition. Cambridge, Massachusetts; London, England: The MIT Press.

Singelnstein, Tobias. 2018. "Predictive Policing: Algorithmenbasierte Straftatprognosen Zur Vorausschauenden Kriminalintervention." *Neue Zeitschrift für Strafrecht* 38, no. 1: 1–9.

Skopek, Jan. 2012. "Methodologie und Daten." In *Partnerwahl im Internet*, by Jan Skopek, 121–143. Wiesbaden: VS Verlag für Sozialwissenschaften. https://doi.org/10.1007/978-3-531-94064-9_6.

Slack Technologies, LLC n.d. "slack." Accessed Mar. 19, 2024. https://slack.com/.

Söbke, Heinrich, Pia Spangenberger, Philipp Müller, and Stefan Göbel, eds. 2022. *Serious Games: Joint International Conference, JCSG 2022, Weimar, Germany, September 22–23, 2022, Proceedings*. Lecture Notes in Computer Science 13476. Cham: Springer. https://doi.org/10.1007/978-3-031-15325-9.

Sørensen, Estrid, and Jan Schank. 2017. "Praxeographie. Einführung." In *Science and Technology Studies. Klassische Positionen und aktuelle Perspektiven*, edited by Susanne Bauer, Torsten Heinemann, and Thomas Lemke, 407–428. Berlin: Suhrkamp.

Srnicek, Nick. 2017. *Platform Capitalism*. Theory Redux. Cambridge; Malden, Massachusetts: Polity.

Star, Susan Leigh. 1989. "Human Beings as Material for Artificial Intelligence: Or, What Computer Science Can't Do." Conference presentation presented at the *American Philosophical Association*. Berkeley, California.

———. 1999. "The Ethnography of Infrastructure." *American Behavioral Scientist* 43, no. 3 (Nov.): 377–391. https://doi.org/10.1177/00027649921955326.

———. 2008. "An Interview with Susan Leigh Star." Interview by Mark Zachry. *Technical Communication Quarterly* 17, no. 4 (Oct.): 435–54. https://doi.org/10.1080/10572250802329563.

———. (2007) 2015. "Living Grounded Theory: Cognitive and Emotional Forms of Pragmatism." In *Boundary Objects and Beyond. Working with Leigh Star*, edited by Geoffrey C. Bowker, Stefan Timmermans, Adele E. Clarke, and Ellen Balka, 121–141. Cambridge, Massachusetts; London, England: The MIT Press.

———. (1991) 2015. "Power, Technology, and the Phenomenology of Conventions: On Being Allergic to Onions." In *Boundary Objects and Beyond. Working with Leigh Star*, edited by Geoffrey C. Bowker, Stefan Timmermans, Adele E. Clarke, and Ellen Balka, 263–289. Cambridge, Massachusetts; London, England: The MIT Press.

———. (1988) 2015. "The Structure of Ill-Structured Solutions: Boundary Objects and Heterogeneous Distributed Problem Solving." In *Boundary Objects and Beyond. Working with Leigh Star*, edited by Geoffrey C Bowker, Stefan Timmermans, Adele E. Clarke, and Ellen Balka, 243–259. Cambridge, Massachusetts; London, England: The MIT Press.

Star, Susan Leigh, and James R. Griesemer. 1989. "Institutional Ecology, 'Translations' and Boundary Objects: Amateurs and Professionals in Berkeley's Museum of Vertebrate Zoology, 1907–39." *Social Studies of Science* 19, no. 3 (Aug.): 387–420. https://www.jstor.org/stable/285080.

Star, Susan Leigh, and Karen Ruhleder. 1996. "Steps Toward an Ecology of Infrastructure: Design and Access for Large Information Spaces." *Information Systems Research* 7, no. 1 (Mar.): 111–134. https://doi.org/10.1287/isre.7.1.111.

Stardust@home. n.d. "Stardust@home." Accessed Mar. 19, 2024. https://web.archive.org/web/20230206173841/https://stardustathome.ssl.berkeley.edu/.

Starider. 2021. "How did you experience catching stalls alongside GAIA?." Forum – Human Computation Institute. May 21, 2021, 11:34 PM. https://forum.hcinst.org/t/how-did-you-experience-catching-stalls-alongside-gaia/1089/6.

Starzmann, Maria Theresa. 2015. "Kommentar Zu Matthias Jung, 'Citizen Science' – eine Programmatik zur Rehabilitierung des Handelns Wissen-Schaftlicher Laiinnen und Laien und ihre Implikationen für die Archäologie." *Forum Kritische Archäologie* 4: 55–58. https://doi.org/10.6105/journal.fka.2015.4.7.

Stevens, Philipp. 1980. "Play and Work: A False Dichotomy?" In *Play and Culture*, edited by Helen B. Schwartzmann, 316–323. New York: Westpoint.

Strasser, Bruno J., Jérôme Baudry, Dana Mahr, Gabriela Sanchez, and Elise Tancoigne. 2018. "'Citizen Science'? Rethinking Science and Public Participation." *Science & Technology Studies* 32, no. 2 (May): 52–76. https://doi.org/10.23987/sts.60425.

Suarez, Pablo. 2015. "Rethinking Engagement: Innovations in How Humanitarians Explore Geoinformation." *ISPRS International Journal of Geo-Information* 4, no. 3 (Sept.): 1729–1749. https://doi.org/10.3390/ijgi4031729.

Suchman, Lucy. 2007a. "Feminist STS and the Sciences of the Artificial." In *New Handbook of Science and Technology Studies*, edited by Edward Hackett, Olga Amsterdamska, Michael Lynch, and Judy Wajcman, 139–163. 3rd edition. Cambridge, Massachusetts; London, England: The MIT Press.

———. 2007b. *Human-Machine Reconfigurations: Plans and Situated Actions*. 2nd edition. Cambridge; New York: Cambridge University Press.

———. 2021. "Talk with Machines, Redux." *Interface Critique*, no. 3 (Sept.): 69–80. https://doi.org/10.11588/ic.2021.3.81328.

Sudmann, Andreas. 2015. "Deep Learning als dokumentarische Praxis." *Sprache und Literatur* 46, no. 1–2 (Jan.): 155–170. https://doi.org/10.30965/25890859-0460102011.

———. 2018. "On the Media-political Dimension of Artificial Intelligence. Deep Learning as a Black Box and OpenAI." *Digital Culture & Society* 4, no. 1 (Oct.): 181–200. https://doi.org/10.14361/dcs-2018-0111.

———, ed. 2019a. *The Democratization of Artificial Intelligence: Net Politics in the Era of Learning Algorithms*. AI Critique 1. Bielefeld: transcript.

———. 2019b. "The Democratization of Artificial Intelligence. Net Politics in the Era of Learning Algorithms." In *The Democratization of Artificial Intelligence: Net Politics in the Era of Learning Algorithms*, 9–31. AI Critique 1. Bielefeld: transcript.

Surowiecki, James. 2005. *The Wisdom of Crowds*. New York: Anchor.

Sutton-Smith, Brian. 2001. *The Ambiguity of Play*. Cambridge, MA: Harvard University Press.

———. 2008. "Play Theory. A Personal Journey and New Thoughts." *American Journal of Play* 1, no. 1: 80–123. https://theplayethic.typepad.com/sutton-smith%20-%20play%20theory.pdf.

Sztompka, Piotr. 2000. *Trust: A Sociological Theory*. Cambridge Cultural Social Studies. Cambridge; New York, NY: Cambridge University Press.

Taenzel, Tobias. 2017. "Measuring Similarity of Artworks Using Multidimensional Data." Master's thesis, LMU Munich. Munich, Germany: LMU Munich. https://www.en.pms.ifi.lmu.de/publications/diplomarbeiten/Tobias.Taenzel/MA_Tobias.Taenzel.pdf.

Tegmark, Max. 2017. *Life 3.0: Being human in the age of Artificial Intelligence*. London: Penguin Books.

Terranova, Tiziana. 2000. "Free Labor. Producing Culture for the Digital Economy." *Social Text 63* 18, no. 2: 33–58. muse.jhu.edu/article/31873.

———. 2012. "Free Labor." In *Digital Labor: The Internet as Playground and Factory*, edited by Trebor Scholz, 33–57. Abingdon, Oxon; New York, NY: Routledge.

Thanner, Sarah, and Libuše Hannah Vepřek. 2023. "Imaginieren – Intraagieren – Rekonfigurieren: Mensch–Technologie-Relationen Im Werden." In *Zeit. Zur Temporalität von Kultur*, edited by Manuel Trummer, Daniel Drascek, Gunther Hirschfelder, Lena Möller, Markus Tauschek, and Claus-Marco Dieterich, 321–338. Münster; New York: Waxmann.

The Folding@Home Consortium (FAHC). n.d. "FOLDING@HOME." Accessed Mar. 19, 2023. https://foldingathome.org/?lng=en.

The Hybrid Intelligence Centre. n.d. "Hybrid Intelligence." Accessed Jun. 1, 2024. https://www.hybrid-intelligence-centre.nl/.

The MathWorks, Inc. n.d. "MathWorks." Accessed Mar. 19, 2024. https://de.mathworks.com/.

Thirion, J.-P. 1998. "Image matching as a diffusion process: An analogy with Maxwell's demons." *Medical Image Analysis* 2, no. 3 (Sept.): 243–260. https://doi.org/10.1016/S1361-8415(98)80022-4.

Tinati, Ramine, Markus Luczak-Roesch, Elena Simperl, and Wendy Hall. 2016. "Because science is awesome: studying participation in a citizen science game." In *Proceedings of the 8th ACM Conference on Web Science*, 45–54. New York, NY: ACM. https://doi.org/10.1145/2908131.2908151.

———. 2017. "An investigation of player motivations in Eyewire, a gamified citizen science Project." *Computers in Human Behavior* 73 (Aug.): 527–540. https://doi.org/10.1016/j.chb.2016.12.074.

Tischberger, Roman. 2020. "Computer sagt Nein. Fehlerkulturen in der Softwarearbeit." In *Vernetzt, entgrenzt, prekär? Kulturwissenschaftliche Perspektiven auf Arbeit im Wandel*, edited by Stefan Groth, Sarah May, Johannes Müske, 113–134. Arbeit und Alltag 17. Frankfurt; New York: Campus.

Touvron, Hugo, Thibaut Lavril, Gautier Izacard, Xavier Martinet, Marie-Anne Lachaux, Timothée Lacroix, Baptiste Rozière, Naman Goyal, Eric Hambro, Faisal Azhar, et al. 2023. "LLaMA: Open and Efficient Foundation Language Models." arXiv. https://doi.org/10.48550/arXiv.2302.13971.

Traweek, Sharon. 1992. *Beamtimes and Lifetimes: The World of High Energy Physicists*. Cambridge, Massachusetts: Harvard University Press.

Tsing, Anna. 2011. "Worlding the Matsutake Diaspora: Or, can Actor-Network Theory Experiment with Holism?" In *Experiments in Holism: Theory and Practice in Contemporary Anthropology*, edited by Ton Otto and Nils Bubandt, 47–66. Oxford, UK: Wiley-Blackwell. https://doi.org/10.1002/9781444324426.ch4.

Turing, Alan M. 1950. "Computing Machinery and Intelligence." *Mind* LIX, no. 236 (Oct.): 433–460. https://doi.org/10.1093/mind/LIX.236.433.

Turkle, Sherry. 2005a. "The New Philosophers of Artificial Intelligence: A Culture with Global Aspirations." In *The Second Self: Computers and the Human Spirit*, 20th anniversary edition, 219–244. Cambridge, Massachusetts; London, England: The MIT Press.

———. 2005b. *The Second Self: Computers and the Human Spirit*. 20th anniversary edition. Cambridge, Massachusetts; London, England: The MIT Press.

Turner, Victor. 1995. *Vom Ritual zum Theater. Der Ernst des menschlichen Spiels*. Frankfurt am Main: Fischer.

Under Secretary of Defense for Acquisition Technology. 1998. "DoD Modeling and Simulation (M&S) Glossary." DOD 5000.59-M. Washington D.C.: Department of Defense. United States of America. Jan. 15, 1998. https://web.archive.org/web/20070710104756/http://www.dtic.mil/whs/directives/corres/pdf/500059m.pdf.

University of California. n.d. "BOINC. Compute for Science." Accessed Mar. 19, 2024. https://boinc.berkeley.edu/.

University of California, Davis. n.d. "Protein Structure Prediction Center." Accessed Mar. 19, 2024. https://predictioncenter.org/.

University of Washington. n.d. "Rosetta@home." Accessed Mar. 19, 2024. https://boinc.bakerlab.org/.

Vaicaityte, Grete. 2021a. "Bots, that are going to play Stall Catchers along humans." Human Computation Institute Blog. Sept. 30, 2021. https://blog.hcinst.org/bots-that-are-going-to-play-stall-catchers-along-humans/.

———. 2021b. "Meet the bot authors – Roman Solovyev and his bot 'ZFTurbo'!" Human Computation Institute Blog. Nov. 9, 2021. https://blog.hcinst.org/meet-the-bot-authors-roman-solovyev-and-his-bot-zfturbo/.

———. 2021c. "Microsoft's annual Giving Campaign – a time of year when volunteering blooms." Human Computation Institute Blog. Dec. 1, 2021. https://blog.hcinst.org/microsofts-giving-month-a-time-of-year-when-volunteering-blooms-out-2/.

———. 2021d. "Meet the bot authors – Kirill Brodt and his bot 'Clsc2'!" Human Computation Institute Blog. Dec. 11, 2021. https://blog.hcinst.org/meet-the-bot-authors-kirill-brodt-and-his-bot-clsc2/.

Vepřek, Libuše Hannah. 2020. "Citizen Scientists Wanted! In the Fight Against the Coronavirus." Transformations. Apr. 1, 2020. https://web.archive.org/web/20220123054956/http://transformations-blog.com/citizen-scientists-wanted-in-the-fight-against-the-coronavirus/.

———. 2021a. "Hello from Libuše!" Human Computation Institute Blog. Aug. 19, 2021. https://blog.hcinst.org/hello-from-libuse/.

———. 2021b. "Multiplicities of meanings: citizen science between knowledge production, gameplay and coping with everyday life." Paper presented at the Breaking the Rules? Power, Participation, Transgression conference 2021 (SIEF2021), Online, Jun. 22.

———. 2022a. "Between Means and Ends: Data Infrastructures in Biomedical Research." Paper presented at the Mobilizing Methods in Medical Anthropology Conference 2022 (RAIMed2022), Online, Jan. 18.

———. 2022b. "Towards More Collaborative and Adaptive Ethical Review Platforms. With the example of the ethical review of human computation-based citizen science projects." Master's thesis, LMU Munich. Munich, Germany: LMU Munich. https://www.en.pms.ifi.lmu.de/publications/diplomarbeiten/Libuse.Veprek/DA_Libuse.Veprek.pdf.

———. 2023a. "Ein Gefühl für die Daten entwickeln. Eine ethnografische Annäherung an große Textdaten am Beispiel digitaler Chats." *Kulturanthropologie Notizen* 85: 167–187. https://doi.org/10.21248/ka-notizen.85.12.

———. 2023b. "Spielerisch helfen, «sinnvoll» zerstreuen?" *Zerstreuung*. das bulletin. Für Alltag und Populäres. https://www.dasbulletin.ch/post/spielerisch-helfen-sinnvoll-zerstreuen.

Vepřek, Libuše Hannah, Patricia Seymour, and Pietro Michelucci. 2020. "Human Computation Requires and Enables a New Approach to Ethics." *Proceedings of the Crowd Science Workshop: Remoteness, Fairness, and Mechanisms as Challenges of Data Supply by Humans for Automation co-located with 34th Conference on Neural Information Processing Systems (NeurIPS 2020)*: 26–33. Vancouver, BC, Canada. http://ceur-ws.org/Vol-2736/paper5.pdf.

Vepřek, Libuše Hannah, Sarah Thanner, Lina Franken, and The Code Ethnography Collective (CECO). 2023. "Computercode in seinen Dimensionen ethnografisch begegnen." *Kulturanthropologie Notizen* 85: 139–166. https://doi.org/10.21248/ka-notizen.85.13.

Verbeek, Peter-Paul. 2001. "Don Ihde: The Technological Lifeworld." In *American Philosophy of Technology: The Empirical Turn*, edited by Hans Achterhuis, 119–146. Indiana Series in the Philosophy of Technology. Bloomington: Indiana University Press.

———. 2005. *What Things Do: Philosophical Reflections on Technology, Agency, and Design*. Pennsylvania: Penn State University Press. https://doi.org/10.1515/9780271033228.

VERBI – Software. Consult. Sozialforschung. GmbH. n.d. "MAXQDA." Accessed Mar. 19, 2024. https://www.maxqda.com/.

Vercauteren, Tom, Xavier Pennec, Aymeric Perchant, and Nicholas Ayache. 2009. "Diffeomorphic demons: Efficient non-parametric image registration." *NeuroImage* 45, no. 1 (Mar.): S61-72. https://doi.org/10.1016/j.neuroimage.2008.10.040.

Verein der Europäischen Bürgerwissenschaften – ECSA e.V. n.d. "ccsa. European Citizen Science Association." Accessed Mar. 19, 2024. https://www.ecsa.ngo/.

Vertesi, Janet. 2014. "Seamful Spaces: Heterogeneous Infrastructures in Interaction." *Science, Technology, & Human Values* 39, no. 2 (Jan.): 264–284. https://doi.org/10.1177/0162243913516012.

Vertesi, Janet, and David Ribes, eds. 2019. *digitalSTS: A Field Guide for Science & Technology Studies*. Princeton, New Jersey: Princeton University Press.

v_mulligan. 2014. "Open Source." *Foldit forum*. Feb. 11, 2014. https://fold.it/forum/suggestions/open-source/page-3.

Vogl, Elisabeth. 2018. *Crowdsourcing-Plattformen als neue Marktplätze für Arbeit: Die Neuorganisation von Arbeit im Informationsraum und ihre Implikationen*. Augsburg: Rainer Hampp.

Vohland, Katrin, Anne Land-Zandstra, Luigi Ceccaroni, Rob Lemmens, Josep Perelló, Marisa Ponti, Roeland Samson, and Katherin Wagenknecht. 2021. "Editorial: The Science of Citizen Science Evolves." In *The Science of Citizen Science*, edited by Katrin Vohland, Anne Land-Zandstra, Luigi Ceccaroni, Rob Lemmens, Josep Perelló, Marisa Ponti, Roeland Samson, and Katherin Wagenknecht, 1–12. Cham: Springer. https://doi.org/10.1007/978-3-030-58278-4_1.

Von Ahn, Luis. 2005. "Human Computation." Ph.D. thesis, Carnegie Mellon University. Pittsburgh, PA: Carnegie Mellon University. http://reports-archive.adm.cs.cmu.edu/anon/2005/CMU-CS-05-193.pdf.

———. 2010. "Human Computation." Paper presented at the Voices From the Future, National Science Foundation, Aug. 26. https://www.nsf.gov/news/special_reports/voices/luis_von_ahn.jsp.

Von Ahn, Luis, and Laura Dabbish. 2008. "Designing games with a purpose." *Communications of the ACM* 51, no. 8 (Aug.): 57. https://doi.org/10.1145/1378704.1378719.

Wahlberg, Ayo. 2022. "Assemblage Ethnography: Configurations Across Scales, Sites, and Practices." In *The Palgrave Handbook of the Anthropology of Technology*, edited by Maja Hojer Bruun, Ayo Wahlberg, Rachel Douglas-Jones, Cathrine Hasse, Klaus Hoeyer, Dorthe Brogård Kristensen, and Brit Ross Winthereik, 125–144. Singapore: Palgrave Macmillan. https://doi.org/10.1007/978-981-16-7084-8_6.

Walmsley, Mike, Lewis Smith, Chris Lintott, Yarin Gal, Steven Bamford, Hugh Dickinson, Lucy Fortson, et al. 2020. "Galaxy Zoo: probabilistic morphology through bayesian CNNs and active learning." *Monthly Notices of the Royal Astronomical Society* 491, no. 2 (Jan.): 1554–1574. https://doi.org/10.1093/mnras/stz2816.

Ward, Dave, and Mog Stapleton. 2012. "Es are good. Cognition as enacted, embodied, embedded, affective and extended." In *Consciousness in Interaction: The role of the natural and social context in shaping consciousness*, edited by Fabio Paglieri, 89–104. Amsterdam: John Benjamins.

Watson, David, and Luciano Floridi. 2018. "Crowdsourced Science: Sociotechnical Epistemology in the e-Research Paradigm." *Synthese* 195, no. 2 (Feb.): 741–764. https://doi.org/10.1007/s11229-016-1238-2.

Watson, Rod. 2014. "Trust in Interpersonal Interaction and Cloud Computing." In *Trust, Computing, and Society*, edited by Richard H. R. Harper, 172–198. Cambridge: Cambridge University Press. https://doi.org/10.1017/CBO9781139828567.012.

Weichselbraun, Anna, Shaila Seshia Galvin, and Ramah McKay. 2023. "Introduction: Technologies and Infrastructures of Trust." *The Cambridge Journal of Anthropology* 41, no. 2 (Sept.): 1–14. https://doi.org/10.3167/cja.2023.410202.

Weingardt, Markus A., ed. 2011. *Vertrauen in der Krise: Zugänge verschiedener Wissenschaften*. Baden-Baden: Nomos.

Welz, Gisela. 2021a. "Assemblage." In *Theoretische Reflexionen. Perspektiven der Europäischen Ethnologie*, edited by Peter Hinrichs, Martina Röthl, and Manfred Seifert, 161–176. Berlin: Reimer.

———. 2021b. "More-than-human Futures: Towards a Relational Anthropology in/of the Anthropocene." *Welt. Wissen. Gestalten. 42. Kongress der Deutschen Gesellschaft für Volkskunde (dgv) 2019*, edited by Gertraud Koch, Johannes Moser, Lara Hansen, and Stefanie Mallon: 36–46. Hamburger Journal für Kulturanthropologie (HJK), no. 13 (Jul.). http://nbn-resolving.de/urn:nbn:de:gbv:18-8-17075.

Westphal, Andrew J., Anna L. Butterworth, Christopher J. Snead, Nahide Craig, David Anderson, Steven M. Jones, Donald E. Brownlee, Richard Farnsworth, and Michael E. Zolensky. 2005. "Stardust@home: A Massively Distributed Public Search for Interstellar Dust in the Stardust Interstellar Dust Collector." In *Lunar and Planetary Science XXXVI*, no. 21 (Jan.). https://ntrs.nasa.gov/citations/20050180792.

Wiederrich, Dave. 2019. "Stall Catchers: Citizen Scientists Speeding Alzheimer's Research." Memory Cafe Directory. Aug. 27, 2019. https://www.memorycafedirectory.com/stall-catchers-citizen-scientists-speeding-alzheimers-research/.

Wietschorke, Jens. 2021. "Zwischen Aushandlungsparadigma und Kontextualismus." In *Theoretische Reflexionen. Perspektiven der Europäischen Ethnologie*, edited by Peter Hinrichs, Martina Röthl, and Manfred Seifert, 51–67. Berlin: Reimer.

Wietschorke, Jens, and Moritz Ege. 2023. "Was sind kulturelle Figuren? Zur Einführung". In *Kulturelle Figuren. Ein empirisch-kulturwissenschaftliches Glossar (Festschrift für Johannes Moser)*, edited by Daniel Habit, Christiane Schwab, Moritz Ege, Laura Gozzer, and Jens Wietschorke, 11–23. Münchner Beiträge zur Volkskunde 49. Münster; New York: Waxmann.

Wiggins, Andrea, and John Wilbanks. 2019. "The Rise of Citizen Science in Health and Biomedical Research." *The American Journal of Bioethics* 19, no. 8 (Jul.): 3–14. https://doi.org/10.1080/15265161.2019.1619859.

Willson, Michele, and Katharina Kinder-Kurlanda. 2021. "Social Gamers' Everyday (in)Visibility Tactics: Playing within Programmed Constraints." *Information, Communication & Society* 24, no. 1 (Jan.): 134–149. https://doi.org/10.1080/1369118X.2019.1635187.

Winnicott, David. 1965. *The Maturational Processes and the Facilitating Environment: Studies in the Theory of Emotional Development*. Madison, CT: International Universities Press.

Woolgar, Steve. 1991. "Configuring the User: The Case of Usability Trials." In *Sociology of Monsters. Essays on Power, Technology and Domination*, edited by John Law, 57–99. Abingdon, Oxon; New York, NY: Routledge.

Worldwide Protein Data Bank (wwPDB). n.d. "RCSB PROTEIN DATA BANK (RCSB PDB)." Accessed Mar. 19, 2024. https://www.rcsb.org/.

Wynn, James. 2017. *Citizen Science in the Digital Age: Rhetoric, Science, and Public Engagement*. Rhetoric, Culture, and Social Critique. Tuscaloosa: The University of Alabama Press.

Wynne, Brian. 1988. "Unruly Technology: Practical Rules, Impractical Discourses and Public Understanding." *Social Studies of Science* 18, no. 1 (Feb): 147–167. https://doi.org/10.1177/030631288018001006.

Yang, Jianyi, Ivan Anishchenko, Hahnbeom Park, Zhenling Peng, Sergey Ovchinnikov, and David Baker. 2020. "Improved protein structure prediction using predicted interresidue orientations." *Proceedings of the National Academy of Sciences* 117, no. 3 (Jan.): 1496–1503. https://doi.org/10.1073/pnas.1914677117.

Y Combinator. n.d. "Startup Directory." Accessed Mar. 19, 2024. https://www.ycombinator.com/companies?batch=W23.

Zimmer, Carl. 2017. "Scientists Are Designing Artisanal Proteins for Your Body." *The New York Times*. Dec. 26, 2017. https://www.nytimes.com/2017/12/26/science/protein-design-david-baker.html.

Zimmerli, Walther Ch. 1990. "Wieviel Akzeptanz erträgt der Mensch? Bemerkungen zu den Hintergründen der Technikfolgenabschätzung." In *Mensch — Gesellschaft Technik*, edited by Ernst Kistler and Dieter Jaufmann, 247–260. Wiesbaden: VS Verlag für Sozialwissenschaften. https://doi.org/10.1007/978-3-322-95524-1_17.

zo3xiaJonWeinberg. 2021a. "Recipe: intro-to-the-AlphaGoStarCraft-A.i.game5.Lua." Foldit. Jan. 13, 2021. https://fold.it/recipes/104332.

———. 2021b. "Recipe: rate1star output anime Minecraft13pub.Lua." Foldit. Aug. 22, 2021. https://fold.it/recipes/105123.

Zooniverse. n.d. "Galaxy Zoo. The Science behind the Site." Accessed, Mar. 19, 2024. https://www.zooniverse.org/projects/zookeeper/galaxy-zoo/about/research.

Zoran. 2009. "Sneak Preview: Custom Tools (Macros)." Foldit forum. Mar. 12, 2009. https://fold.it/forum/blog/sneak-preview-custom-tools-macros.

www.ingramcontent.com/pod-product-compliance
Lightning Source LLC
Jackson TN
JSHW052132131224
75386JS00037B/1254